THE EUROPEAN ENVIRONMENT AND CAP REFORM

Policies and Prospects for Conservation

The European Environment and CAP Reform
Policies and Prospects for Conservation

Edited by
Martin Whitby
Professor of Countryside Management
Centre for Rural Economy
The University of Newcastle upon Tyne
UK

CAB INTERNATIONAL

CAB INTERNATIONAL
Wallingford
Oxon OX10 8DE
UK

Tel: +44 (0)1491 832111
Fax: +44 (0)1491 833508
E-mail: cabi@cabi.org
Telex: 847964 (COMAGG G)

A catalogue record for this book is available from the British Library.

ISBN 0 85199 106 8

Typeset in Photina by Solidus (Bristol) Limited
Printed and bound in the UK by Biddles Ltd, Guildford

Contents

Contributors

Mr David Baldock is Deputy Director of the Institute for European Environmental Policy, London, UK.

Dr Eric Bignal is an agricultural ecologist with the UK Joint Nature Conservation Committee and is an Executive Committee Member of the European Forum on Nature Conservation and Pastoralism.

Gert van der Bijl is International Coordinator of the Centre for Agriculture and Environment (CLM), Utrecht, the Netherlands.

Jean-Marie Boisson is Professor of Economics and Jean Monet Professor of European Integration, Université de Montpellier I, France.

Dr François Bonnieux is Director of Research, INRA, Rennes, France.

Dr Henry Buller is ESRC 'Global Environmental Change' programme research fellow at the Department of Geography, King's College London and Lecturer at the Department of Geography, University of Paris, France.

Fernando Garrido is a PhD student in rural sociology, in the Instituto de Estudios Sociales Avanzados – Andalucia (IESA-CSIC), Cordoba, Spain.

Andreas Höll is a senior research fellow in the Institut für Ländliche Strukturforschung, Frankfurt, Germany.

Philip Lowe is Duke of Northumberland Professor of Rural Economy and director of the Centre for Rural Economy in the Department of Agricultural Economics and Food Marketing, the University of Newcastle upon Tyne, UK.

Dr David McCracken is an agricultural ecologist in the Department of Environmental Sciences, Scottish Agricultural College, Auchincruive, UK.

Dr Heino von Meyer is economist, researcher and consultant for studies on rural development and environment, Hamburg, Germany.

Dr Eduardo Moyano is Doctor in Rural Sociology and Professor in the Instituto de Estudios Sociales Avanzados – Andalucia (IESA-CSIC), Cordoba, Spain.

Ernst Oosterveld is staff member on nature management on the farm at the Centre for Agriculture and Environment (CLM), Utrecht, the Netherlands.

Dr Andrea Povellato is a senior research officer in the Instituto Nazionale di Economia Agraria (INEA), Padua, Italy.

Dr Jørgen Primdahl is Associate Professor of Countryside Planning and Landscape Management in the Department of Economics and Natural Resources, The Royal Veterinary and Agricultural University Copenhagen, Denmark.

Dr Bengt Rundqvist is Principal Officer in the Natural Resources Department of the Swedish Environmental Protection Agency, Stockholm, Sweden.

Dr Martin Scheele is the administrator dealing with environmental policy issues in the field of agriculture in the Commission of the EU, Brussels, Belgium.

Robert Weaver is Professor, Pennsylvania State University and was on sabbatical leave with INRA, Rennes, France.

Martin Whitby is Professor of Countryside Management and co-director of the Centre for Rural Economy, Department of Agricultural Economics and Food Marketing, University of Newcastle upon Tyne, UK.

Preface

The saga of the Common Agricultural Policy (CAP) has been an enduring feature of the EC since it began in 1957. Its early productivist decades were terminated by events in 1984 and 1985 which, taken together and viewed with hindsight, amount to a substantial turning point of policy. The two events were, first, the imposition of milk quotas and, second, the introduction of formal environmental policy mechanisms within the CAP. The, now familiar, Article 19 of Regulation 797/85 merely permitted member states to make payments for the provision of public goods by farmers. Later, support for these payments was provided through Regulation 1760/87, to the limit of 25% of costs. The level of these payments, significantly, was based on the concept of income forgone rather than reflecting direct demand for the public goods. Contracts for production of 'desired environmental attributes' were already familiar in Northern Europe, where they had been used in one or two countries for some years. In these countries the new Regulation was taken up quickly once a financial contribution from Brussels was assured but the rest of the Community was slower to respond.

The next step emerged from the pressures for CAP reform, arising from the combination of continuing budgetary crises and international trade pressures, emanating from the Uruguay round of GATT negotiations. These pressures led to the now famous MacSharry Reforms which culminated in a commitment to reduce producer prices towards world prices whilst supporting farmers' incomes with measures 'decoupled' from agricultural production. Accompanying the MacSharry Reforms, the European Union introduced a special regulation with the twin aims of protecting the environment while supporting farm incomes. These measures, defined in Regulation 2078/92, form the subject of this volume.

The theme is developed through a series of essays by academics, administrators and policy analysts from NGOs. This heterogeneity of origins of contributors is mirrored, too, in authors' disciplinary backgrounds. The contents page thus includes sociologists, geographers, political scientists, economists, ecologists and a landscape architect. Although such a mixture would not produce a homogenous treatment of the subject, the authors' common interest in the Agri-Environment Regulation and a willingness to collaborate across conventional disciplinary boundaries has generated valuable synergy in assembling the text. The result is a multi-disciplinary review of the introduction of a new environmental regulation by the EU. Some chapters treat particular issues – administration, politics, ecological resources, the value of benefits, as issues across the EU; but the majority present studies of the response to the regulation by individual member states.

The work was initiated in 1994 when the University of Newcastle upon Tyne made a generous grant to fund a meeting at which chapters could be debated. That meeting took place in Brighton, in March 1995, and the papers presented there have been edited to produce this volume.

The Editor of such a book is unlikely to emerge from his work without debts to others. In this case I am happy to acknowledge the contribution of many colleagues at Newcastle and elsewhere. To Hilary Talbot for organizing the Brighton meeting, to Eileen Curry, Margaret Hall and Lisa Trewitt for diligent typing of text, to Richard Hill for competent editorial assistance, to Philip Lowe, Caroline Saunders and Lionel Hubbard for carefully reading my own chapters and finally, to all the authors for their patient response to my many editorial demands, I am profoundly grateful. This acknowledgement in no way absolves me from errors remaining in the volume.

Martin Whitby
Newcastle upon Tyne
October 1995

Exchange Rates Against the ECU

	Currency unit	1992	1993	1994
Denmark	DKr	7.81	8.39	8.23
France	FFr	6.85	7.32	7.18
Germany	DM	2.02	2.14	2.10
Italy	Lit	1981.9	2033.6	2086.7
Netherlands	G	2.27	2.40	2.35
Spain	Ptas	132.5	164.6	173.4
Sweden	SKr	7.53	10.07	9.98
UK	£	0.74	0.86	0.84

INTRODUCTION

1 The Agri-environmental Measures in the Context of the CAP Reform

Martin Scheele

Agriculture has – by definition – many effects on the state of the environment. The very nature of agriculture is to modify the original status of natural sites while using land and animal rearing to produce food and raw materials. Some of the environmental effects of agriculture are beneficial, others are negative.

The beneficial effects are produced because over hundreds of years agricultural production has created cultivated landscapes that correspond to today's perception of desired landscapes and which provide a level of bio-diversity that is often much richer than that of the original status. Once cultivated landscapes have been established, continuation of agricultural production is a central precondition for avoiding land erosion and degradation of soils. Negative effects of agricultural production have also increasingly become an issue of public concern. These include contamination of waters by pesticides and fertilizers, deterioration of the bio-diversity once created and destruction of the historical form of cultivated landscapes.

Neither avoidance of negative effects nor stimulation of positive ones can be achieved at a satisfactory level on the basis of market activities, although the former are side-effects of the latter. Environmental goods and avoidance of environmental damage most often have the character of so-called public goods which are consumed jointly while the individual contribution to their supply is difficult to enforce. Therefore, environmental effects of agriculture have become the subject of both environmental and agricultural policies.

The European Commission underlined in its 'Green Book' (EC, 1985) that environmental policies must set the framework within which agricultural production takes place. As regards regulatory activities aiming at the abatement of negative environmental effects, the 'Polluter Pays Principle' was emphasized as a basic principle.

As regards the provision of public goods in the field of the environment, the Green Book underlines the important role of agriculture as a steward of the environment and cultivated landscapes, and states that this stewardship should be remunerated. In this context the Green Book already drew a clear line between giving financial incentives for the provision of environmental goods and services by agriculture and the necessity to diversify agricultural income sources and to combine extensification of agricultural production with the need to control agricultural production.

The idea of combining environmental policies with agricultural market and income policies in a mutually beneficial way was an important pillar of the 1992 reform of the Common Agricultural Policy (CAP). It was recognized that enough farmers must be kept on the land in order to preserve the natural environment and traditional landscapes.

A central element of the CAP reform was the shift from price support to direct payments, aiming to combine control of agricultural markets with extensification of agricultural production. Direct payments foreseen in the Market Organizations for cereals and oilseeds are linked to obligatory set-aside of land leading to a general extensification of crop rotations. In the case of the Beef Premium, eligibility is linked to maximal stocking rates. A supplementary extensification premium is granted if the stocking rate does not exceed 1.4 livestock units (LU) per hectare.

The CAP reform reflected, furthermore, that farmers fulfil or can fulfil an important function as stewards of the environment and the countryside by introducing an agri-environmental scheme which was established under Council Regulation (EEC) 2078/92 as a part of the so-called accompanying measures. The accompanying measures which include also afforestation of agricultural land and an early retirement system should – together with the rural policy under the reformed structural funds – improve the standard of land use and land conservation and ensure a balanced development of the countryside.

A basic consideration underlying the introduction of agri-environmental measures under Regulation (EEC) 2078/92 was to combine the encouragement of extensification in the context of market organization with specific measures tailored to the situation in individual Member States. The measures are first and foremost land-use related; the environmental impact must be significant and must go beyond what could be seen as being 'good agricultural practice'.

The agri-environmental scheme had to be implemented by zonal programmes at the Member State level. Co-financing from the Community budget is provided for 75% of the payments in Objective 1 areas and 50% in all others. The scheme is aiming for supplementary action. National measures should not be replaced by funds from the EU budget which should provide some leverage for additional funds to be raised by the Member States.

Under the agri-environmental scheme, payments are provided to encourage farmers to use environmentally sound production methods. This involves a significant reduction of polluting inputs like fertilizers, pesticides and herbicides in crop production. Maintenance and introduction of organic farming is also promoted. As regards livestock farming, reduction of livestock density is sought where damage is being caused by the overstocking of sheep and cattle.

In addition to the extensification of input use, environmentally friendly management of farmed land is promoted with respect to conserving or re-establishing the diversity and quality of the natural environment involving the flora, the fauna and scenic features of cultivated landscapes. Promotion of farming practices compatible with the requirements of protecting the environment and natural resources includes payments for the rearing of local breeds in danger of extinction and cultivation of plants endangered by genetic erosion.

Payments are also provided to ensure the upkeep of abandoned farmland in rural areas and the management of land for public access and leisure activities. The agri-environmental scheme is completed by a scheme providing for 20-year set-aside for ecological purposes, like the constitution of conservation reserves, the creation of biotopes or natural parks and so on. In addition to the promotion of environmentally sound farming activities, the agri-environmental scheme provides for measures to improve the training of farmers with regard to farming or forestry practices compatible with the environment.

The agri-environmental measures are established on a voluntary and contractual basis. Farmers participating in the zonal programmes are paid in return for associated income losses. The scheme requires that farmers engage themselves for at least 5 years, except for the set-aside scheme where the time period is 20 years. The upper limits of premia which are granted on an annual basis are defined by the regulation. Specific premia set up on a regional basis must be justified by calculations provided in the context of zonal programmes.

Since 1993 altogether 163 zonal programmes have been notified for adoption by the Commission and 152 programmes have been adopted. The budget engaged for co-financing the agri-environmental measures already approved is 3711.6 million ECU for the 1993–97 period. The programmes adopted vary enormously as regards environmental priorities and ways of implementation. The great variety of approaches reflect certainly not only differences in objectives but also differences in natural conditions and site-specific problems.

Some of the Member States felt that it was important to provide aid to as many farmers as possible which means that the environmental rules were relatively light. Other Member States thought that it was vital to focus the aids on specific problems and have earmarked very high premia with very strict rules. Some Member States opted for à la carte selections of measures while others introduced full package solutions.

The agri-environmental scheme is a new instrument in the context of agricultural policy. The voluntary, contractual basis of the scheme is following the idea of so-called economic instruments which are suggested by the EU's fifth Action Programme for environmental policies.

The system has the potential for efficient solutions: the premia are calculated according to opportunity costs of farmers due to participation in zonal programmes. As a consequence, the efficiency condition at the micro level is met where the marginal costs of using factors of production for the provision of environmental goods and services equal those of alternative use in agricultural production.

Efficiency of the policy-mix as a whole is also possible, because the reform of the market organizations together with the agri-environmental scheme meets the efficiency condition of Tinbergen (1964) requiring that the number of policy instruments at least equals the number of objectives. This condition allows for fine-tuning of the policy system in engaging a specific instrument for each objective while taking into account positive or negative side-effects caused by other combinations of objectives and instruments.

Efficiency of the Tinbergen type is possible, because both the replacement of price-support by direct payments and the agri-environmental scheme pursue different objectives while having mutually positive side-effects. Direct payments aim first to improve agricultural incomes while having at the same time – due to decreasing price support and eco-conditionality of premia – a positive environmental side-effect. The agri-environmental scheme is pursuing environmental objectives as a priority, while having positive income effects due to the diversification of income sources, at the same time.

In practice, there is, at first glance, a very positive response as regards the implementation of measures under Regulation (EEC) 2078/92 from the farmers, the public and environmental groups. Whether the agri-environmental scheme can prove its practicability as a tool of agricultural policy, which might even be worth extending, will be subject to thorough investigation. Analysis of economic and ecological effects could help to improve the scheme in the future.

One of the open questions is whether the premia are set correctly. In a situation where the CAP reform was being implemented and given the innovative character of the agri-environmental approach, calculation of premia had to a certain degree to be based on assumptions about the costs of participation. Actual participation in zonal programmes and comparison with the number of participants originally foreseen by the Member States will give some indication of the appropriateness of the levels of premia set.

As regards the ecological effects, similarly, new terrain has been entered. Given the necessity of tight control of agri-environmental measures, it is sometimes difficult to find the zero point, the bottom line against which improvement of the environment can be evaluated. Furthermore, it must be taken into consideration that the agri-environmental scheme is set-up in the spirit of the principle of subsidiarity. This implies on the one hand a broad range of objectives and measures which are very difficult to compare according to a uniform set of criteria. On the other hand, it is left explicitly to the Member States to define measures according to their environmental objectives and priorities.

Concerns have been raised whether the agri-environmental scheme providing financial incentives for environmentally sound behaviour complies with the polluter pays principle. This can be assured as long as maintenance or improvement of the state of the environment is not compulsory under national law. In other words, as long as using natural resources is covered by private property rights, supply of environmental goods and services is a private economic activity. Society is willing to pay for such activities as long as the provision of public goods in the field of the environment is desired. In the light of these considerations, analysis of specific zonal programmes might be necessary in order to clarify whether or not compatibility with the polluter pays principle has to be concerned.

Open questions arise also from the fact that the financial engagement of Member States as measured by percentage of agricultural surface involved and premia per hectare vary considerably. In the light of these differences, analysis has to be made of whether the general framework of Regulation 2078/92 is sufficiently flexible to allow for implementation under different natural and economic conditions. It has to be borne in mind, however, that funds for the agri-environmental scheme should be spent according to environmental needs and not in proportional ratios among Member States or regions.

With the agri-environmental scheme new terrain has been covered. Whether or not this is an approach that is worth maintaining or even extending will be the subject of thorough investigation. Analysis of the scheme is even compulsory according to Article 10 of Regulation 2078/92. Related work has already started in various contexts, the results of which will certainly help with future stages of policy making in this field.

2

The Development of European Agri-environment Policy

David Baldock and Philip Lowe

Introduction

European agriculture has undergone a profound technological revolution in the postwar period. While the social consequences have aroused criticism in much of Europe for most of this period, in recent years there has been mounting concern over the environmental consequences as well. Awareness of specific environmental problems has been accompanied, or even superseded, by a more general sense of an ecological crisis surrounding modern agriculture which indicts the predominant 'industrial' model of high input/high output farming.

Undoubtedly, agriculture across Europe has experienced more rapid change during the past 40 years than ever before. It is worth reflecting, therefore, why it is only so recently that the impact of agricultural change has become one of the most prominent environmental issues in European countries. After all, the contemporary environmental movement first emerged in most countries in the 1960s.

There are a number of reasons for the partial shift in focus that has happened since then and which has brought agricultural environmental problems to prominence. First, as regulatory mechanisms have developed in relation to urban and industrial pressures on the environment, then the comparative absence of controls, or at least effective controls, over agriculture has become more and more apparent. This lack stems in part from traditional perceptions of farming as a naturally conserving land use, seemingly set apart from extractive and industrial activities that cause pollution and despoliation. It is also a function of the dominant way in which environmental controls have been formulated and promulgated – largely through the setting of specific standards and their administration by urban-based regulatory agencies and inspectorates. These systems are geared up to dealing with large industrial concerns, concentrated and acute forms of pollution and point sources. Diffuse sources (not just in agriculture but also noise, vehicular emissions, and the transport, storage, use and disposal of hazardous chemicals) create much more intractable problems of regulation, and it is only recently that serious consideration has been given by governments and officials to how these problems might be overcome.

Second, the recession of the late 1970s and early 1980s alleviated some of the well recognized pressures on regional environments, while those from agriculture intensified or were recognized for the first time. The closure of plant, especially in traditional heavy industries, though leaving a legacy of extensive dereliction, eliminated some long-standing sources of air and water pollution. The oil shocks of 1973 and 1979 also gave impetus and urgency to resource conservation generally, either because drives to save or recover energy facilitated other forms of recycling, or because cost-conscious manufacturers were induced to look for other ways to reduce their raw material costs. In turn, these efforts alleviated problems of pollution and waste disposal. Through all this, though, agricultural production, with its protected economic status, continued to expand.

A third factor in the increased salience of agricultural problems amongst environmentalists has been recognition of the Common Agricultural Policy as an engine for the intensification of farming and thus for the destruction of the rural environment. Unlike many other sources of environmental degradation in which market forces and technological change are the dominant factors, agricultural intensification is also strongly driven by government support policies. It has therefore been an obvious target for lobbyists. But equally, given the supranational nature of the CAP, it was evident that, in this sector, satisfactory environmental reform could not be achieved in one country alone. Concerted multilateral pressure was required, and it has taken considerable time and effort to build up the necessary transnational infrastructure and to mobilize support for reform across the Member States. Action at the European level has also had to contend with separate national perspectives on the nature of the rural environment problem.

At least three environmental agendas can be discerned in the rural policy domain of the 1970s and 1980s in the European Community. The first of these is associated particularly with France, the Alpine countries, Scandinavia and parts of southern Europe. This is a concern best summarized by the French term 'desertification'. As farming in marginal areas, particularly in the mountains and high uplands became less viable in both economic and social terms, substantial areas of land became vulnerable to abandonment or a decline in the intensity of management. In some areas, this had resulted in the total abandonment of villages, terraces, fields and areas of common grazing land, with important ecological, as well as sociocultural implications. In such areas, it was seen as the priority to maintain agriculture, usually some form of pastoralism, partly for environmental reasons as well as for wider social, economic and cultural reasons. The maintenance of farming through appropriate subsidies is thus the primary objective of this environmental agenda. At Community level, it is best expressed through the Less Favoured Areas Directive (Beaufoy *et al.*, 1994).

A second environmental debate emerged in the intensively farmed areas of Europe in the 1970s. This centred on the damage caused to the environment by contemporary agriculture, with a particular focus on pollution by livestock wastes, inorganic fertilizers and pesticides. In many areas, it was compounded by the often severe environmental consequences of land consolidation schemes which often entailed the wholesale removal of ditches, woodland and other landscape features. This critique was clearly associated with a newly emerging and largely urban based environmental movement, the prime target of which was pollution and environmental stress stemming from industrialization. Agriculture became a focus of criticism, particularly in the Netherlands, Germany, Denmark, the UK and northern Italy. For the first time, agriculture was depicted as a despoiler rather than a guardian of the environment. Many of the proposals from these new critics focused on pollution control, the promotion of organic farming and the production of healthy food, coupled with a general preference for smaller-scale, less intensive family farming (Baldock and Bennett, 1991; Lowe, 1992).

In the UK, a distinctive variant of this movement arose in the late 1970s. This grew out of the traditional British concern with protection of the landscape and wildlife. It applied the analysis of the destructive effects of intensive agriculture to wildlife and landscape issues, rather than to pollution, soil erosion and large-scale production. Agriculture was depicted as the main engine for destroying nature in the lowlands especially, primarily through intensification. The traditional farmed landscape was seen to be 'under sentence of death' from modern agricultural practices (Shoard, 1980). This perception was greatly reinforced by national statistics showing the large-scale loss of semi-natural habitats throughout the country and clearly identifying agriculture as the main cause of this process (Lowe *et al.*, 1986).

European Decision-making Processes

In principle, the development of an agri-environment policy should represent the coming together of agricultural and environmental concerns. However, decision making in the European Union, as in most Member States, is sectoralized. The development of European agricultural policy and of European environmental policy have been largely separate. Each has had its own momentum within its own network of established policy actors.

Agricultural policy is the most mature and well established policy field in the European Community (Marsh, 1991). Agricultural interests are deeply embedded constitutionally and administratively within the Community's institutions. Agricultural policy making is also set apart somewhat with its own separate decision-making structures. The policy network includes Agricultural Ministries and the major farming unions, as well as the Agricultural Directorate (DGVI) of the European Commission, and it has not been easy for non-agricultural interests to penetrate this network (Neville-Rolfe, 1983). Environmental policy is a more recent development and was only explicitly recognized in the basic treaties in 1987, although in fact the development of European environmental policy dates back to the early 1970s (Liefferink *et al.*, 1993). This has been a much more fluid and open policy field (Mazey and Richardson, 1992 and 1993).

Much of the public pressure for change in agriculture on environmental grounds has come from environmental groups, mainly in northern Europe and mainly concerned with the intensification of agriculture and the ecological consequences. To a certain extent, such concerns have also been taken up by some of the northern Member States' governments. However, while the development of European environmental policy has been fairly accommodating of the environmental movement, European agricultural policy making has resisted the direct involvement of environmental interests (Baldock and Beaufoy, 1993).

Indeed, the closed nature of national and European agricultural policy communities, which typically exclude from decision making organizations not directly involved in implementing policy, and the complex and interlocking interdependencies between state agencies and agricultural organizations make for structures which are usually highly resistant to change, but also which seldom generate reform from within (Hervieu and Lagrave, 1992). Major change tends to arise from external pressures, particularly under conditions of crisis in which the management of the agricultural sector can no longer contain the problems it generates, thus politicizing policy making (Lowe *et al.*, 1994).

In this case, the conditions of crisis arose when European agriculture's efforts to cope with its own structural overproduction spilled over into budgetary crises and international trade conflicts, brought to a head in the General Agreement on Tariffs and Trade (GATT) negotiations. This undoubtedly created the opportunity for significant changes to agricultural policy, including the development of agri-environment policy.

Agri-environment policy, though, remains very much an initiative of, and in the realm of, the agricultural policy community and not of the environmental policy network. The environmental pressures upon agricultural policy makers have been indirect ones. In particular, the development of EC environmental policy has gradually impinged upon agriculture in ways which have increasingly raised questions concerning the interactions of agricultural and environmental policies and in turn have stimulated demands for policy integration which the agricultural policy community has been unable to ignore (Delpeuch, 1991).

The EC's very first Action Programme for the Environment, published in 1973, recognized the need to tackle emerging problems of agricultural pollution (O.J. C112, 20 December 1973) – a sentiment reiterated four years later in the Second Action Programme (O.J. C139, 13 June 1977). More ambitiously, the EC's Third Action Programme for the Environment stated the need to:

> ... promote the creation of an overall strategy, making environmental policy a part of economic and social development, (resulting) in a greater awareness of the environmental dimension, notably in the field of agriculture (and) ... enhance the positive and reduce the negative effects on the environment of agriculture
>
> (O.J. C46, 17 February 1983)

These intentions were echoed in the Commission's 1985 Green Paper on the future of the CAP which – following interventions by the Environment Directorate in response to drafts from the Agricultural Directorate – departed from previous such documents in including a section which proposed that agricultural policy should 'take account of environmental policy, both as regards the control of harmful practices and the promotion of practices friendly to the environment' (European Commission, 1985).

The legal requirement to integrate environmental protection into other EC policy areas was established in 1987 by the Single European Act. The need for 'environmental integration' is also a main theme of the Fifth Environmental Action Programme and is given a more comprehensive legal basis in the Maastricht Treaty. At the time of agreeing the MacSharry reform package in May 1992, the EC Agricultural Council declared its commitment to 'make environmental protection an integral part of the Common Agricultural Policy'.

It would be wrong, however, to see in this and subsequent policy initiatives the triumph of environmental interests alone. Instead, environmental arguments have coincided with other powerful arguments for reform and together these have induced notable changes. The chronic funding problems of the EC, the strain placed on the already overstretched budgets by the accession of the southern European states, the mounting costs and public scandal of burgeoning agricultural surpluses, and rising international opposition to the dumping of surpluses on world markets, have demanded consideration of means of curbing overproduction and the public costs of farming supports. Thus, some agricultural policy makers have responded to environmental concerns, not necessarily through any deep convictions, but because of a perceived coincidence between the aims of environmental improvement and the need to reduce agricultural output, thereby contributing to the alleviation of surplus and budgetary problems. At the same time, in northern Europe farming leaders, in a context of chronic oversupply of staple products and falling farm incomes, have begun to look to the provision by farmers of environmental 'products', in order to underpin or renew their claims for public support.

In the Commission's proposals for the package of CAP reforms agreed in 1992 the first objective of the CAP was thus reshaped as follows:

> Sufficient numbers of farmers must be kept on the land. There is no other way to preserve the natural environment, traditional landscapes and a model of agriculture based on the family farm as favoured by the society generally
> (European Commission, 1991, pp. 9–10).

The Roots of European Agri-environment Policy

The first agri-environment measure at the European level is generally taken to be Article 19 of Council Regulation 797/85 on Improving the Efficiency of Agricultural Structures. This authorized Member States to introduce 'special national schemes in environmentally sensitive areas' to subsidize farming practices favourable to the environment.

This amendment to the EC's Structures Directive was promoted by the British Government and in terms of the comparative political economy of agri-environment policy it is worth reflecting why. Britain is not renowned as a lead state in the development of EU environmental policy. On the contrary, by the mid-1980s, it had earned quite a reputation for being one of the lag states, repeatedly resisting the development of a more progressive environmental policy at the European level.

Of course, Article 19 was a development not in environmental policy but in agricultural policy, and during the 1980s the British Government was leading pressure for the reform of European agricultural policy. The Government's main concern was the costs of the CAP, and in 1984 its opposition to the escalating budget had been an important factor leading to the introduction of milk quotas.

By the mid-1980s agrarian corporatism had weakened in the UK. Amongst European countries, the agricultural sector was making the smallest percentage contribution to GDP and to employment. Moreover, in a highly urbanized nation, in which the peasantry had long since vanished and in which agriculture was decidedly capitalistic, agrarian ideologies did not have such a hold on society. The Ministry of Agriculture was very much a sectoral ministry and not a territorial ministry, and therefore there was more scope to raise issues concerning the wider environmental costs of agricultural development on the departmental agenda (Lowe and Buller, 1990).

Campaigns to protect rural landscapes and habitats from agricultural intensification had begun in the UK in the early 1960s. Britain's entry into the EC in 1972 stimulated yet more intensification, particularly the conversion of grassland to arable in response to higher cereal prices induced by the CAP. Increased agricultural land prices also encouraged landowners to remove small habitats in favour of productive land. The Government's response to mounting calls to protect important threatened sites and landscapes was to allow conservation authorities an opportunity to object to farmers' plans that might damage key sites but to require the authorities then to compensate the farmer for any income foregone, including the loss of any agricultural capital grants (i.e. investment aid) for which the farmer would be eligible. This resolution, which was enacted in domestic legislation in 1981, disturbed environmentalists. They particularly objected to the fact that compensation to farmers for not destroying wildlife habitat had to be paid by relatively impecunious conservation authorities rather than by the Ministry of Agriculture which had helped create the problem by promoting agricultural productivity and land reclamation. Environmental lobby groups pressed the Ministry to make its supports to farmers conditional on environmental criteria. However, the Ministry argued that it was bound by European rules and had no mandate to support farmers other than in producing food.

As political pressure mounted on the Ministry to resolve a series of heavily publicized local conflicts over ploughing of environmentally valuable wet grassland, it became clear that their preferred option was to pay farmers to maintain the existing land use, rather than to rely on the relatively unattractive and cumbersome compensation measures provided under the 1981 legislation. There was no effective domestic legislation under which farmers could have been required to desist from ploughing and the Government had little enthusiasm for amending the 1981 Act which had been agreed only after an epic battle in Parliament (Lowe *et al.*, 1986).

Consequently, it became attractive to press for a new incentive scheme for farmers. Initially, the Ministry of Agriculture claimed that this was not permissible under the prevailing rules of the CAP. Some of the force of this argument was removed when it became apparent that the policy used for implementing the Less Favoured Areas Directive in the Netherlands had been effectively providing incentives for farmers in wet grassland areas to continue with traditional systems, rather than to drain the land. This was a close parallel with what the Ministry of Agriculture was seeking to achieve in England. Nevertheless, it was concluded that a change in EC legislation was essential to permit the proposed incentive schemes in an explicit way.

The UK Government pursued this option with some vigour and was assisted in no small measure by environmental groups who mounted a European campaign. Many Member States and Commission officials were irritated by this initiative, partly because it had arisen late in the negotiations over a major new Regulation, 797/85, partly because it seemed to reflect predominantly British preoccupations and partly because of wider suspicions about the British agenda for the future of the CAP. Some Member States were already addressing the kind of problems Britain was encountering either (as with the Netherlands) through the creative interpretation of existing European rules or (as with Denmark) through domestic land-use planning controls over the conversion of pasture land. Other Member States saw the British initiative as an attempt to introduce a new subsidy for northern Member States at the very time when southern Member States, including the recent entrants – Spain, Portugal and Greece – were attempting to secure a larger portion of the Community market and the CAP budget, and when Britain was pressing for limits on farm expenditure. In the event, the new Article 19 merely permitted governments to introduce environmental incentive schemes without providing any aid from the Community budget.

In the light of these objections it is somewhat surprising that, 2 years later, in Regulation 1760/87, it was agreed that, up to a certain ceiling, ESA payment schemes could be eligible for a maximum of 25% reimbursement from the European Agricultural Guidance and Guarantee Fund (EAGGF =FEOGA). This development must be seen in the context of the mounting budgetary crisis of the CAP, caused through overproduction. It marked the initial acceptance that supporting farmers to conserve the countryside might also help, albeit in a modest way, to curb overproduction.

The 1987 change brought Article 19 into line with other elements of Regulation 797/85 concerned with extensification and set-aside, introduced at the initiative of the German Government. The purpose of these had been the reduction of market surpluses rather than environmental protection. Article 1a provided for premia to farmers who set aside at least 20% of their arable land and Article 1b provided premia for the extensification of production that resulted in a 20% reduction in output.

Regulation 1760/87 formally required the Agricultural Council to re-examine the extensification and ESA schemes within 3 years. The Commission took the opportunity to review the three measures (including the voluntary arable set-aside provision) and came up with proposals for 'reinforcing the relationship between agriculture and the environment' (European Commission, 1990) which led to a proposed regulation 'on the introduction and the maintenance of agricultural production methods compatible with the requirements of the protection of the environment and the maintenance of the countryside' (European Commission, 1990). This eventually became the core of the Agri-environment Regulation that accompanied the MacSharry reforms of the CAP (European Commission, 1992).

A number of considerations seem to have influenced the Agricultural Directorate of the Commission (Interview with Bertrand Delpeuch, 26 June 1995). The first of these was the need to give some substance to the formal commitments made in various policy documents to integrate environmental considerations into agricultural policy. The 1985 Green Paper *Perspectives for the CAP* (European Commission, 1985) had recognized that the role of agriculture is not only to produce food but also that it has an important contribution as the main economic activity required for the management of the countryside and conservation of the environment. With its 1988 statement on *The Future of Rural Society* (European Commission, 1988a) and on *Environment and Agriculture* (European Commission, 1988b) the Commission conceded the need to adapt agriculture to the requirements of protecting the environment and maintenance of the countryside. At the very least this suggested adjustments to the extensification and voluntary set-aside schemes to make them environmentally beneficial.

A second and related consideration was the need to respond to specific problems emerging from the implementation of EU environmental policy. The most pressing of these concerned agricultural pollution which came to public recognition in many parts of the European Community in the late 1980s, in part through the implementation of the Drinking Water Directive. Although this directive did not entail specific restrictions on agriculture, they were envisaged in the proposed Nitrates Directive which was intended to address one of the most serious problems thrown up by the Drinking Water Directive – nitrate pollution from agricultural sources. The Nitrates Directive was intended to promote a preventative approach by reducing leaching from farm land. The Agricultural Directorate was concerned over the consequences for farmers' livelihoods.

A third consideration was the very limited scope and impact of the three existing measures. In part this was seen to be due to inadequacies in their design and in the incentives available. More generally, the Commission felt that such schemes were oriented too narrowly, reflecting specific national concerns rather than Community-wide problems. The extensification scheme had only been taken up by four countries (Germany, Belgium, France and Italy); and the rules were considered difficult to apply and excessively complex. Article 19 also was implemented by only four countries (UK, Germany, the Netherlands and Denmark). It, in particular, was seen to suffer from a northern European bias and in its existing form was thought to be of little relevance to the rural problems of Southern Europe. The voluntary arable set-aside scheme was likewise seen to neglect French and southern European worries over desertification.

Ever since the instigation of Less Favoured Area (LFA) policy, the Agricultural Directorate had shown itself sensitive to this more socially and agrarian oriented definition of the rural environment problem. But the Commission is also driven by an integrationist logic and while it was evident that some Member States (notably the UK) were pushing rural environmental concerns as part of their general opposition to the CAP, the Commission was keen to see the development of agri-environment policy within a strong, Community-wide framework. It was important therefore in devising the Agri-Environment regulation to address specifically French and southern European concerns.

Some Member States had simply ignored the 1985 measures. The

Portuguese version of Regulation 797/85 did not even include Article 19 (Vieira, 1992). In general, the southern Member States and Ireland had shown little interest in the three measures (there had been a small uptake of the set-aside scheme in Italy). Essentially, in these countries the major concern was support for farming populations on social, not environmental, grounds and there was only limited public pressure for an environmental scheme. Southern European governments saw the priority as further intensification to close the gap with northern Europe, rather than environmental enhancement. However, they could not ignore the possibility of financial aid and between 1989 and 1991 Italy, France, Luxembourg, Ireland and Spain started small experimental schemes under Article 19 (Dixon, 1992). The one introduced by France was, in particular, a pointer to the widening agenda (Métais, 1992).

Initially the French Government had set its face against implementing Article 19 out of concern at the budgetary consequences of introducing not only a new type of farm subsidy but also one that was seen to be a means of supporting uncompetitive farms (Delpeuch, 1992). However, in 1989 it came forward with a proposal for four categories of experimental schemes which addressed certain national concerns. Two of these categories were accepted as eligible by the Commission under Article 19: aid for areas of low farming productivity, where farms are becoming abandoned; and support for the protection of scarce habitats. However, two categories were rejected as ineligible: aid for areas affected by water pollution by nitrates or pesticides from intensive farming; and management payments for areas in the Mediterranean zone, threatened by forest fires, where grazing of the scrub layer should be encouraged. In what was accepted and what was rejected, the French preoccupation with land abandonment was apparent, echoed also in national criticisms of set-aside. The other rejected category – water pollution from intensive farming – confirmed the arrival, on the agenda of French agricultural policy, of an important element of what we have termed the northern environmental agenda. This was stimulated to a considerable degree by the implementation of the Drinking Water Directive and the proposed Nitrates Directive.

A final consideration for the Agricultural Directorate was the market situation facing farmers. In devising an agri-environment regulation, that consideration could not be put to one side. A necessary condition of any initiative was that it should help to ease overproduction. On the one hand, this meant that the Commission wished to see Article 19 schemes of a kind which contributed not only to conservation but also to the reduction of surpluses; on the other hand, it impelled the Commission to look at a considerably expanded agri-environment programme that might make a significant contribution to the control of surpluses. Insofar as the new agri-environmental schemes would provide an additional source of income for farmers in an era of price restraint, this would complement the new compensatory payments being introduced within the MacSharry package.

The Commission's proposals, set out in Com (90) 366, contained a number of elements to address these considerations (European Commission, 1990). First of all it was proposed to turn the former extensification scheme into a pollution reduction measure and to make its introduction obligatory on Member States. This compulsory element reflected the Commission's attempt to impose its own Community-wide logic. Significantly, it also represented a convergence of separate national agri-environment agendas around a set of ecological and public health concerns to do with agricultural pollution.

Second, significant additions were proposed to what had been Article 19, covering the positive management of the countryside. This would remain an optional measure but a palette of schemes was proposed to make it more widely relevant. The Commission was critical of the existing requirement to demarcate environmentally sensitive areas. 'In consequence of this limitation,' it argued, 'the measures adopted by the Member States often aim, in the first place, at the protection of the flora and fauna or of specific biotopes in numerous small or minuscule areas demarcated with a view to nature protection and with a direct link to existing national legislation in the field'. While wisely not tampering with the existing option of limiting measures to sensitive areas, the Commission proposed additional options to take account of a wider range of landscape types and of potential problems, including schemes targeted on particular categories of land (e.g. grassland or field edges), and schemes of management or afforestation for areas threatened by natural hazards or fire, due to land abandonment.

Finally, the Commission proposed adjustments to the voluntary set-aside scheme to encourage more afforestation of set-aside land and to allow for long-term set-aside for ecological objectives. Com (90) 366 became swept up in the debate on the MacSharry proposals. Crucially it became, as Regulation 2079/92, an integral part of CAP reform. It was included by those who designed the reform as another possible compensatory measure for farmers for production and price cuts.

The Agri-environmental Regulation

The three 'accompanying measures' to the principal CAP reform measures were agreed in May 1992. None was entirely new; each was constructed from earlier measures. However, they were presented as a package which extended the range of the CAP into zones of policy which hitherto had been regarded as ancillary, rather than central, to the main agricultural support structures. The three measures were:

1. the agri-environment programme (Regulation 2078/92);
2. the early retirement scheme (Regulation 2079/92);

3. the forestry aid scheme (Regulation 2080/92).

(O.J. L215, 30 July 1992)

It is not surprising that one of the most contentious issues during the debate over these measures in the Agricultural Council was over the decision to provide the FEOGA funding through the Guarantee, rather than the Guidance Fund. Since the Guarantee Fund is the mechanism for supporting the CAP market measures, such as export refunds and intervention purchase, this signalled the incorporation of the accompanying measures within the core of the CAP. More significantly perhaps, the Guarantee Fund is not subject to the same budgetary restrictions as the Guidance Fund and so the three accompanying measures were not limited from the beginning by a fixed annual ceiling on Community expenditure. Whereas the European Commission had dispensed 10 million ECU co-financing agri-environmental payments in 1990, it was expected that the budget would reach 1.3 billion by 1997 under the new CAP (Delpeuch, 1992; European Commission, 1992).

The forestry aid scheme and early retirement scheme are both optional for Member States. They may choose whether to implement measures of the kind permitted by the Regulations. Where they do, they are able to reclaim 50% of the eligible cost from the Community budget or 75% of the cost in Objective 1 regions – those officially designated as lagging behind in economic terms. Most of the Mediterranean countries, Ireland, the Eastern Länder of Germany and the Highlands of Scotland currently are designated as Objective 1.

The same rates of FEOGA reimbursement apply to the agri-environment programme as to the other two Regulations. However, by contrast, it is obligatory on Member States to implement a national agri-environment programme and to include within it all the individual categories of measures listed in Article 2, unless there is a clear reason why these should not apply. Thus, Member States that had not taken any agri-environmental initiatives were to be obliged not only to do so for the first time, but also to develop a programme with a variety of different objectives. By making implementation of the Regulation obligatory, it was hoped to prevent a repetition of experience with Article 19, which was implemented only by a few Member States in Northern Europe. Furthermore, it was hoped that implementation of the Regulation might lead to a reduction in the intensity of agriculture over a significant area of land and help thereby to stabilize or reduce production and contribute to the wider goals of the MacSharry reforms.

The types of voluntary incentive scheme which Member States may introduce under the Regulation are set out in the Box 2.1. Measures (a), (b) and (c) are all concerned with reducing the intensity of agriculture. Taken as a group, they cover both the crop and livestock sectors and make explicit reference to the conversion of arable land into extensive grassland. Measure (a) is concerned directly with reducing inputs of fertilizers and pesticides and

Box 2.1. Aid schemes possible under regulation 2078/92.

1. Subject to positive effects on the environment and the countryside, the scheme may include aid for farmers who undertake:

(a) to reduce substantially their use of fertilizers and/or plant protection products, or to keep the reductions already made, or to introduce or continue with organic farming methods

(b) to change, by means other than those referred to in (a), to more extensive forms of crop, including forage, production, or to maintain extensive production methods introduced in the past, or to convert arable land into extensive grassland

(c) to reduce the proportion of sheep and cattle per forage area

(d) to use other farming practices compatible with the requirements of protection of the environment and natural resources, as well as maintenance of the countryside and the landscape, or to rear animals of local breeds in danger of extinction

(e) to ensure the upkeep of abandoned farmland or woodlands

(f) to set aside farmland for at least 20 years with a view to its use for purposes connected with the environment, in particular for the establishment of biotope reserves or natural parks or for the protection of hydrological systems

(g) to manage land for public access and leisure activities.

2. In addition, the scheme may include measures to improve the training of farmers with regard to farming or forestry practices compatible with the environment.

* There is also provision for aid under Article 4 of the Regulation for the cultivation and propagation of 'useful plants adapted to local conditions and threatened by genetic erosion'.

See Appendix for the full text.

could be seen as a means of reducing pollution arising from agriculture, as well as promoting extensification. There has been some debate about whether incentive payments should be made to farmers who reduce their inputs of fertilizers and pesticides in order to meet pollution control targets, since this can be construed as a breach to the Polluter Pays Principle. In this sense, measure (a) can be distinguished from measures (d) to (g) where farmers are being offered incentives to undertake activities which maintain or enhance the countryside.

None the less, measure (a) has been popular with several Member States. There is explicit reference to funding for farmers converting to, or continuing with, organic forms of production. Several Member States had already implemented organic conversion schemes before 1992 and several more have done so since the Regulation was put into place. Incentives for organic farming are available to all producers – both existing organic producers and new entrants – in Germany, the Netherlands, Denmark and Spain but in the UK payments are restricted to those converting to organic farming. In France the national scheme covers only new entrants, but certain regional programmes are offering payments to existing organic producers.

The second set of measures, (d) to (g), is concerned primarily with the landscape, nature conservation and public access to the countryside, as well as a rather vague reference to 'protection of the environment and natural resources' in measure (d). These measures are rather broadly framed and cover a potentially large range of different schemes, an impression confirmed by the plethora of proposals transmitted to the Commission since 1992.

Measure (d) is a development of the wording in Article 19, later Article 21. It also includes the option of providing farmers with incentives to rear breeds of livestock 'in danger of extinction'. The intention is to preserve a genetic inheritance in the European Community which is in danger of being eroded as certain breeds disappear. Although aid for such breeds already existed on a modest scale in Portugal and Germany, the initiative to include it in the Agri-environment Regulation arose from Commission staff who had prepared Com (90) 366 and who felt that rare local breeds could be a focus, especially in the southern Member States, for the public recognition of the regional distinctiveness of traditional farming systems that also conserve wildlife and the landscape.

Option (e), paying farmers for maintaining abandoned farmland or woodland, is a significant addition and permits aids of the kind put forward by the French Government under the old Article 19 scheme but rejected because they fell outside the scope of the scheme, which was confined exclusively to existing farmland. There has been continued sensitivity about this option because it allows stock to be reintroduced into areas from where they have been withdrawn and some officials in the Commission have been concerned that this may be an invitation to allow increases in production, rather than to limit it.

Option (g) was introduced on a proposal from the British Government. It reflects a concern to be able to pay farmers and other landowners for making land available for walking and other forms of public access in return for payment. Few other Member States have been interested in this option.

The education and training option, spelled out in Article 6, allows Member States to reclaim from the Community budget part of the cost of environmental training schemes. This might be thought an important desideratum if agri-environmental measures are to succeed in practice. However, eligible schemes have to be new ones related to the aims of the Regulation and it is not possible simply to claim reimbursement for existing schemes. In some Member States this has limited the applicability of the option considerably.

A further option not referred to in Article 2, but mentioned in Article 4, is the possibility of providing incentive payments for the cultivation and propagation of threatened varieties of agricultural plants, which includes fruit trees and varieties of vegetable. Both this and the training scheme lie outside the group of quasi-obligatory measures listed in Article 2.

The measures in Article 2 should be presented within 'multi-annual zonal programmes' with a duration of at least 5 years. The programmes are to 'reflect the diversity of environmental priorities'. This underlines the intention that Member States should generate schemes that are sensitive to local circumstances, rather than simply introducing standard national schemes. The reference to environmental priorities presumably refers to Community measures such as the nitrates, birds and habitats Directives. In practice, a number of Member States have designated areas affected by these directives for incentive schemes introduced in response to the Regulation.

The schemes submitted do not have to be new. However, they do have to respect a number of rules and there is a maximum payment per hectare which is eligible for Community reimbursement, e.g. there is a limit of 250 ECU/ha on schemes applying to pasture. Many Member States have put forward programmes which contain a mixture of both old and new measures. Arguably, some of the countries that have come to the matter for the first time, such as Ireland, Spain and Portugal, have made more ambitious but perhaps less precisely targeted use of the Regulation than countries, such as the UK and the Netherlands, with established agri-environmental programmes.

Following on what was initially a German idea, the Regulation suggests that Member States should put forward zonal programmes covering areas reasonably homogeneous in environmental terms. This suggests a series of small-scale programmes tailored to local conditions. There is no requirement that these programmes address specific local objectives or be administered by local or regional authorities. Furthermore, there is the option of establishing a 'general regulatory framework' instead of a series of independent zonal programmes. In practice, many Member States have developed national programmes or frameworks complemented by schemes focused on individual areas of high environmental priority. In others, regional authorities have taken the lead in putting forward programmes (Baldock and Mitchell, 1995). The tendency has been to fall back on administrative units rather than to tailor schemes to coherent geographical areas. The option of presenting a cluster of local schemes with no over-arching national element has not been popular.

The FEOGA budget for the three accompanying measures originally was set at 4.04 billion ECU over 5 years. Of this, 2.16 billion ECU was earmarked initially for the Agri-environment Regulation. However, as national schemes have been proposed it has been necessary to revise expenditure requirements upwards and a more recent estimate suggests that 3.16 billion ECU is a more likely total for the period up to 1997 (de Putter, 1995). In 1994 the Commission decided to set national allocations negotiated with each member state in order to keep this part of the FEOGA budget under control.

The timetable for implementation was extended further than originally intended. Member States were to have submitted their proposals by the end of July 1993 for approval by the Commission on the advice of the 'STAR'

Committee composed of Member State representatives. Some programmes were ready on time but many trickled in after the deadline, several had to be revised following negotiations with the Commission, and the new Member States also joined the process. In consequence, the approval process continued into 1995. By May, the Commission had approved at least 140 programmes for co-financing and more were to follow (de Putter, 1995).

Conclusions

The Agri-environment Regulation (2078/92) has its roots in earlier European measures. These include voluntary set-aside, experimental extensification and, most significantly, the initial provisions permitting Member States to establish Environmentally Sensitive Areas (ESAs). The Agri-environment Regulation, though, differs in crucial respects from these earlier measures. First, it contains a wider range of measures intended to address the rural environmental concerns of all Member States and to avoid what came to be seen as a northern European bias in the applicability of the earlier measures. Second, it is obligatory on all member states. Third, it is co-funded from the Guarantee rather than the Guidance, section of FEOGA, thus opening access to a much larger and more flexible budget and bringing the agri-environment measures nearer to the core of the CAP.

The Regulation is an attempt to encompass many of the strands that had come to prominence in the debate on agriculture and the environment in the late 1980s. Measures to 'extensify' agriculture remain central to the Regulation, although it is less specific about the degree of de-intensification that is required than the previous measure for this purpose. There is an expectation that the Regulation will contribute to a scaling back of production but there are no clearly defined mechanisms for ensuring that this occurs. Ironically, some of the more imaginative schemes proposed by Member States have involved the introduction of grazing animals, such as sheep, into habitats where they have previously disappeared. This can be seen as increasing, rather than reducing, livestock production and several schemes have encountered objections from the Commission on these grounds.

The structure of programmes proposed by the Member States has been highly variable and they do not appear to have felt constrained to give special weight to the concept of Zonal Programmes. They have submitted a range of General Regulatory Frameworks, National Programmes, Regional Programmes and Local Programmes, only some of which could be categorized as truly Zonal Programmes (de Putter, 1995). The scale of these proposals has ranged from relatively modest National Programmes, such as the one in the Netherlands in which a training scheme is the largest element, to the voluminous group of measures put forward by the Länder and Federal Government in Germany at a total cost of approximately 1.7 billion ECU (de

Putter, 1995). While the Commission has reviewed these proposals and caused many of them to be altered substantially, the picture that emerges is far from uniform. Member States have ample opportunity to exercise their freedom to design schemes to suit their own circumstances and conditions, and both the environmental rules imposed and the rates of payment for farmers exhibit considerable variation. The Regulation can be seen as one of the most important of several recent initiatives allowing more regional differences in the otherwise rather rigid CAP. Whether it will give rise to serious concern about distortion of the market and unfair competition between farmers receiving different levels of aid remains to be seen.

The principle that farmers can be paid both for reducing the intensity of production and for supplying environmental services is established unequivocally in the Regulation. However, the long-term implications of providing support for this purpose are not yet entirely clear. In many Member States, current institutions have little or no experience of designing and implementing agri-environmental policies and have had scant opportunity to debate either the principles or the practical mechanisms with their own farming and environmental constituencies. Many of the schemes are unavoidably experimental and the need to review their operation at an early date is already being recognized. Incentives are being deployed on a large scale to address environmental issues, which have been defined partly at a European level and partly by local administrations, particularly agriculture ministries. The standard procedure of offering farmers a form of management agreement for a fixed period may not be the most appropriate response to some of these issues and there are already questions about whether some schemes will generate any significant environmental benefit. Other questions are also emerging. What will happen after the management agreements expire? How far are national and regional administrations becoming committed to incentive payments as a permanent feature of rural policy?

It is safe to predict that tensions between environmental and income support objectives will be a recurring theme in the implementation of the Regulation. For the most part, agriculture ministries will be responsible for resolving this issue but there will be continual pressures to allow other players into the ring and to legitimize these schemes through broader participation in the policy community.

In terms of expenditure, the Agri-environment Regulation remains a relatively small element in the CAP. There are conflicts with the major commodity regimes, not least when farmers are eligible for larger payments for arable set-aside than for entering environmental schemes. The organizations that represent large-scale productivist agriculture continue to be a powerful force in directing the CAP and there is a widespread perception that the Regulation is an adornment to the CAP rather than a more far-reaching attempt to integrate environmental concerns into the heart of the policy. Environmental interests may still find it difficult to penetrate the fabric of

European agricultural decision making but they can no longer be marginalized entirely. The architecture of the CAP is changing, not only through the incorporation of environmental objectives but also because of the proliferation of regional and local schemes which it now embodies. The concentration of power within a tightly knit negotiating community at EU level remains pronounced but it is being eroded at the edges and, given their past record, environmental interests will not be slow to exploit new opportunities.

3

The Ecological Resources of European Farmland

Eric Bignal and David McCracken

Background and Introduction

The agri-environment programme (Regulation 2078/92) is wholly agriculturally based and is structured such that payments are given to farmers for farmland management. Its objectives are clearly stated in the formal title *Council Regulation on Agricultural Production Methods Compatible with the Requirements of the Protection of the Environment and the Maintenance of the Countryside*. Funded partly by the European Union (EU) and partly by the government of the individual Member States, the aim of all 2078 schemes is to encourage farming practices which are, in the broadest sense, of ecological or nature conservation value. The Regulation is directed at farmland with the objective of modifying (or in some cases sustaining) farming operations which would otherwise be driven entirely by the economic signals from the market and from other CAP support mechanisms. But it has another objective in that the 'extensification' of production, which is deemed to have environmental benefits, is also intended to reduce the overall production of agricultural commodities within the EU.

The ecological resources targeted by this Regulation are therefore by definition limited to currently productive farmland. While this might suggest that the task of defining the ecological resources is easier, this is not the case. Historically throughout Europe the classification of biotopes and the identification of areas of nature conservation value has tended not only to concentrate on the identification of the small proportion of 'best' areas but has also focused on the concept of 'naturalness'. As a result, virtually all listed sites of high nature conservation value, which have some form of designation or protection, do not generally occur on the managed component of farmland. Although in recent years it has gradually become more widely appreciated that mobile species (particularly birds but also mammals and insects) cannot be protected by these site-based measures alone, the various legislative measures aimed at protecting individual species do not in themselves influence land management outside of 'special' designated areas.

The upshot of this is that for any wider assessment of the ecological resources of farmland there is a fundamental problem: existing data and information collected at the national or European level avoids the wider countryside and has certainly not been focused on farmland. Few approaches to reconciling this problem are open to us, but the two most obvious are discussed below.

Collate and review information on the extent and the distribution of semi-natural vegetation and then to relate this to farmland

This has the immediate disadvantage that the data collected have not been focused on the matrix of farmland. Although this may be less of a problem in intensively managed farmland, where farming operations are mostly destructive and where there is little intrinsic nature conservation value on the managed land, in low-intensity farming systems, where wildlife of high nature conservation value survives in association with farming operations, it is a major problem.

This problem is confounded by the fact that there is in any case very little satisfactory data available on the extent and distribution of semi-natural vegetation, land cover or land-use types in the EU. There are major inconsistencies in the way that each Member State collects information. For example, the EUROSTAT categories of land use, which are reported on a regular basis for each Member State, do not even use standard definitions of permanent pasture and permanent meadow nor do they consistently differentiate between improved and unimproved grasslands across all the Member States. As such these statistics reveal little about the extent of semi-natural vegetation or the ecological potential of farmland throughout the EU.

The EU's CORINE (Co-ordination of Information on the Environment) programme, which started in 1985, contains three inventories of data: air quality, biotopes and land cover. The biotopes inventory, which is specifically geared to semi-natural vegetation, should in theory provide a systematic basis for implementing the EU Habitats and Species Directive (CEC, 1992). It could also be used to estimate the proportion of biotopes of European importance which are agricultural land. Table 3.1 gives an indication of those Directive 43/92 habitats occurring in the United Kingdom which are (or are potentially) affected by agriculture. It is clear from this list that much of the ecological resources of United Kingdom farmland (and the real target of Regulation 2078/92) is not regarded as a 'habitat' and that in any case most of these habitats need specific conservation management. Most will by

Table 3.1. Habitats on Annex 1 of the Habitats And Species Directive which occur in the United Kingdom and an indication of which are (or are potentially) affected by agriculture. (Source: CEC, 1992.)

Habitat	
Sandbanks which are lightly covered by sea water all the time	
Estuaries	X
Mudflats and sandflats not covered by seawater at low tide	
Lagoons	
Large shallow inlets and bays	
Reefs	
Annual vegetation of drift lines	
Perennial vegetation of stony banks	
Vegetated sea cliffs of the Atlantic and Baltic coasts	X
Salicornia and other annuals colonizing mud and sand	
Spartina swards (Spartinion)	
Atlantic salt meadows (Glauco-Puccinellietalia)	X
Continental salt meadows (Puccinellietalia distantis)	X
Mediterranean salt meadows (Juncetalia maritimi)	
Mediterranean and thermo-Atlantic halophilous scrubs (Arthocnemetalia fructicosae)	
Embryonic shifting dunes	
Shifting dunes along the shoreline with *Ammophila arenaria* (white dunes)	
Fixed dunes with herbaceous vegetation (grey dunes)	X
Decalcified fixed dunes with *Empetrum nigrum*	X
Eu-atlantic decalcified fixed dunes (Calluno-Ulicetea)	X-h
Dunes with *Hippophae rhamnoides*	
Dunes with *Salix arenaria*	X
Wooded dunes of the Atlantic coast	
Humid dune slacks	X
Machair	X
Dune juniper thickets (*Juniperus* spp.)	X
Open grassland with *Corynephorus* and *Agrostis* of continental dunes	X-h
Oligotrophic waters containing very few minerals of Atlantic sandy plains with amphibious vegetation: *Lobellia, Littorella* and *Isoetes*	
Oligotrophic waters in medio-European and perialpine areas with amphibious vegetation: *Littorella* or *Isoetes* or annual vegetation on exposed banks	
Hard oligo-mesotrophic waters with benthic vegetation of *chara* formations	
Natural eutrophic lakes with Magnopotamion or Hydrocharition-type vegetation	
Dystrophic lakes	
Mediterranean temporary pools	X-h
Floating vegetation of *Ranunculus* of plain and submountainous rivers	
Northern Atlantic wet heaths with *Erica tetralix*	X
Southern Atlantic wet heaths with *Erica ciliaris* and *Erica tetralix*	X
Dry heaths (all subtypes)	X
Dry coastal heaths with *Erica vagans* and *Ulex maritimus*	X
Alpine and subalpine heaths	X
Sub-Arctic willow scrub	
Stable *Buxus sempervirens* formations on calcareous rock slopes (Berberidion p.)	

Juniperus communis formations on calcareous heaths and grasslands	X
Calaminarian grasslands	
Siliceous alpine and boreal grasslands	X
Alpine calcareous grasslands	X
Semi-natural dry grasslands and scrubland facies on calcareous substrates (Festuco-Brometalia)	X
Species-rich *Nardus* grasslands in siliceous substrates in mountain areas and submountain areas in continental Europe	X
Molinia meadows on chalk and clay (Eu-Molinion)	X
Eutrophic tall herbs	X
Lowland hay meadows (*Alopecurus pratensis, Sanguisorba officinalis*)	X
Mountain hay meadows (British types with *Geranium sylvaticum*)	X
Active raised bogs	
Degraded raised bogs (still capable of natural regeneration)	
Blanket bog	X
Transition mires and quaking bogs	X
Depressions on peat substrates (Rhynchosporion)	X-h
Calcareous fens with *Cladium mariscus* and *Carex davaillianae*	
Petrifying springs with tufa formations (Cratoneurion)	X
Alkaline fens	X
Alpine pioneer formations of Caricion bicoloris-atrofuscae	
Siliceous scree	
Eutric scree	
Calcareous subtypes of chasmophytic vegetation on rocky slopes	
Siliceous subtypes of chasmophytic vegetation on rocky slopes	
Limestone pavements	X
Beech forests with *Ilex* and *Taxus* rich in epiphytes (Ilici-Fagion)	X
Asperulo-Fagetum beech forests	
Stellario-Carpinetum oak-hornbeam forests	
Tilio-Acerion ravine forests	
Old acidophilous oak woods with *Quercus robur* on sandy plains	X
Old oak woods with *Ilex* and *Blechnum* in the British Isles	X
Caledonian forest	
Bog woodland	
Residual alluvial forests (Alnion glutinoso-incanae)	
Taxa baccata woods	

X = currently managed/exploited by farmers at least at some site.
X-h = historically managed/exploited by farmers.

definition occur in Special Areas of Conservation (SAC) and will not be the target for 2078 schemes. The same is true for other Member States.

Focus on farming systems of intrinsically high ecological value

This is our preferred approach for describing the ecological resources of farmland. While this inevitably suffers from many of the same problems as basing the description on semi-natural vegetation (e.g. no pan-European classifications, poor or variable availability of data etc.), it does have the distinct advantage of describing the real target of Regulation 2078: the farmland. Because of the dearth of information available on the extent and distribution of farming systems of perceived high natural value, in 1993 the Joint Nature Conservation Committee (JNCC) and the World Wide Fund for Nature (WWF) commissioned a study of low-intensity farming systems in nine European countries (France, Greece, Hungary, Ireland, Italy, Poland, Portugal, Spain and the United Kingdom). The results of this work are presented in a summary report called *The Nature of Farming* (Beaufoy *et al.*, 1994) and a booklet/poster of the same name (Bignal *et al.*, 1994b).

In this study we concentrated on low-intensity farming systems partly because we believed these to be of highest nature conservation value, but also because a large volume of work has now been undertaken, or is currently in progress, by many organizations on the environmental aspects of intensive farming. In contrast, work on the nature conservation aspects of less intensive systems is much sparser, despite their greater importance for wildlife. It is ironic that many environmental initiatives on farmland tend to concentrate (often with little prospect of success) on reversing actions that have been destructive, yet tend to ignore practices that are currently benign and could be sustained. In this respect, most environmental initiatives tend to reward some farmers for their previously destructive activities but not others for their contribution to the maintenance of biodiversity. In addition, many (indeed most) low-intensity farming systems are changing rapidly through intensification or abandonment (Curtis and Bignal, 1991; Curtis *et al.*, 1991; Bignal and McCracken, 1992) and there is an urgent need to highlight their importance for nature conservation.

Methods and Rationale

In the study a number of characteristics were chosen as good indicators of low-intensity farming systems (Table 3.2). These are generally indicators of systems that have tended to adapt their management techniques to the natural environmental constraints of the region rather than adapting the environment (and in many cases the livestock breeds and crop varieties) to meet a standardized, often industrialized, production system. Although our interpretation is subjective, these management techniques could generally be described as 'practices which have been out of fashion for many years and techniques which are no longer part of modern agriculture' (Beaufoy *et al.*, 1994). Clearly these labour-intensive and more traditional management practices vary across Europe, not only because of regional differences but also because of the more rapid rate of agricultural intensification in some countries. The question is whether a classification, or typology, which is useful for the targeting of 2078 schemes or other environmentally-oriented policies, can be developed.

The relevance of low-intensity farming systems to Regulation 2078 is that to date survival of these more integrated and mixed management systems

Table 3.2. Characteristics of low-intensity livestock and crop-based farming systems. (Source: Beaufoy *et al.*, 1994.)

Low nutrient inputs, low output per hectare	
Livestock systems	Crop systems
Low nutrient input; predominantly organic	Low nutrient input; predominantly organic
Low stocking density	Low yield per hectare
Low agrochemical input	Low agrochemical input (usually no growth regulators)
Little investment in land drainage	Little investment in land drainage
Relatively high percentage of semi-natural vegetation	Crops and varieties suited to specific regional conditions
Relatively high species composition of sward	More traditional crop varieties
Low degree of mechanization	Low degree of mechanization
Often hardier, regional breeds of livestock	Use of fallow in the crop rotation
Survival of long-established management practices, e.g. hay-making, transhumance	More traditional harvesting methods
Reliance on natural suckling	Tree crops tall rather than dwarf
Limited use of concentrate feeds	Absence of irrigation

has generally been by default. In many cases severe environmental constraints have limited the degree to which farming practices could be intensified and mechanized. If we can characterize the systems that survive, and the areas where they still occur intact, there is then the opportunity to maintain them by design. It should be possible to begin to place an environmental value on systems of management that never had the conservation of biological resources as an objective but which we now recognize as the producers and managers of ecological conditions of high natural value.

The Extent and Location of Low-intensity Farming Systems

The approximate distribution of low-intensity farming systems across the nine countries studied is indicated in Fig. 3.1, and an estimate of the amount of such farmland in each country is provided in Table 3.3. Low-intensity farmland now mostly survives in upland and remote areas (especially in the context of distance and difficulty of transport to markets) where there are considerable physical constraints on the development and modernization (especially mechanization) of agriculture. Southern Europe has both the most types and the greatest area of land under low-intensity farming, with Spain, Portugal and Greece in particular all having over 60% of their utilized agricultural area under such systems. Although the areas shown on the map and the figures in the table are preliminary and indicative, it is estimated that

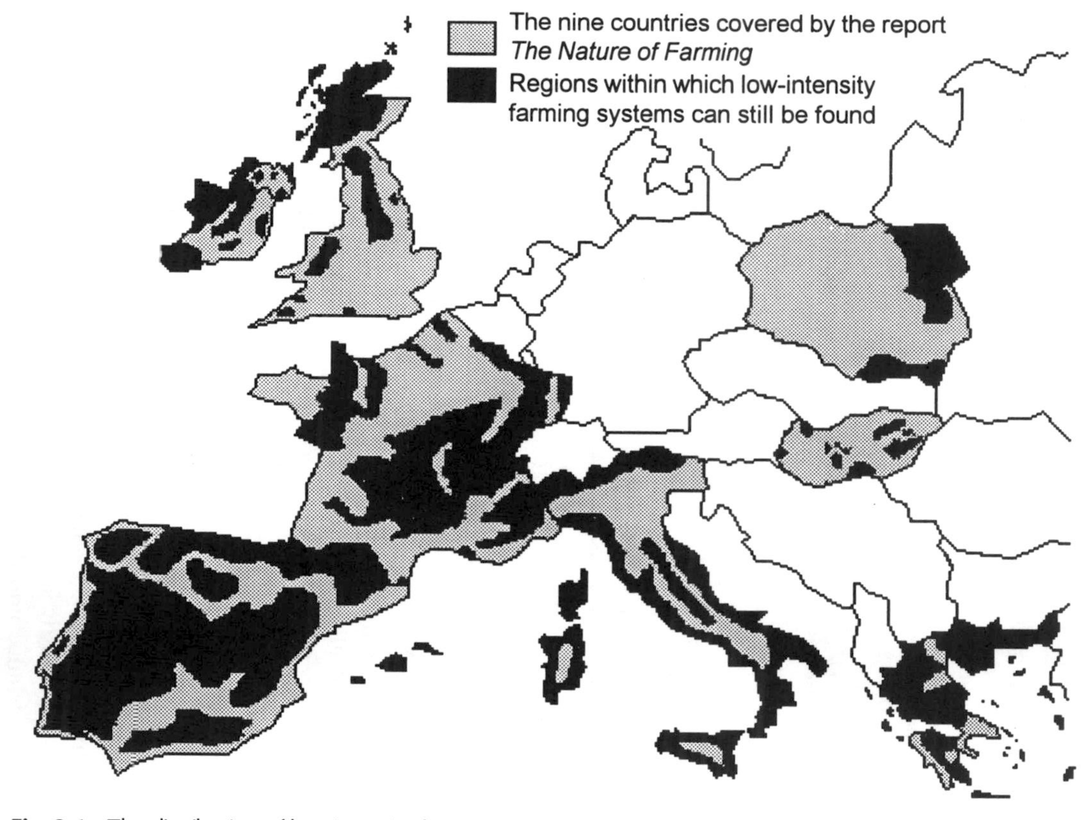

Fig. 3.1. The distribution of low-intensity farming systems.

Table 3.3. Estimated area of farmland under low-intensity farming systems in each of the nine countries considered by Beaufoy *et al.* (1994). (Source: Burrell *et al.*, 1990; Economist Intelligence Unit, 1994.)

	Land surface ('000 ha)	Land surface area under agriculture ('000 ha)	Agricultural area under low-intensity systems ('000 ha)
France	54,702.6	31,016.4	7,754
Greece	13,194.4	9,183.3	5,600
Hungary	9,303	6,493.5	1,500
Italy	30,122.5	22,650	7,100
Poland	31,267.7	19,135.8	2,735
Portugal	9,208.2	4,558.1	2,735
Republic of Ireland	7,028.3	5,650.8	2,000
Spain	50,478.2	30,589.8	25,000
UK	24,413.9	18,425.5	2,000
Total	229,718.8	147,703.2	56,424

across the nine study countries there are more than 55 million hectares of land under these systems, 30 million hectares of which are associated with livestock systems alone. The Iberian peninsula contains approximately half of the low-intensity farmland resource occurring across these nine countries.

Typology and Distribution of Low-intensity Farming Systems

It is possible to identify the broad similarities in agricultural practices that exist across the nine study countries. These are grouped according to livestock, arable, mixed and permanent crop systems and, although simplified, they form quite robust groups describing broad types of low-intensity farming found in western and central Europe (Table 3.4). Using this typology, links have been established between apparently very different systems by identifying common themes, for instance, in the livestock systems there are many common issues, such as foddering practices, livestock breeds and grazing and pasture management techniques.

Livestock Systems

Low-intensity livestock systems are now mainly practised in the upland and mountainous regions of Europe and in the arid zones in the south. Such systems are variable and range from semi-wild and largely unmanaged cattle and horses in remote regions of Spain, to dairy farms producing specialist cheeses and incorporating closely managed hay meadows in the French Jura. Low-intensity sheep systems are the most widespread livestock type covering

Table 3.4 Typology of low-intensity livestock, arable, permanent crop and mixed farming systems. (Source: Beaufoy *et al.*, 1994.)

Livestock systems	Arable and permanent crop systems	Mixed systems
Low-intensity livestock raising in upland and mountain areas	Low-intensity dryland arable cultivation in Mediterranean regions	Low-intensity mixed Mediterranean cropping
Low-intensity livestock raising in Mediterranean regions	Low-intensity arable cultivation in temperate regions	Low-intensity, small-scale, traditional mixed farming
Low-intensity livestock raising in wooded pastures	Low-intensity rice cultivation	
Low-intensity livestock raising in temperate lowland regions	Low-intensity tree crops	
	Low-intensity vineyards	

large areas of upland, mountain or dry pasture of grassland and scrub. *Transhumance* and other seasonal movements of livestock between grazing areas are also an important characteristic of these systems in many areas of southern Europe. Long distance movements between lowland (winter) and upland (summer) regions now usually only involve sheep and goats, however, short movements within a region can also still involve cattle and horses.

Central to all the livestock systems is a long continuity of sustainable use of large areas of grasslands, heaths and woodlands. In Spain, there are about 14 million hectares of such biotopes, including 3.5 million hectares of wood pastures, where livestock is still raised traditionally, while in Greece 5 million hectares are used each year for seasonal grazing by one million sheep and goats. Even in intensively managed Hungary there are still 1.5 million hectares of low-intensity farmland, mostly *puszta* (unimproved natural grassland) grazed by sheep cattle and working horses. Virtually all the remaining high nature conservation value grasslands within the nine study countries are associated with low-intensity livestock systems, and the greatest threat to these biotopes comes from agricultural intensification or abandonment.

Low-intensity livestock raising in upland and mountain areas

This typically involves sheep, beef cattle and/or horses grazing rough grassland, moorland, heaths and forests. Such land is often in communal or public ownership, and in some cases grazing is supplemented with hay grown in meadows or other forage crops. This system occurs over large areas of the uplands and mountains in Scotland and northern England, western and central Ireland, southern and eastern France, northern Spain, central and northern Portugal and northern Italy. Smaller areas are found in the mountains of western Hungary and south-east Poland.

Low-intensity livestock raising in Mediterranean regions

This predominantly consists of sheep and goats grazing dry grassland and rough grazing, including *maquis* and *garrigue* scrub. Grazing land is often rented or in communal ownership, and supplementary forage is provided by arable crop residues and fallow. This system is practised over large areas in southern France, Mediterranean Spain, Italy, Portugal and northern Greece.

Low-intensity livestock raising in wooded pastures

This system mainly involves sheep, pigs and cattle grazing on permanent pasture with dispersed tree cover. In some areas, supplementary forage is provided by shifting cultivation of forage cereals. The largest area occurs in the *dehesas* (Spain) and *montados* (Portugal) in the south-west of the Iberian Peninsula. Small areas also occur in north-east Hungary, south-east Poland and central Italy. In addition, in some upland and mountain regions (northern Spain, central Italy, southern France, south-west Hungary, Greece and Portugal), the livestock are allowed to graze and browse in woodlands and forest.

Low-intensity livestock raising in temperate lowland regions

This predominantly entails beef, sheep and/or dairy production on permanent meadows and/or pastures. In some areas the livestock also graze in salt or freshwater marshes. This system is now restricted to remnants of chalk grassland (southern England, northern France and central Hungary), and to isolated patches of grazing marsh (e.g. the Camargue in southern France).

Arable Systems

Once common across all of Europe, low-intensity arable systems are now mainly confined to the Mediterranean regions, where the dryland (non-irrigated) systems in Spain, Portugal, southern Italy and Greece are particularly significant. These systems are low yielding and use fallowing (in association with grazing) to maintain soil fertility and organic content. Traditionally, 30–80% of this land is left uncultivated or fallow each year, with some fallows lasting for 5 years or more. This management creates and maintains a *pseudo-steppe* landscape of great importance for nature conservation (e.g. Goriup *et al.*, 1991). In Spain there are still 13 million hectares of dryland arable, including 4 million hectares of dryland fallow.

Low-intensity dryland arable cultivation in Mediterranean regions

This typically comprises cereal production, often in combination with seasonal grazing by sheep of stubbles and fallows. Input levels (especially of herbicides) are low, and a large proportion of the land is usually left fallow each year. Large areas occur in Spain (especially in Castilla y León, Aragón, Extremadura, Andalucía and Almería), Portugal (in the interior regions of Alentejo and Tras-os-Montes) and southern Italy (in the Apennines). Small areas also remain in Greece, southern France and in parts of the Great Plain in central Hungary.

Low-intensity arable cultivation in temperate regions

This system mainly involves cereal production which is sometimes practised in combination with sheep, beef or dairy cattle grazing grassland and feeding on forage crops. Input levels of artificial fertilizers and pesticides are low. This system is nowhere widespread and mostly occurs in north-west Europe.

Low-intensity rice cultivation

This system consists of flood-irrigated rice production with very low inputs of pesticides. Remnants of traditional systems are found in Mondego valley in central Portugal, and some organic rice production occurs in Catalonia in north-eastern Spain.

Permanent Crop Systems

Permanent crops (such as olives, fruit and vines) are an important component of the Mediterranean lands. Much of this cultivation has been intensified in recent years and the surviving low-intensity systems are generally in the poorer areas where farming is less and inter-cropping (for example of olives, almonds, carobs and cereal with livestock grazing) is still practised. In Greece, over 0.6 million hectares (95%) of olive groves are still managed in a traditional way.

Low-intensity tree crops

This system is mostly associated with fruit orchards or olive groves where there is a low input of chemicals and where harvesting often occurs by hand. Many of these orchards and olive groves are also regularly grazed and/or used for hay production. Such orchards occur in parts of Spain (especially Asturias, Galicia and Cantabria), France (Normandy) and Hungary, with isolated remnants in southern England. Low-intensity olive groves are common in Greece (particularly in the Peloponese, Crete and the Aegean and Ionian islands), the interior regions of Portugal, the Calabria region of Italy, and in the vicinity of villages in Mediterranean Spain and France.

Low-intensity vineyards

This typically involves the long-established cultivation of old varieties of grape with low (or no) input of pesticides and fungicides. Such vineyards often occur in a mosaic with arable cultivation and tree crops, e.g. in Italy or as part of the *tanya* mixed system in Hungary.

Mixed Systems

There are still several areas of Europe where truly small scale mixed systems using far less than conventional inputs still survive. Some are virtually subsistence farming and most are in remote areas where farming is often combined with other occupations such as fishing, forestry or other paid work outside agriculture.

In some places the value of these systems as a component of *pluriactivity*, which could help to maintain viable rural communities, is being recognized by rural planners (Rennie, 1991).

Low-intensity mixed Mediterranean cropping

This incorporates a mosaic of low-intensity cereal and permanent crop (e.g. orchards, olive groves or vineyards) production. This system is practised in many areas of Spain, Portugal, Italy, Greece and southern France where crop production has not been highly rationalized nor intensified.

Low-intensity, small-scale, traditional mixed farming

This system usually consists of subsistence or part-time farming involving integrated crop and livestock production. There is normally little use of external inputs, but (as in many of the above systems) labour input may be high. There are many examples throughout Europe and most are highly localized and declining, such as *crofting* in northern Scotland, the *tanya* system in central Hungary, *coltura promiscua* in Italy, and *minifundia* in central and northern Portugal.

Examples of the Character and Nature Conservation Value of Low-intensity Farming Systems

The majority of low-intensity farming systems have been in existence for hundreds and in some cases thousands of years. Although these systems have developed and evolved during that time, changes have mostly occurred slowly over long periods. As a result, the annual farming cycle practised in these systems has created and maintained ecological conditions, often with great temporal stability, within which many plants and animals can find the resources that they require for some or all of the stages of their own annual cycle. Indeed for many of the landscapes, biotopes and wildlife communities upon which we now place high nature conservation value, the only practical, socially acceptable and sustainable management involves the continuation of low-intensity farming. A selection of examples of some of these systems from across Europe are given below.

Greek transhumance

The most common agricultural land use in Greece is low-intensity rearing of sheep and goats for meat and milk, with the livestock often herded in large mixed flocks. This system covers much of the mainland and is especially significant in maintaining the nature conservation value of mountainous areas. In total, around 5 million hectares are utilized as seasonal grazings, with the vast majority occurring above 600 m on the mainland. In summer, the largest concentration of migrant animals is on the high alpine pastures of the Pindos mountains, and the flocks descend to the surrounding foothills and lowland plains of Thessaly and Epiros in the autumn. Many of the mountain pastures have great botanical interest, and high grazing pressure by livestock for a relatively short period each year is essential to prevent scrub encroachment and maintain floristic diversity. In addition, grazing by livestock and the presence of carcasses maintains the open areas with carrion that are essential for foraging by scavengers (such as the Egyptian vulture (*Neophron percnopterus*) and the griffon vulture (*Gyps fulvus*)) and predators (such as the golden eagle (*Aquila chrysaetos*) and Bonelli's eagle (*Hieraaetus pennatus*)).

Portuguese montados

Cork oak (*Quercus suber*) *montados* currently occupy 700,000 ha in the higher rainfall areas close to the coast and on higher ground in the interior, while the more drought resistant holm oak (*Q. ilex rotundifolia*) *montados* cover 600,000 ha of the interior provinces of Alentejo, Algarve and Beira Baixa. The traditional farming system practised beneath the oak trees combines extensive cereal cropping and grassland management with livestock production. Inputs are low, outputs are low (between 1.5 and 2.5 tonnes of wheat per hectare) and stocking densities are low (2–4 hectares per cow and 1–4 sheep per hectare). The trees provide shelter for the livestock and crops, but are also essential productive components in their own right. Acorns and cut branches are used as extra fodder for the livestock and cork is harvested from the bark of the cork oak trees. Traditional *montado* management maintains a characteristic landscape containing elements of woodland, cropped and fallow land, species-rich grassland and scrub. This mixture of habitats contains a remarkable abundance and variety of wildlife, e.g. mammals such as the wolf (*Canis lupus*) and Iberian lynx (*Lynx pardina*) and especially a large number of bird species.

Montados are essential to the survival of threatened species such as the imperial eagle (*Aquila heliaca*), the black vulture (*Aegypius monachus*) and the black stork (*Ciconia nigra*), and provide important wintering areas for millions of northern European migrants such as the common crane (*Grus grus*), black-cap (*Sylvia atricapilla*) and chiffchaff (*Phylloscopus collybita*).

Spanish dryland arable production

The great plains of the Duero, Tajo and Ebro river basins and of La Mancha, La Campiña in Andalucia and a large part of the south-east (Murcia and Alicante) have been cultivated for centuries. Where the land is irrigated, these arable production systems have come under quite intensive management in recent years. However, the high proportion of non-irrigated arable land (4 million hectares) which is left fallow annually indicates that this dryland cultivation is predominantly low-intensity in character compared with northern and central Europe, where fallows have now all but disappeared. This fallow land is of considerable conservation interest for their flora (especially arable weeds) which have been eliminated from more intensive arable systems. Grazing of stubbles and fallows by sheep and goats is still an important feature of much arable cultivation, and these livestock play an important role in fertilizing the ground and dispersing the arable weed seeds. The combination of low-intensity cultivation, fallow land and patches of permanent pasture also creates a habitat of high value for many rare birds, such as the great bustard (*Otis tarda*), sandgrouse (*Pterocles alchata* and *Pterocles orientalis*) and stone curlew (*Burhinus oedicnemus*). Habitats of this sort in Spain harbour the most important populations in Europe of steppe bird and plant species.

Italian olive groves

Olive groves extend over 1,033,600 ha in the areas characterized by a Mediterranean climate (mainly Tuscany, Latium, Umbria, Liguria, Apulia, Sardinia and Sicily). The majority only receive small quantities of pesticides and fertilizers, and under such conditions the olive trees support a high diversity and density of insects. Many different varieties of trees are grown and the associated insects and fruit are exploited by many passerine birds (mainly thrushes, warblers, finches and starlings (*Sturnus vulgaris*)) during autumn and winter. In addition, olive groves provide important nesting areas for many birds such as the roller (*Coracias garrulus*), Scops owl (*Otus scops*) and hoopoe (*Upupa epops*).

Traditional management involves winter pruning, ploughing the ground during the summer (to control competitive vegetation) and manual harvest in the autumn. The olives are either picked direct from the tree by hand (Tuscany, Latium, Umbria) or by shaking and collecting the fruit in nets (Apulia, Calabria).

Hungarian mixed farming

Tanyas are small-scale, usually privately owned, mixed holdings which survive on predominantly sandy soils in the central parts of Bács-Kiskun county in southern Hungary. In total, around 200,000 ha are farmed in this way. Typically, each *tanya* consists of a house, outbuildings and between 2 and 25 ha of land. Rye, wheat and maize are the main products from the arable land. Intercropping is also common, typically involving maize, beans and marrows or oats, vetch, lucerne and barley. Organic manures are used and pesticides are seldom applied. Horses still play an important part in cultivation and transport of agricultural produce. On most holdings, two or three cattle and a small flock of eight to ten sheep are kept. The heavier livestock can erode the fragile sandy soils through trampling and so are often kept within defined areas and are not moved unnecessarily. Much of the grasslands consist of unimproved meadows from which hay is taken for winter fodder. The mosaic of different habitats and land uses contained within the *tanya* landscape contains a remarkably rich variety of insect, plant and vertebrate species. The survival of some globally endangered vertebrates (such as Ursini's viper (*Vipera ursinie*), great bustard, imperial eagle, slender-billed curlew (*Numenius tenuirostris*), lesser white-fronted goose (*Anser crythropus*) and mole rat (*Spalax leucodon*)) depends on unimproved grasslands and the continuation of low-intensity farming systems.

The Need for Better Descriptions of the Ecological Resources

Several observers have commented on the doubtful success to date of some 2078 schemes (Baldock and Beaufoy, 1992; Baldock, 1993; Dixon, 1995; Ganzert, 1995). There continues to be a fundamental problem in devising schemes that have real ecological value, despite a growing awareness and recognition of the European importance of farmland managed in a traditional, long-established way (Baldock *et al.*, 1993; Dixon *et al.*, 1993; Bignal *et al.*, 1994a; McCracken *et al.*, 1995).

We have suggested some reasons for this (Bignal *et al.*, 1995) but one which is fundamental must be the absence of a good scientific base of information that describes the functional and habitat value of regional farming systems. A key concept which we need to develop is that of the farmland biotope or farmland matrix. We need to focus attention on the ecological importance of managed farmland and the way this creates a biotope within which a variety of habitat conditions for plants and animals are created by seasonal management operations. In low-intensity farming systems we must think less about 'remnants of habitat left amongst farmland' and more of a farmland biotope for which the optimum management practices need to be developed.

Finally, if we regard man the farmer as part of farmland's ecological resources we cannot afford to ignore the human dimension (Bignal *et al.*, 1988). There is a need to understand why certain traditional forms of management have survived in some areas and to discover what incentives are needed to perpetuate them. In a series of case studies commissioned by the JNCC we have identified that there are factors that are common to many of the systems despite strong regional differences in farming operations on the ground (Institute for European Environmental Policy, 1994 unpublished results). Since these factors are now linked both with the historical reasons for the survival of these systems and the reasons for their current demise, these also need to be addressed by policy. Although there is little doubt that Regulation 2078 is an important step on the road to developing a more environmentally sensitive European agricultural policy, it must be said that widespread progress will only be made when the environment becomes a central element of the CAP. Having said this, it is clear from this brief review that throughout Europe there is also a need for clear conservation priorities for existing farmland (both intensive and low-intensity) and also for areas where, without continued support, agriculture will be abandoned altogether.

Acknowledgements

The study into the extent, distribution and characteristics of low-intensity farming systems in Europe was conducted by the Institute for European Environmental Policy and funded by the Joint Nature Conservation Committee and World Wide Fund for Nature International. Thanks are due to David Baldock, Guy Beaufoy, Julian Clark, Heather Corrie, Kate Partridge, Mike Pienkowski and the many others in all the different countries who contributed valuable assistance and encouragement during the study.

The meetings of the European Forum on Nature Conservation and Pastoralism have also enabled us to consider and discuss the future of low-intensity farming systems with many of our colleagues from throughout Europe. We are therefore grateful to all those attending the various meetings for their useful contributions towards the development of many of the views presented here.

The Response of the Member States

Denmark

Jørgen Primdahl

Introduction

When the proposed national legislation needed to implement Article 19 of Regulation 797/85 was discussed in the Danish Parliament in 1989, proposals for the afforestation and the voluntary set aside schemes were on the agenda as well. As it appears from the parliamentary record afforestation and set-aside were the controversial issues. No member of parliament presented critical views on the agri-environmental scheme.

From the beginning, Article 19 has been viewed as an essentially positive policy innovation by the various political parties, as well as by the farming communities and the different environmental groups. This enthusiasm has continued with the agri-environmental policies under Regulation 2078/92, although there have been disagreements on concrete policy design and distribution of duties.

In this chapter the agri-environmental policies in Denmark are presented and discussed. The policies are seen against the background of the present structure and developments of agriculture. Traditions and the status of environmental regulations are described to provide the policy context. The main part of the chapter pictures the policy design and implementation of Article 19 and the new scheme under Regulation 2078/92. The future of agri-environmental schemes and discussions on the experiences so far conclude this chapter.

Agriculture and Environmental Conservation

A rural country

Agriculture is a key factor influencing the Danish landscape, and all Danish landscapes are essentially agricultural fens.

No other European countries are farmed to so great an extent; approximately 60% of the total area is used for agricultural purposes of which 90% is arable land (Hansen, 1994). The prominent position of agriculture is reflected by its impact on the total national output. In 1992 the agricultural sector was exporting about DKr 50 million, equivalent to 20% of the total export value.

Historically Denmark can be considered a rural society. Industrialization came relatively late compared with other North European countries. Thus, it was not until 1960 that the export value from the industrial sector exceeded the agricultural export, with twentieth century literature and works of art bearing the imprint of rural life. There is not an explicit urban view or appreciation of the countryside, as for example is found in England. The concept of 'countryside' is non-existent in the Danish language, other words, such as 'the country' and 'the open country' are used to denote rural landscapes. Nor is there any centrally designated, protected national landscapes or national parks of the type found in most other European countries.

Agricultural developments

Danish agriculture is characterized by relatively large, farmer-owned holdings, predominantly intensive livestock production, and a highly developed cooperative tradition. This structure has evolved through three major transitions:

1. The comprehensive land reforms in the late eighteenth century are of the utmost importance in this context, as village communities were changed into individual holdings with outlying farms (Dombernowsky, 1988). The ownership structure then established remains a dominant feature of the present structure, one reason being that the rules of succession did not allow the parcelling out of farms into smaller units, e.g. for male heirs.
2. The second major transformation phase, in which Denmark changed from being a grain exporting agricultural nation into one based on husbandry, took place in the second half of the nineteenth century (Rasmussen *et al.*, 1988). During the agricultural crises that started when the European market was opened to grain imports from North America, agriculture was thrown into an enormous phase of adjustment. The process also triggered several political reforms, which in turn led to the creation of a large number of small holdings as well as technological innovations.

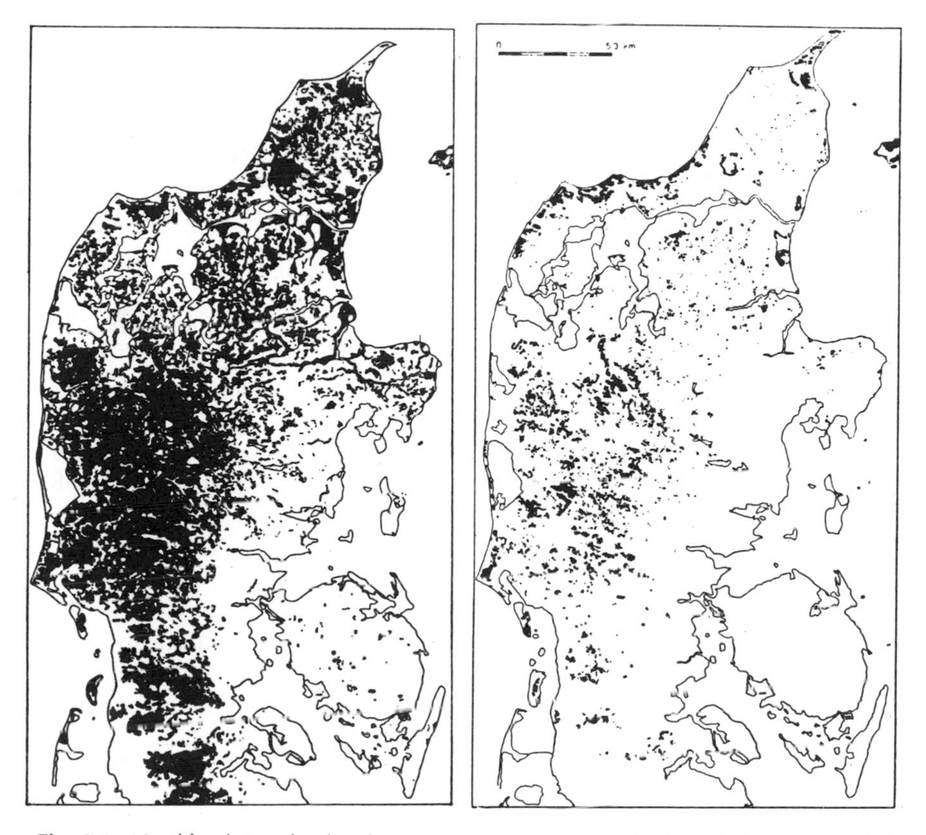

Fig. 4.1. Heathlands in Jutland and Funen in 1800 (left) and 1940 (right). (From Trap 1958, p. 68.)

The organizational structure was altered in several ways to meet changing needs and numerous co-operative societies for crop production, dairy production, and slaughterhouses were set up. The establishment of the Danish Folk High Schools was another driving force of the era. Their express aim was to offer general education to rural youth, and they thus came to play a significant role in the general progress taking place in rural districts.

The transformation is illustrated by:

- The reclamation of heathland, meadows, pastures, etc. resulted in an increase in farmland acreage from 63% (1861) to 74% (1912) of the total national acreage, while the share of farmland actually in use as arable land increased from 82% to 89% (Fig. 4.1). The agricultural development of the period was strongly influenced by the heathland reclamation visions, and the public debates on the topic have been documented by Olwig (1984).
- The total number of livestock increased from 850,000 (1837) to

2,460,000 (1914), approximately 50% being dairy cattle, and pigs showed an increase from 240,000 to 2,500,000.

- Crop output, measured in fodder value, doubled in the period 1860 to 1914, while animal production quadrupled. Around 1914 agricultural exports amounted to about 90% of the total national export value.

3. The third major transformation started with agricultural industrialization in this century. Unlike the changes mentioned under (1) and (2), the resulting changes were comparable to those taking place in other European countries. Agricultural industrialization is characterized by the following features:

- A general intensification of production, reflected in increased chemical input per hectare. In Denmark, the use of fertilizers and pesticides underwent a dramatic increase, especially after 1950, and with major environmental consequences.
- Concentration, through amalgamation of small holdings, and land purchase in general, increased the average acreage and output of the holdings. This applies especially to pig production, which is today concentrated on relatively few large units.
- Specialization: in 1967, mixed holdings with both dairy cows and pigs comprised more than 75% of all farms, and less than 10% of the holdings were purely arable farms. By 1982, 15 years later, about 25% were mixed holdings, a similar fraction had no animal husbandry, while some 20% raised pigs or dairy cows. Specialization has also had considerable regional impact, as dairy farms are now concentrated in West and North Jutland, while cash crop units are mainly found in the eastern parts of the country. This trend has had some major landscape impacts: today grazing livestock required for managing the semi-natural grasslands, are in short supply in the eastern parts of the country, especially for the grazed salt marshes along the coasts. Furthermore, the regional concentration of dairy cattle has created imbalances due to an over-supply of manure in some local areas. Recently imposed counter-measures include amendments to agricultural legislation requiring a balance of animal units and land on the individual farm. Large-scale biogas plants which are being constructed in livestock-intensive regions are also reducing the balance problem.

In conclusion Danish agriculture can be characterized as a livestock intensive, highly industrialized sector that has been subjected to rapid, major changes. Since late 1994, there has been a broad debate in the media about organic farming, and for the first time, the farming organizations decided to make organic farming a priority issue for the future development of Danish agriculture. The biggest dairy cooperative, MD Foods, even decided to add 15% to the state subsidy during the 2-year conversion period.

Environmental and landscape developments

Environmental issues, landscape structures and dynamics relating to agriculture are described in this section.

Glacial landforms

Glacial landforms are the dominant elements in Danish landscapes. Some areas were created during the most recent Ice Age (the Weichsel Period), when the entire country except West Jutland was covered by the ice cap, others developed during the earlier Saale Ice Age (Schou, 1949). Ground or terminal moraines with fertile soils are the prevalent landforms of the islands and East Jutland.

The western parts of Jutland are composed of mainly outwash plains and old moraines, which makes sandy soils the dominant type in this part of the country. Jutland and the major islands are intersected by tunnel and glacial water valleys. Raised seabed stretches are found in the northern part of the country, while the south eastern isles form a unique 'drowned' moraine landscape with extensive, shallow and fertile coastal areas. Rather large diked farmlands are found in this region. South-west Jutland forms the utmost part of a continuous marsh area stretching north from Germany and The Netherlands.

Climate

The climate is Atlantic, with rather cool summers and mild winters. Average precipitation is 640 mm, and apart from a minor water deficit during the growing season (April–October), precipitation and general climate offer favourable conditions for farming. In western Jutland, most of the farmland is irrigated, due to the sandy soils with a low root zone capacity. Practically all water for irrigation is extracted from the ground water, with permits for using lake and river resources granted in exceptional cases.

Ground water resources

Due to the precipitation surplus and the mostly morainic composition of the geological material, Danish ground water reserves are plentiful and relatively well protected. Thus, ground water is the primary source for drinking water all over the country, surface water is used to a limited extent in the Copenhagen Metropolitan area. However, in recent years, problems associated with nitrate and pesticide contaminated ground water resources are becoming more frequent. Pesticide contamination has so far been limited to 'point sources' for instance leaching from buried barrels to the ground water, but the problem of nitrate leaching is on the increase. At present, the nitrate load is a problem mainly associated with the farmlands of West and North Jutland. In these parts, 50,000 ha have been designated as ground water reserves under protection, and starting in the spring of 1995, farmers are being offered management agreements, with the aim of having arable land converted into extensively farmed grasslands, afforestation or nature areas. A similar increase in agriculturally related ground water problems is anticipated for the remainder of the country, so major initiatives in this field are foreseeable.

Surface waters

All Danish watercourses, lakes and coastal waters are affected by agriculture. Regarding watercourses, the most frequent types of intervention are physical measures such as straightening and piping to facilitate drainage. In addition, the land is often tilled right down to the riverbanks, and the combined impacts lead to a deterioration of the ecological quality of these watercourses. Agricultural practices also affect lakes and coastal waters, when the nitrate load combined with phosphate loads, mainly from urban areas, gives rise to eutrophication processes. An action plan for the aquatic environment has been implemented since 1987, its main objective being to harness the pollution of lakes and coastal waters resulting from nutrient-input from towns and farmlands.

Landscape dynamics

Agriculture, combined with urban development, is the driving force behind Danish landscape dynamics. Since the turn of the century, agricultural development has been highly influenced by industrialization, which has led to major landscape changes. Heathland and wetlands have been reclaimed, and hedgerows have been planted on the sandy soils of West Denmark: for several years at a rate of approximately 1000 km a year, and at present the annual rate is on the increase.

The small uncultivated elements found in the cultural landscape, in Denmark called small biotopes, have been comprehensively monitored, initially on the islands and later, since 1986, in eastern Jutland and on Bornholm (Agger and Brandt, 1988; Brandt *et al.*, 1994). Developments in small biotopes are fundamental factors for visual and ecological functioning of the landscape. In addition they have proved valuable as indicators for agricultural land use change.

From Table 4.1 it appears that the decrease in most types of biotopes in the period of 1981–1986 has been followed by an increase in dry linear types (road verges, hedgerows, stone walls etc.) and dry areas (barrows, game plantings, thickets, etc.), while in Jutland all types are increasing. The research will be updated in 1996, and will then cover the entire country, including western Jutland. Expectations are that the contractive agricultural aspects have now penetrated completely, meaning increases for most or all types of small biotopes throughout the country.

An interview survey of a representative sample of Danish farmers conducted recently revealed an increase in permanent grasslands, including

Table 4.1. Development of small biotopes in Denmark in 1981–91. (Source: Brandt *et al.*, 1994.)

	1981–86* percentage change per year	1986–91* percentage change per year
13 sample areas in Denmark and the islands (4 km^2 each)		
Wet linear biotopes (ditch, canal, brook etc.)	–0.1	–1.1
Dry linear biotopes (road, verge, hedgerow, stone wall etc.)	–0.1	+0.2
Wet patch biotopes (pit, pond, bog, swamp etc.)	–1.8	–0.8
Dry patch biotopes (barrow, thicket, game planting)	+0.9	+2.0
10 sample areas in eastern Jutland (4 km^2 each)		
Wet linear		+3.2
Dry linear		0.0
Wet patch		+2.4
Dry patch		+4.7

*Indicated as a percentage annual change of the average for all sample areas; the linear biotopes in percentage of length; the patch biotopes in percentage of number.

grazed heathland, salt marshes, meadows, and pastures. Extended to the whole country it was found that 16,000 ha was brought into rotation during the period of 1982–1992 (most of this was permanent grassland) whereas 29,000 ha was changed from arable to forest and grasslands (Andersen, undated). However, it remains to be analysed to what extent the new grasslands are actually being grazed, or have just been abandoned.

Accompanying the growing number of biotopes have been increases in landscape heterogeneity in most regions – a development reinforced by increased set-aside associated with the CAP reforms. In the growing seasons 1992–1993 the acreage covered by set-aside contracts amounted to 205,000 ha, that is 7.5% of the total agricultural area (Landbrugsministeriet, 1994).

Policy Background and Administrative Context

Danish legislation directed towards agriculture, environment and rural landscapes has shown a marked increase in the last 10 years. This is true of general regulations addressing environmental and landscape protection issues, and even more so regarding pro-active plans and policies with associated subsidy schemes. EU co-funded agri-environmental policies have had increasing importance in this development.

Generally speaking, the overall regulation of the open countryside could be seen as an aggregate of discrete, uncoordinated schemes, involving several authorities and levels. The following section will give a summary of (i) national strategies including action plans, (ii) current programmes including agri-environmental measures, and (iii) the physical planning system including different designations. A presentation of regulations and planning for the countryside is found in Wulff (1991) and Primdahl (1991).

National strategies

In the late 1960s, following several years of deliberation and public debate, it was decided that Danish nature conservation was to be based on general regulation, including planning regulation and the so-called conservation orders (the latter being to some extent comparable with the Sites of Special Scientific Interest found in the UK).

Proposals for establishing a national park system were thus dismissed, based on the argument that the entire country should be viewed as a National Park. Conservation orders hold provisions for the protection of specifically designated areas against undesired changes not subject to regulation and are invariably followed by once-and-for-all compensations. In addition, they normally include landscape management directions, entitling county councils to manage specific areas regardless of the owner's wishes. Most conservation orders also provide for wider recreational access for the public.

By 1993 about 5% of the total surface area (190,000 ha), distributed in sites of varying sizes, ranging from a few to several thousand hectares, had been placed under conservation orders. Only a few conservation orders had been initiated by the state, the majority being brought by The Nature Conservation Association, a private body with a special entitlement to bring conservation orders – a right they share with county councils and municipal councils. Most areas under conservation orders are private property, which supplement the forests and natural areas owned and managed by The National Forest and Nature Agency (about 120,000 ha).

Since the late 1980s a succession of national strategies and action plans directed towards agricultural landscapes have been adopted. The major ones are cited in the following list, based on a comprehensive overview (Ministry of the Environment, 1994).

Action plan on the aquatic environment

After several years of inventories and public debates, the Action Plan on the aquatic environment was presented by the Government in 1987. The aim was to reduce the nitrogen and phosphorus loads to the aquatic environment by reducing discharges of nitrogen by 50% and of phosphorus by 80%. Increased storage capacity in agriculture, aiming for better utilization of manure and large investment in municipal treatment plants, have been the most significant results of the plan. Investments in storage capacity at farm level have been more than DKr 2 billion during the period 1985–1989.

Action plan for a sustainable development in agriculture
This plan, submitted to Parliament in April 1991 targeted strategies for the development of agriculture. The plan regulation aims at reducing the negative impacts from fertilizer, manure and pesticide application, the targets are for a 50% reduction in nitrogen by the end of the century and a 50% reduction in the use of pesticides before 1997, compared with the average consumption in the 5-year period 1981–1985.

Pesticide reduction plan
This plan was published in 1986 by the Minister of Environment and aims at reducing pesticide usage in agriculture. Protection of users, food, drinking water, and of flora and fauna are the main purposes of the plan which should reduce the use of pesticides by 25% before 1990 and by a further 25% before 1997. In addition to new rules on re-evaluation and approval of pesticide products the following means should be exercised to reach the goals: increasing the advisory service; training the users; improving information; and increasing research within the pesticide sector. So far the goal of a 25% reduction has not been reached and the introduction of new regulations is being considered – including the introduction of pesticide taxes.

Strategies for sustainable forestry and natural forest
With reference to the Forest Declaration of the Rio Conference a number of measures concerning sustainable forest and protection of natural forest and traditional management systems have been undertaken. Government plans from 1989 aim to double the woodland from 12 to 24% of the total area of Denmark within a forest rotation (80–100 years). In addition, measures to be taken until the turn of the century include the identification and protection of a minimum of 5000 ha of 'untouched' forest, the protection of at least 4000 ha of traditional management systems, the preparation of a strategy for genetic resources, the implementation of specific programmes for conservation and biodiversity within forest ecosystems; and the provision of information to private forest owners.

Strategy for nature management
The former marginal land strategy implemented between 1987 and 1993 has been replaced by the strategy of nature management, according to which 10% of the wetlands (20,000 ha) should be subject to nature restoration projects. New conservation orders for areas requiring special protection are expected to increase the protected areas by 20,000–30,000 ha in the coming 5–10 years. The goal of doubling the afforested area mentioned above is also part of this strategy.

A common feature of these strategies and action plans (apart from being relatively new) is that they are rather explicit in their objectives, they are followed by monitoring systems, and they continue to spur legal amendments and new financial programmes, unfortunately they are also quite uncoordinated.

Current programmes and agri-environmental measures

Less than a decade ago few measures addressing countryside issues existed. Among the current schemes cited in Table 4.2, only three existed 10 years ago, namely the afforestation scheme, the shelterbelts scheme and the conservation order compensation scheme. The present picture is somewhat different. Annual budgets total approximately DKr 250 million and the amount is increasing. In addition expenditure under the CAP reform for set-aside amounted to DKr 422 million in 1993 (Danmarks Statistik, 1994). The CAP set-aside is developing, so in the future the measure will be associated with more direct objectives relative to environment and landscape.

Though several of the schemes cited in Table 4.2 address afforestation and habitat restoration, the vast majority are still related to the agricultural landscape. Two of these are EU co-financed agri-environmental policies which will be described in detail later.

Planning and legislation

In Table 4.2, the important schemes oriented towards landscape protection and/or enhancement are shown. Schemes without direct bearing on land-

Table 4.2. Current programmes concerning agricultural land use and agri-environmental policies.

	Budget 1994 in million DKr
The National Forest and Nature Agency (Ministry of the Environment)	
1. Afforestation aid for private land	10.4
2. Forest improvement aid	17.0
3. Deciduous forest improvement and forest management	
4. Shelterbelts and hedgerow scheme	23.4
5. Game planning scheme	2.8
6. Nature management scheme, afforestation of state owned land, major nature restoration projects, and outdoor recreation initiatives	131.4
7. New conservation orders	13.5
Environmental Protection Agency (Ministry of Environment)	
8. Water course restoration scheme	3.0
9. Reduction of ochre problems in watercourses	4.3
Danish Directorate for Development in Agriculture and Fisheries (Ministry of Agriculture and Fisheries)	
10. ESA measures	20.0
11. New ESA measures	21.0

scape issues, such as set-aside under the CAP reforms and some environmental schemes are not mentioned even though they may have substantial indirect influence on landscape dynamics.

The schemes shown are administered under three different bodies, two of which are part of the Ministry of the Environment, while the third belongs to the Ministry of Agriculture and Fisheries. Except for three schemes (4, 5 and 6) they were all introduced since the late 1980s. The pro-active part of landscape and land-use oriented regulation has grown dramatically, which means that the needs for coordination and continuity of the schemes, that is for planning, have grown proportionally. However, the schemes are only coordinated to a limited degree and several of them have been changed or replaced by others within recent years.

Table 4.3 shows the overall Danish planning system referring to agricultural areas. Several sectors are involved in planning activities affecting the rural areas, mainly at the regional level. Regional plans prepared by the county administrations are the coordination instrument for inputs from the various sectors, and regional planning provisions must be followed by other sectors and local authorities.

Agri-environmental Policies

The first Danish agreements under Regulation 797/85 (1760/87) were signed in 1990. However, at that time, public grants for land use and landscape improvements were already a long-established tradition. Viewed by today's standards, several of those projects would not be considered improvements but, in the past, they earned credit for their originators, and monuments to commemorate the achievements of the individuals and organizations involved were raised. This was also true of many heath and wetland reclamation projects mentioned earlier.

Public grants for hedgerow plantings are the epitome of a long-standing public policy which is still under development, with considerable consequences for Danish countryside. In recent years, and with grants under the Shelterbelts Act, three-row hedgerows have been established at a rate of about 800 km/year under the collective agreement, whereas hedgerows covered by individual agreements total about 250 km/year. Since 1993, about half the planted hedgerows have had six to seven rows and grant aid for hedgerow plantings in Denmark has been organized as part of Regulation 2080/92. Management agreements between farmers and public authorities are another long-standing practice. These include agreements between farmers and the National Forest Service for grazing and other activities, as well as similar agreements between farmers and their respective counties regarding private land subject to nature conservation orders.

Thus, there is a well established tradition for cooperation regarding land

Table 4.3. The overall Danish planning system referring to agricultural areas. (Zoning and designations are marked with *.)

Sector/level	Water catchment	Surface waters	Agriculture	Conservation	Comprehensive physical planning
Central (State)	Guidance/approval Test cases	Guidance/approval Test cases	Soil classification Strategy for sustainable agriculture Agri-environmental policies CAP-reform	Guidance/approval *EU-bird and habitat protection *Conservation plan for the sea territory *Strategy for sustainable forestry including designation of nature forests Nature management projects	Guidance/approval *National planning directives including coastal protection zones
Regional (County)	Water catchment pl. Designation of: *Groundwater protection zones	Water quality pl. Water course restoration and management	Agricultural planning Designation of: *Afforestation zones and minus zones *ESAs and buffer zones *Areas for environmental set-aside	Nature conservation pl. *Identification of protected habitats Nature conservation orders	Regional planning including different zones for: *Rural areas *Summerhouse areas *Urban development
Local (Municipal)	Water supply pl. Administration	Waste water pl. Administration			Municipal planning including different zones for: *Rural areas *Urban development *Urban renewal Development planning
Legislation	Water Supply Act	Environmental Protection Act Watercourses Act Ochre Act	Agricultural Holdings Act Land Consolidation Act Agricultural Structure Act	Nature Protection Act Forestry Act	Planning Act

use and landscape issues – presumably one reason why the ESAs (Sensitive Agricultural Areas, henceforward referred to as ESAs even though they are no longer termed so in Denmark) measures were undertaken expeditiously and with considerable involvement.

The Danish ESA measures under 797/85

Danish legislation on grants for agriculture in Environmentally Sensitive Areas was put on the agenda of the Danish parliament in spring 1989, together with a motion for grant aid for set-aside and private afforestation. The entire debate focused on the two last measures, and the members of the Danish parliament expressed reservations regarding the introduction of efficient set-aside and afforestation schemes. By contrast, grant aid for management agreements in ESAs did not give rise to major debate, and the political speeches all seconded the scheme without reservations.

The scheme was to be based on a 1987 report – the Committee on Implementation of EC Socio-structural Measures in Denmark. According to the report, the scheme would be applicable only to areas designated by the counties according to a set of national guidelines, and the counties would co-finance part of the scheme (Struktur-og Planudvalget, 1987). The bill was passed in 1989, and largely followed the recommendations of the committee. The considerable delay, between the presentation of the Agricultural Committee report and the passing of the required legislation was due to opposition, not to the ESA measure, but to reservations regarding the two other schemes mentioned earlier.

The ESA measure was designed to make the county authorities the primary public authority. They were to designate ESAs in addition to entering individual agreements. The Directorate of Agriculture was charged with approving the designated ESAs and was also the responsible/liable Danish authority for the EC. The following sections on the ESA measure in Denmark are based mainly on a research project concerning the implementation and effects of the Danish measure, carried out by researchers at the Royal Veterinary and Agricultural University (Hansen and Primdahl, 1991; Primdahl, 1992; Primdahl and Hansen, 1993). The project material consisted of detailed analyses of the ESAs together with the agreements signed in 1990, 3042 agreements in total, a detailed study on 126 agreements in five counties, including interviews with involved farmers and registration of land use and landscape appearance in relevant areas. Consequently 970 agreements signed in 1991 were not included in the study. The data in Tables 4.4–4.6 do not cover all ESA-agreements signed under Regulations 797/85 and 1760/87.

Designation of Environmentally Sensitive Areas

The designation of ESAs should be viewed in the light of the fact that Denmark has no national parks or other prioritized national landscapes. The counties started designating ESAs in 1989, based on the guidelines of the Agricultural Ministry. The guidelines left comparatively wide scope for the specific strategies of each county, and the counties made optimal use of their discretion. Nation-wide, the counties designated a total of 915 areas, and Table 4.4 shows that counties relied on highly different strategies for their area designation. There was uncertainty concerning the designation of areas protected with conservation orders. By including such areas, the scheme could also be applied to the protection of grazing and other management practices on land where management programmes were already in operation. Some counties targeted their designations on such 'conservation areas', others left them out to designate other valuable areas instead. About one-third of the total ESA area is partly or totally covered by conservation orders.

Table 4.4, indicates that counties selected different strategies, notably to do with the size of ESAs. South Jutland and Frederiksborg (to the north of

Table 4.4. Number of ESAs and agreements.

					Number of ESAs with various degrees of uptake (%) the first column showing number of ESAs that have no agreements		
County	Number of ESAs	Mean ESA area ha*	Number of agreements	Mean agreement (ha)	0%	0–50%	50–100%
Copenhagen	4	116	8	11.8	2	2	0
Frederiksborg	64	32	85	8.3	31	16	17
Roskilde	36	55	31	8.5	24	9	3
Vertsjaelland	70	119	123	8.6	38	26	6
Storstrøm	31	306	140	8.8	8	18	5
Bornholm	11	133	56	5.8	1	7	3
Fyn	170	74	155	9.0	99	56	15
Sønderjylland[†]	46	–	412	6.4	20	23	3
Ribe	50	171	229	8.8	21	19	10
Vejle	13	625	326	5.6	0	13	0
Ringkøbing	32	414	382	9.2	5	23	4
Aarhus	38	344	404	6.6	3	30	5
Viborg	21	755	276	9.2	7	8	6
Nordjylland	38	557	415	9.5	7	27	4
All	624	–	3042	7.9	266	277	87

*The fact that some ESAs consist of several sub-areas it is not included in the mean.
[†]For practical and statistical reasons the 339 designated ESAs are grouped and consequently reduced to 46.

Copenhagen) targeted their designation towards numerous, minor areas, while other counties selected few, relatively extensive areas. It is also seen from Table 4.4 that no agreements (or rather few) were made for some ESAs – which obviously weakens the effects of the scheme. In virtually all counties, the local branches of the Danish Nature Conservation Association were active in the designation process, as well as the local Farmers' Unions. One county actually ran an advertising campaign with the purpose of ascertaining farmers' immediate interests before making the first designation proposal. That is, the county authorities knew the areas for which agreements were desirable and they also knew about the farmers already interested. For the new designations, many counties chose to adopt the principle of involving farmers (via their organizations and newspaper advertisements) before the actual designation processes.

Whether 'conservation areas' or not, counties focused their area designations on landscapes with a comparatively high priority, i.e. extensively farmed areas and areas with a major component of semi-natural areas. Many designations also included considerations related to protection of the environment.

This is seen in Table 4.5, showing the objectives of the individual designations, for instance, the protection of surface waters (watercourses, lakes, and coastal areas) is to be included for several areas, whereas the protection of ground water and protection against ochre pollution are explicit objectives for rather few areas. However, agreement uptake for ground water areas is comparatively low. The reason is that these areas were predominantly well-drained for which payments were too low, while a major part of the remaining areas were wetlands with relatively inferior farming conditions.

Table 4.5. Agreements in relation to ESA objectives. Each ESA may include several objectives.

ESA objective	(1) Number of ESAs	(2) ESA area (ha)	(3) Number of agreements	(4) Agreement area (ha)	Column (4) as percentage of (2)
Protection of habitats	574	112,984	2,949	23,419	21
Conservation of landscape quality	254	39,731	904	6,018	15
Protection of surface waters	286	57,268	1,342	9,776	17
Protection of groundwater	45	20,357	210	1,828	9
Protection against ochre pollution	22	16,787	409	3,042	18

For practical and statistical reasons, the 915 designated ESAs have been grouped and reduced to 624 for the purposes of this study.

Table 4.5 also shows the classic nature conservation issues (i.e. habitat and landscape protection) to be the most frequent objectives. Generally these objectives were in conformity with the published guidelines.

The agreements

By the spring of 1990 all designations were settled and approved by the Ministry of Agriculture. Signing of agreements commenced in April, and after 8 months, by the end of the year just over than 3000 agreements had been signed. This was an achievement made possible by two factors, (i) agreements were made on a de-centralized level, and (ii) most agreements had only a limited impact on present management practices making the conditions relatively attractive to farmers. All the agreements run for 5 years and payments vary from DKr 400–900/ha/year. Five counties have used the maximum rate for all agreements, whereas with others, payments vary according to the degree of restrictions. The average payment for all agreements signed in 1990 is DKr 790/ha/year.

Table 4.6 shows the agreements grouped according to 'main purpose' – that is the substance of each agreement relative to present practice – clearly preservation of existent permanent grasslands is the most important outcome of the agreements. As many as 86% of agreements comprise areas for which a continued function as permanent grasslands is found desirable, by means of grazing and/or mowing, involving different prescriptions regarding management of the areas. Of the areas covered by the agreements 8% are converted from arable land into grasslands, and another 8% remain in rotation, typically contingent on (i) green winter fields, (ii) spring harrowing, succeeded by a spring crop. The major restrictions, apart from the prescription that permanent grasslands should be grazed/mown, are also shown in Table 4.6.

Thus, most agreements include restrictions on grazing (maximum and/or minimum number of livestock units per hectare, time of turn-out etc.), pesticides (most often a ban), soil preparation (usually a ban on re-sowing or other types of soil preparation), and fertilizers. It is noteworthy that practically all agreements feature a limitation or a ban on the use of fertilizers and manure.

Effects

Preservation of semi-natural grasslands is the main effect of the Danish ESA measure. The uptake of agreements has been particularly successful in ESAs with salt marshes in which more than on third is covered by 1990-agreements. Relatively few agreements are concerned with land that is to remain arable, and few agreements lead to landscape enhancement in the form of conversion from arable lands into permanent grasslands. A relevant question is: What protective effects does the ESA measure have?, or to be more specific: How effective has the ESA scheme been as a means to maintain

Table 4.6. ESA-agreements and their content.
A. Main purpose

Number of agreements*	Agreement area (ha)	Preservation of existing permanent grassland	Conversion to permanent grassland (%)	Remains in rotation
1,634	14,138	86	8	6

B. Detailed restrictions

	Number of agreements	Agreement area (ha)	Agreement area (%)
Grazing			
No restriction	728	5,126	21
Restriction	2,313	18,848	79
Sum	3,041	23,974	100
Use of pesticides			
No restriction	799	7,371	31
Restriction	2,240	16,587	69
Sum	3,039	23,958	100
Tillage including re-seeding			
No restriction	727	6,231	26
Restriction	2,314	17,743	74
Sum	3,041	23,974	100
Fertilization and manuring			
No restriction	217	1,705	7
Restriction	2,824	22,269	93
Sum	3,041	23,974	100

*Only from 54% of the analysed agreements is it possible to see the status of the agreement area before the agreement becomes effective.

permanent grasslands that would otherwise have changed (e.g. been abandoned, with subsequent overgrowing due to natural succession)? For obvious reasons a conclusive answer to these questions cannot be given until after the end of the 5-year agreement term. As part of the detailed study involving the 126 agreements, land use in fields bordering areas was also recorded. Thus, the planned ex-post evaluation will show how their development compares with the development of areas under agreement.

The immediate effects of the agreements are indicated with the 126 farmers interviewed in the detailed study being asked two key questions: (A) 'Does the agreement mean a change in management of the area compared with 1989?', (B) 'Would you have changed the use of the area, for instance abandoned grazing or mowing if you had not signed an agreement?'.

Responses are seen in Table 4.7. It appears that 42 of the 93 respondents managing existing permanent grassland have changed their management practices as a consequence of the agreement. Reduced fertilization and a cessation of re-seeding (during the 5-year period) were typical changes in management practices. However, a major number (51 of the 93 farmers) stated the agreement had entailed no management changes, and that they had been able to continue the way they had operated during the previous year. On the face of it, an agreement must have seemed quite attractive to these farmers, despite the rather modest compensation – a maximum of DKr 900. When the two questions are viewed in combination, Table 4.7 shows that almost half replied to both questions in the negative.

For them an agreement did not mean changes in management; nor did it secure the continuation of grazing or mowing otherwise discontinued. However, not being clairvoyant, the farmers would not have been able to predict the future. A fractured leg or other unlooked-for event could easily put a stop to grazing on the most outlying or poorest grasslands, not to mention a fall in the prices of fodder and other crops. The latter actually happened with the introduction of the CAP reforms in 1992. Although Table 4.7 would seem to indicate a considerable number of 'free riders', their actual share is probably lower – based on the many applications the Ministry of Agriculture received from farmers wishing to cancel their agreements within the 5-year term.

Finally, one important side-effect of the Danish ESA measure should be emphasized – new and closer connections between county landscape officials, farmers and agricultural advisers was a positive result. Dealing with positive issues (from a farmer's viewpoint, as opposed to communications regarding general restrictions and prohibitions) has surely provided the counties with experience and relations that have already proved themselves useful concerning the new schemes described in the next section.

Agricultural Policies under 2078/92

Agri-environmental policies under Regulation 2078/92 were proposed and brought for Commission approval in 1993. Implementation of the policies

Table 4.7. Effects of ESA agreements for existing grassland areas.

Does the agreement mean a change in management of the area compared with 1989?	Would you have changed the use of the area if you had not signed an agreement?		
	Yes	No	Sum
Yes	6	36	42
No	6	45	51
Sum	12	81	93

started in 1994, and the entering of agreements will be continued in 1995. The policies are for new comprehensive schemes, and not merely extensions of the former ESA measure. However, though differently organized, they still share a number of fundamental characteristics with the former ESA measure.

The new schemes differ from the earlier ones in that they are framework organized to a limited extent. Their administration lies with the Ministry of Agriculture (i.e. the Directory of Development Agriculture and Fisheries) and also includes the signing of agreements. These agreements can now be made both inside and outside the newly designated ESAs. A higher grant aid rate is paid within the designated areas, however, few schemes apply to ESAs only (Table 4.8). At present the role of the county councils is limited to designating areas eligible for the new schemes. Whenever agreement areas are part of the new ESAs county officials will confirm this, but they are otherwise no longer involved with the entering of agreements. However, quite a few counties have chosen an active approach to inform the farmers and others of the new schemes.

With the exception of the county schemes, the new schemes were launched with an initial application round in April 1994 – before their final approval in Brussels and before the new ESAs had been designated. The new measure is composed of the seven schemes listed in Table 4.8. It is seen that they imply an enlargement of the agreement issues and also a raised grant aid rate for converting arable lands into grasslands. New issues include 20 year set-aside schemes and unsprayed buffer zones. The Italian ryegrass and nitrogen reduction options were offered in few counties in the first ESA measure but are now offered everywhere in the country. In Table 4.8 it appears that a proportion of agreements consist of nitrogen reduction, namely 15.7% of the new agreement area.

Three-quarters of the agreement area is existing grasslands (Table 4.8), indicating that the primary effects of the new schemes will be preserving values associated with grassland, as were the former ones. Nevertheless it is interesting that 'conversion agreements' make up more than 6% of the agreement area (approximately 2000 ha) – given that the conversion of land in rotation to grassland has some competition from the reform set-aside subsidy, the grant aid rate was DKr 2700 throughout the country in 1994.

The new schemes were first offered to the farmers in 1994, when short periods were set for entering agreements. Despite the rather limited information campaign as well as a rather brief time span (7 weeks), and that the new ESAs remained to be identified, agreements were signed for more than 28,000 ha. The area was practically the same size as the area covered 5 years previously when the time for entering agreements amounted to 8 months. The new ESAs were not designated before the end of 1994, so the farmers were not sure that they would be entitled to the raised grant aid. Only a minor (unknown) fraction of the approximately 28,000 ha could have been a continuation of the former ESA agreements.

Table 4.8. The new agri-environmental schemes under Reg. 2078/92. The right columns show the area for agreements signed during 7 weeks of spring/summer 1994.

Schemes	1994 grant aid rates DKr/ha Outside new ESAs	Inside new ESAs	Agreement area (%)
Applicable to all areas			
40% reduction in N-supply level for land in rotation; amount of aid depends on yields of areas concerned, application must be countersigned by agricultural advisor to certify that the actual and stated yields agree.	650	700–900	15.7
Conversion of arable to grassland. The grant aid rate depends on past yields with confirmation by an agricultural advisor. Minimum of 1 ha to be covered. Two levels can be applied for:			
Level 1 – Fertilization to continue at reduced rate, at a maximum of 80 kg N/ha.	Level 1: 1000	1200–1800	4.1
Level 2 – Fertilization banned, maximum of 1 LU/ha, no cutting before 15 July	Level 2: 1700	1750–2300	2.2
Maintenance of permanent grassland* Measures are the same as the programme for conversion of arable land to grassland above. (Areas affected by Article 3 of the Nature Protection Act must be entered for Level 2.)	Level1: 400 Level 2: 600	650 900	39.1 32.5
Italian ryegrass as a catch crop on land in rotation, to be sown no later than 15 May as undersown grass.[†] After harvest, the area is not to be fertilized or sprayed and the land should be tilled no sooner than 15 February. A minimum of 5 ha should be entered	300	450 (Sum 100% = 28,193 ha)	6.4
Maintenance of organic farming for arable land only.	750	950	
Applicable to designated areas only			
20-year set-aside of arable land in nitrate affected groundwater catchment areas for conversion to			

grassland or woodland, or for the establishment of lakes or waterholes. Grazing of stock is allowed but without supplementary feeding. Minimum of 5 ha to be entered.	Not applicable	2300
12-m wide unsprayed buffer zones along watercourse, lakes and valuable habitats. Either all designated areas should be entered, or a minimum of 500 m.	Not applicable	DKr 3 per m

*Under Article 3 of the Nature Protection Act, fertilizer use is banned in areas to which fertilizers have not previously been applied. Such areas are not eligible for agreements under this programme. Withdrawal of this limitation is being considered (spring 1995) so agreements may be signed for all 'Article 3-Areas'.

† The numbers of agreement areas under the 'rye-grass scheme' overlap with other schemes and are consequently not shown.

New Schemes and Expected Future Developments

In spring 1995, new schemes were announced and a third phase of the agreements started. Compared with the first schemes under Regulation 2078, the current schemes differ primarily with respect to grassland agreements. The previous difference between 'conversion-agreements' and agreements for existing grasslands has not been continued in the new schemes.

No matter how the area has been used before the agreement payments concerning grasslands are the same, namely DKr 815–1415/ha/year depending on previous yields at level 1 (maximum 80 kgN/ha) and DKr 1300–1950/ha/year at level 2 (no nitrogen). Another substantial difference from the last scheme is that the agreement for permanent grassland can only be signed within the ESAs. For existing permanent grasslands there has been a considerable increase in payments compared with the schemes offered in 1994 (Table 4.8).

The main reason for these changes in grassland schemes has been to avoid renewals of previous 'conversion agreements' and to concentrate agreements on designated areas in order to maximize the environmental and landscape effects. Another change concerns the ryegrass scheme. In the new scheme all ryegrass types are permitted, not just Italian ryegrass. Except for existing permanent grasslands the general payment level has been reduced compared with the 1994 schemes.

Suggested changes

A modification of the 1994 schemes has been suggested by the counties, agricultural organizations, and conservation organizations. The proposed modification concerns the limitation of grassland agreements saying that no agreements under these schemes can be made for areas on which fertilization or manuring is already prohibited by existing nature conservation legislation. This limitation is causing problems with renewal of many first generation ESA agreements and many farmers cannot understand why agreements signed in 1990 and 1991 cannot be continued. So far the Ministry of Agriculture has not been in favour of removal of the limitation.

The 20-year agreements are restricted to ground-water protection zones designated in three counties in Northern Jutland. It has been considered, however, to make it possible to sign such long term agreements in all counties and it is expected to become a new scheme in the future. Especially in connection with current considerations to re-establish wetlands (wet meadows and pastures), such a new scheme is being appraised.

County schemes

While it is uncertain how the national agri-environmental policies under the Ministry of the Environment will change in the future, specific county schemes have been proposed and submitted to the Commission of the EU.

The county association has applied for EU co-financing of four schemes:

A Management of grasslands (DKr 500–1000/ha/year).
Management of grasslands and the clearance of trees and shrubs (DKr 2300/ha/year).

B Reduced drainage/raised water level (DKr 1500/ha/year).

C 12 m buffer zones (DKr 1500/ha/year).

D Recreational access (DKr 1500/ha/year).

The payments under Schemes A and B may be combined. Applications will be considered individually by the county council in question and may be accepted for areas outside the designated ESAs. According to the programmes submitted to the commission it is not decided whether it is possible to combine agreements under these county schemes with agreements under the national schemes.

Conclusion and Discussion

Agri-environmental policies with reference to Regulation 1760/87 and Regulation 2078/92 have been implemented in Denmark. The policies have been quite successful in terms of uptake: 4200 agreements covering an area of 31,300 ha were signed in 1990–1991 and about 2550 agreements covering a total area of about 28,000 ha were signed under the second phase in 1994. A small (but unknown) number of these are renewed agreements from the first phase. More than 75% of the agreements concern permanent grasslands. The introduction of agri-environmental policies has given new directions to Danish countryside regulations, including changes in agricultural as well as conservation policies.

First, the agricultural policy has been broadened to include non-productive aspects like environmental protection and landscape management. With the signing of several thousand agreements since 1990, farmers have shown their interest in the new schemes. In this way agriculture has been given new functions and consequently, agricultural land has changed in meaning and value.

Traditional marginal lands like semi-natural meadows and pastures have in many areas been designated Environmentally Sensitive Areas in which farmers are offered payments (or higher payments than outside these areas) to introduce or maintain certain management practices. Thus, the marginality of such areas has changed, a tendency which is intensified by the relatively high set-aside payments under the CAP reforms given to arable land in areas with poor soils. Finally, it should be mentioned that agri-environmental policies have introduced a differentiated subsidy form, which is a major change in traditional Danish agricultural policy. Usually agricultural schemes have been available for all farmers irrespective of the location of their land.

Second, agri-environmental policies have changed and challenged the traditional policies for environmental protection and nature conservation. The polluter pays principle which was introduced by the Environmental Protection Act in 1972 is to some degree neglected when farmers are offered payments to meet certain management practices within ground-water protection zones. In nature conservation policy there are similar conflicts between regulations which, without compensation, protect certain landscape elements against undesirable changes and the agri-environmental schemes which are based on a compensatory principle. So far, there have been no examples of payments given to farmers for carrying out activities they were not originally allowed to do, but with most management agreements there are restrictions which are part of the general regulations. In addition to such restrictions many agreements would have restrictions on changes which are not regulated today. Having established a 'tradition' for compensating such restrictions it may be difficult at a later stage to introduce general protection rules. In conclusion one could say that compensatory principles, introduced for all EU-countries, may cause some problems when they meet environmental and nature conservation regulations which are strictly national and therefore vary significantly across the Member States.

Paying farmers for actively introducing or continuing certain management practices is an important characteristic of agri-environmental policies, often linked to the management of permanent grasslands. Clearly, this 'active' aspect of agri-environmental policies has been integrated in nature conservation policies at the regional level, the county councils have used the agri-environmental policies to implement their nature conservation objectives. This has been particularly true for the agreements signed in 1990 and 1991 when the counties were responsible for implementing the scheme.

Landscape protection (against agricultural abandonment) has been the most significant function of the agri-environmental policies so far.

Third, agri-environmental policies have affected the traditional spheres and divisions of labour between central institutions. In the first phase, the counties were given the primary tasks of designating ESAs, designing of specific programmes, and signing of agreements. The Ministry of the Environment together with the Ministry of Agriculture was playing an active role when the guidance material was made and passed on to the counties. In the second phase under Regulation 2078/92 the Ministry of Agriculture had been taking the lead in programme design and the signing of agreements, while the duties of the counties have been reduced to the designation of the new ESAs. However, in the second phase the counties have applied for EU co-financing for specific county programmes under Regulation 2078/92. In Denmark this second phase has resulted in many disagreements and conflicts between the Ministry of Agriculture and the counties, conflicts which do not yet seem to be solved. It could be said that the introduction and later changes in the agri-environmental policies challenged the traditional systems.

Finally, a fourth important dimension of agri-environmental policies should be mentioned, namely problems of continuity (in time) and connectivity (in space). Five years may be considered a long time from the farmers' point of view, because they are used to quick changes in market prices (and in policies for that matter), but in terms of ecological processes like natural succession and evolution of species 5 years (even 20 years) is a short period. Continuation of grazing or mowing practices is therefore crucial for the protection of old grasslands and especially in Denmark, where there are no clear national priorities of landscape values, this is a problem. The agri-environmental policies are used for managing areas of major as well as minor importance. The connectivity problem refers to the situation in which less than half of the designated ESAs are actually under agreement. In most cases the agreement areas are scattered pieces of land in a matrix of dynamic agricultural land uses. This represents a problem for the ecological qualities of each ESA as a whole, simply because biodiversity within a given region is dependent on the size of each patch (each ecosystem) and the interactions between these patches. The lack of connection between the agreement areas is, of course, also a problem when the agri-environmental policies are utilized to fulfil environmental goals with specific areas like ground water protection zones and in buffer zones between intensively used agricultural land and sensitive ecosystems like water courses and lakes.

To summarize, it could be argued that the agri-environmental policies have given new functions to agriculture and at the same time been sources of conflict between established policy spheres. Better use of agricultural policies seems to require more integration of objectives, priorities and substantive means within the spheres of agriculture, environmental protection and nature conservation than is the case today.

Germany

Andreas Höll and Heino von Meyer

Introduction

Over the last two decades 'environment' has become a central issue in the public debate in Germany. Non-governmental, environmental organizations and the Green Party are today established forces in German policy. Several reasons may explain this increasing awareness:

- obvious problems of pollution and degradation in a densely populated, highly industrialized country;
- an increasing scarcity of the 'free' commodity, environment;
- high and rising standards of living, promoting concern with more global issues.

In the early 1980s the environmental debate, previously limited to issues such as nuclear power and industrial pollution, started to focus also on the interdependencies between agriculture, environmental protection and the conservation of wildlife and landscape. Key issues in this agri-environmental debate are:

1. Water pollution: Leaching of nitrate and pesticides into ground water, leading often to the contamination of drinking water supplies; eutrophication of surface waters, also responsible for algae formation in the North and Baltic Seas.
2. Loss of biodiversity: Reduction of habitats and extinction of plant and animal species caused both by the extension of field units and drainage as well as by the intensification of animal husbandry, fertilizer and pesticide use.

In 1985 a major scientific report by the Council of Environmental Advisors to the Federal Government described environmental problems related to modern agriculture in a comprehensive manner. The resulting environmental protection needs were ranked according to the following, in descending order of priority (Der Rat von Sachverständigen (SRU) 1985):

1. biotopes and species;
2. ground water;
3. land and soils;
4. surface waters;
5. food quality.

Concentration, intensification and specialization at farm level, re-enforced by territorial polarization into marginal and intensive zones, are causing severe environmental problems:

1. Fertilizer application levels in Germany are among the highest in Europe. In 1986, for the former West German part of the country, the annual nitrogen surplus (defined as all nitrogen added to the soil minus withdrawals in crops and livestock products) in agriculture amounted to 167 kg N/ha or a total of 2 million tonnes. Of this 41% went into the atmosphere another 31% into the hydrosphere, i.e. into ground and surface water (Isermann, 1989, p.4).
2. Over the last 10 years, nitrate concentration in ground water increased by 1–1.5 mg/l a year (Focken in DLG, 1988, p. 7). There have also been observations of 'nitrate eruptions' – increases by 10 mg/l and more due to the exhaustion of the denitrification capacity of the soils (Hellekes and Perdelwitz, 1986, p. 15). The European limit of 50 mg NO_3/l for drinking water is now exceeded by about 10% of all water supply units (UBA, 1994, p. 35): this tendency is rising.
3. A survey by the German 'Industrieverband Pflanzenschutz' shows that approximately 10% of West German ground water units have quantities of pesticide residues above the European limit of 0.0005 mg/l as the sum of individual substances.
4. According to the 'Red Data Books', 30–50% of all animal and plant species in West Germany are threatened with extinction. In 70% of all cases of endangered plant species agriculture has been identified as the main cause of decline. According to Kaule (1988, p. 17) the main reasons are the destruction of their natural environment. The cause being not only physical destruction but also changes in soil and water ecology due to drainage, fertilizer and pesticide application.

Many additions could be made to the lists of research results documenting the negative environmental effects of agriculture in Germany. Problems are particularly evident in regions with intensive arable or livestock production.

Any serious attempt to systematically reduce environmental damage caused by modern agriculture requires substantial changes in the Common Agricultural Policy of the European Union (CAP). The 1992 CAP reform brought about the first steps towards a reduction in intensity and specialization. However, a thorough evaluation of the environmental impacts of the

1992 CAP reform has not yet been carried out. Some figures suggest that environmental stress has been reduced. For example, German statistics on nitrogen fertilizer use show a reduction of about 15% compared with the level of the mid-1980s. However, taking into account that the set-aside element of the CAP reform brought about a reduction in the cereal and oil seed area of about the same order of magnitude, the extensification effect cannot be considered very high. Other elements of the reform seem to have even encouraged the continuation of unsustainable farming practices.

Basic Problems for an Ecological Agriculture Policy

Despite a strong environmental awareness and the inclusion of agricultural issues in the environmental policy debate, gaps in the German discussion and implementation of agri-environmental policies are still substantial. Although the necessity of environmental protection and conservation is officially recognized, realization of an effective agri-environmental policy has long been hindered by political and legal constraints.

The role of agri-political actors

The Deutscher Bauernverband (DBV) is the most important German farmers' organization representing agricultural interests in the national and international agricultural policy debate. The Arbeitsgemeinschaft bäuerliche Landwirtschaft (AbL) and the Deutscher Bundesverband der Landwirte im Nebenberuf (DBN) are of minor importance. AbL represents a growing group of mostly young farmers opposing the agri-business orientation of the DBV and promoting especially small and medium size family farming. Activities of the DBN are focused in particular on part-time pluriactive farming which is common in about 50% of all German farms. The ArbeitsGemeinschaft Ökologischer Landbau (AGÖL), combining eight different organic farming associations, deals with issues of common interest to organic farms (e.g. production and processing rules; promoting organic farming as a whole; marketing activities). Since the end of the 1970s when the discussion on water pollution, soil erosion and loss of biodiversity began, environmental and nature conservation organizations also began to take a more active part in the German debate on agricultural policy.

From the beginning, DBV and German Agricultural Ministers always jointly pushed for an income oriented CAP. In most instances positions of the DBV were also supported by the agri-food and the agri-chemical industry. Together they form an influential lobby that still dominates the relevant policy discourse. DBV, often being perceived as 'the' German farmers union, has long denied or played down environmental impacts of farming. For many years, arguments concerning income and competition losses were at the centre of the agricultural policy debate. Environmental measures became acceptable for DBV and other agricultural mainstream organizations under two conditions only:

- voluntary participation in environmental protection and nature conservation; and
- obligatory compensation for the resulting agricultural income losses (Höll, 1996).

On the basis of such a combination of income and protection goals the DBV began to change its mind and even started to promote the idea that farmers play an important role in nature conservation and that the respective public expenditures should be increased (Ganzert, 1988).

On the other hand, nature conservation NGOs, initially also supported by the Federal Ministry of the Environment, always argued for a better integration of environmental concerns into agricultural production and policy in general. They argued that protecting natural resources and biodiversity is a task for all societal groups, and that it should therefore be a social duty to all land users (Sozialpflichtigkeit). Although this perception is widely shared by the public, environmental groups were unable to gain sufficient policy support, in particular since the legal framework is still strengthening the position of the farmers' union (see below). Agreement on agri-environmental schemes based on voluntary participation and obligatory compensation was the logical outcome. However, nature conservation NGOs are very much aware of the fact that general agricultural economics and policies continue to encourage environmentally detrimental farming methods and systems.

Legal obstacles

German law assigns land use rights basically to the persons operating the land. Some restrictions on the use of inputs (fertilizers, pesticides) were introduced for the protection of surface and drinking water. Besides that, 'orderly agriculture' is considered to comply with the aims of most of the environmental protection and nature conservation laws:

1. In so called 'agriculture clauses' the notion of 'orderly agriculture' is introduced into nature conservation and water management laws in such a way that almost any farming practice is perceived as not causing any harm to the environment. According to these 'clauses' agriculture, as a rule, even contributes to nature conservation and landscape protection.
2. The term 'orderly', however, has never been clearly defined. In 1987 the German Conference of Federal and Länder Agricultural Ministers agreed on some general principles (Agrarbericht 1988, Materialband p. 142), but no further specifications nor any concrete restrictions were defined.

3. As a consequence, contrary to scientific evidence, German agriculture is in legal terms generally considered compatible with the environment. In turn, environmental protection and nature conservation are in an extremely weak legal position. The 'PPP' (Polluter Pays Principle) has not been implemented to any large extent.

Efforts to change or cancel the 'agriculture clauses' have failed several times during the last 15 years. Even attempts to develop a more precise definition of the rules for 'orderly' agriculture have not been successful up to now.

Implementing Agri-environmental Policies in Germany

Agri-environmental schemes in Germany before Regulation 2078/92

In Germany, agricultural structures policy as well as nature conservation are tasks assigned to the Länder. In the early 1980s several Länder introduced programmes aiming at landscape management and nature conservation on the basis of management agreement contracts with farmers.

Initially these programmes were very site-specific focusing on the protection of certain species or habitats. In Bavaria the first schemes for the protection of meadow birds were established in 1983 with a total budget of about DM 4 million for 1983 and 1984 (ByStMLU, 1986). After 1985, in response to the new Article 19 of the EC Regulation 797/85, several other Länder launched similar programmes. Since 1987 most of these Länder schemes were also co-financed by the EC (under Regulation 1760/87). In 1988, already more than 50 regional programmes were in place targeting about 3.5% of the total agricultural area (ARUM, 1989, p. 95 ff.). Today the Bavarian scheme alone is in the order of DM 200 million annually.

Implementing Regulation 2078/92 in Germany

With the 1992 CAP reform and its accompanying measures under Regulation 2078/92 agri-environmental policy became an issue also for the German Federal Government. The EU-scheme is now implemented primarily through the so called 'Gemeinschaftsaufgabe Verbesserung der Agrarstruktur und des Küstenschutzes' (GAK) (Common Task of Improving Agricultural Structures and Coastal Protection) as a common task for both the Länder and the Federal government.

Common Tasks were introduced in German policy in the early 1970s in policy fields such as

- agricultural structures policy;
- regional policy;
- university infrastructure.

By the Constitutional Basic Act (Grundgesetz) these subjects were originally assigned to the Länder. Due to their importance and urgency, it appeared necessary, however, to establish a common planning framework and to ensure an adequate financial contribution from the federal level. GAK budgets are usually split into a Federal contribution of 60% and a 40% share financed by the Länder. The GAK has become the main mechanism for implementing agricultural structures policy in Germany. It includes, for example, support for land consolidation, farm investment, improvement of market structures.

Since support for environmentally-sensitive farming was not covered by the GAK law, it had to be amended in order to provide a basis for implementation of Regulation 2078/92. The new formulation, however, is rather restrictive, limiting GAK contributions only to those measures that support farming adapted to market and locational conditions (markt- und standortangepaßt). Consequently not all measures foreseen under Regulation 2078/92 are implemented under the GAK. Measures such as long term set-aside or site-specific ecological schemes, remain exclusive Länder tasks and are not co-financed by the Federal budget.

The EU contribution to Regulation 2078/92 schemes in Germany differs from East to West. Since the five New German Länder (Brandenburg, Mecklenburg-Vorpommern, Sachsen, Sachsen-Anhalt, Thüringen) are part of the Objective 1 regions covered by the EU Structural Funds, the EU can co-finance measures with up to 75%. In all West German Länder the maximum EU contribution is limited to 50%.

Measures covered by GAK

GAK support is provided for three types of farming practices:

- extensive arable and permanent crop production;
- extensive grass land management;
- organic farming.

Details are shown in Table 5.1 which also provides information on the premia foreseen by the GAK. The Länder are allowed to either raise (up to 20% points) or reduce (up to 40% points) these premia.

The main characteristics of the GAK implementation are:

- the scheme is applied to all agricultural land in the Federal Republic;
- participating farms have to bring all their arable and/or grassland into the scheme;
- the maximum stocking rate is set at 2.0 LU/ha AA (agricultural area);
- permanent grassland may not be converted into arable land.

Thus, GAK implementation covers only measures (a), (b) and (c) of Article 2 Regulation 2078/92 which had previously been subject to extensification schemes under Regulation 2328/91. The support for maintaining

Table 5.1. Regulation 2078/92 measures under the GAK and premia foreseen. (Source: BMELF, 1994b.)

A. Extensive arable and permanent crop production

	Support in DM/ha	
	Arable land	Permanent crops
No sludge to be applied		
Renunciation of chemical/industrial fertilizers and pesticides	250	1200
Renunciation of chemical/industrial fertilizers	150	250
Renunciation of herbicides in arable and permanent crops	150	350
Renunciation of herbicides in fruit-growing		150

B. Extensive grassland management

	Support (DM/ha) Permanent grassland
Stocking rate: minimum of 0.3 up to a maximum of 1.4 rough grazing LU/ha forage	
Renunciation of chemical/industrial fertilizers and pesticides	
Slurry and farm yard manure to be applied up to a maximum of 1.4 LU/ha	
No melioration	
Reduction of stocking rate to a maximum of 1.4 rough grazing LU/ha forage	
per reduced rough grazing LU/ha	450
minimum per ha permanent grassland	250
Extension of permanent grassland and/or maintenance of extensive grassland use	250
Conversion of arable into extensive grassland, per ha arable land	600

C. Organic farming

	Support in DM/ha	
	Arable and grassland	Permanent crops
Farming in accordance with Reg 2092/91		
Farm is subject to regular checks (of compliance)		
Introduction of organic farming	250	1200
Maintenance of organic farming	250	1200

extensive and/or organic farming is left up to the Länder (all Länder have implemented this support scheme except Schleswig-Holstein, see Table 5.3). GAK implementation had been notified to the EU in 1993 but support schemes are available to farmers only on the basis of specific Länder programmes.

Implementation of Regulation 2078/92 at Länder level

Since implementation is different in the 16 Bundesländer, it is difficult at this stage to provide a comprehensive and complete picture of the present situation in Germany. Länder have used different strategies and approaches for implementing Regulation 2078/92. Up to now some have merely taken over the GAK scheme, in addition to that others have developed zonal and/or site-specific programmes which refer to measues (d), (e) and (f) of Article 2 Regulation 2078/92. Land management for public access and leisure activities [Article 2 g] is not yet subject to Länder programmes in Germany.

The characteristics of Regulation 2078/92 implementation at Länder level depend partly on their individual administrative structure. In some Länder the responsibility for landscape management and nature conservation lies with the agricultural administration, whereas in others the site-specific parts of agri-environmental schemes are conducted under the authority of the environmental ministry.

The present state of implementation of Regulation 2078/92 in the various Länder is very different, partly due to the fact that some Länder had already implemented agri-environmental schemes under the previous Regulations 797/85 and 1760/87 (such as the Bavarian Cultural Landscape Programme). The figures provided in Table 5.2 are based on responses to a survey which was sent to all Länder ministries. They are provisional and not fully comparable. Final results may differ substantially. In some cases figures include payments for nature conservation and site-specific schemes, in others not.

The Federal Agricultural Ministry is presently working on a first analysis of the implementation of Regulation 2078/92 in Germany. According to their data, in the year 1993/94 a total of DM 451 million was spent on agri-environmental schemes (BMELF, 1995b) in German Länder.

Table 5.2 shows the differential take up of Regulation 2078/92 amongst the Länder. In 1993, programmes of Baden-Württemberg, Bavaria, Hessia, Rhineland-Palatinate and Thuringia were approved by the European Commission. For Mecklenburg-West Pomerania and the city Land Bremen programmes are not yet notified by the EU. The programmes shown in Table 5.2 implement all measures foreseen under GAK with the exception of Mecklenburg-West Pomerania where only grassland extensification is included. Other Regulation 2078/92 tasks, such as site-specific measures; rearing of local breeds, or training and education, are sometimes included in the countryside programmes, sometimes approved as additional Länder-programmes. Additional programmes are still planned in some Länder.

The enormous differences in implementation of Regulation 2078/92 are

Table 5.2. Implementation of Regulation 2078/92 and Länder budgets. (Source: Survey sent to Länder administrations in charge of implementing 2078 schemes, February, 1995.)

Länder	Main programmes [name (abbreviated)/date of notification]	Budgets			Total area		Intensity of support
		1992 TDM	1993 TDM	1994 TDM	1000 ha, AA	Total = 100*	Average DM/ha AA†
Baden-Württemberg	MEKA/19.10.93	140,000	140,000	140,000	1,460	8.5	95.9
Bavaria	KULAPI/1.8.93	–	109,000	208,000	3,400	19.7	61.2
Brandenburg	Countryside Protection Programme/1.10.94	–	–	30,200	1,298	7.5	23.3
Berlin	Countryside Protection Programme/Nov. 1994	–	–	180	2	0.01	100.0
Bremen	Countryside Protection Programme/1995	–	–	–	26	0.15	0
Hamburg	Countryside Protection Programme/14.12.94	–	–	–	14	0.08	0
Hessia	HEKUL/4.11.1993	–	17,200	36,400	786	4.6	46.3
Mecklenburg-W. Pomerania	Grassland extensification under 2078: since Feb. 95	10,000	14,000	19,300	1,313	7.6	14.7
Lower Saxony	Countryside Protection Programme/1994	–	–	15,000	2,714	15.7	5.5
North Rhine-Westfalia	Countryside Protection Programme/1994	n.a.	n.a.	n.a.	1,567	9.1	n.a.
Rhineland-Palatinate	F.U.L./3.4.93	–	–	11,200	714	4.1	15.7
Saarland	Countryside Protection Programme/1995	465	500	800	74	0.4	10.9
Saxony	KULAPI/1994	–	–	76,800	898	5.2	85.5
Saxony-Anhalt	Countryside Protection Programme/11.10.94	–	11,320	6,329	1,138	6.6	5.6
Schleswig-Holstein	Countryside Protection Programme/11.10.94	–	–	411‡	1,056	6.1	0.4
Thuringia	KULAP/20.7.93	–	–	38,000	789	4.6	48.2
Total				582,620	17,249	100	

*Länder share in Germany's total AA.

† 1994 Budget for 2078 schemes per hectare of the total AA of each Länder.

‡ Without budget for nature conservation schemes conducted by the ministry for the environment.

n.a. No information provided.

– No budgets provided under Reg 2078/92.

TDM = DM 1000.

AA = agricultural area.

even more obvious in examining the budgetary commitments related to the total land area of each individual Länder. Whereas for Bavaria, Saxonia and Baden-Württemberg payments per ha AA are in the order of DM 60–95 /ha AA, in several other Länder they do not even reach DM 10 /ha AA (Schleswig-Holstein, Lower Saxony and Saxony-Anhalt). Such disparities clearly reveal differences in the sensitivity to agri-environmental concerns and corresponding policy priorities at Länder level. Partly these differences are because those Länder programmes which were first agreed by the EU were not yet confronted with a budgetary ceiling, as originally funding for those schemes was meant to be open ended. In 1994, however, it was agreed that for the period 1993–1997 Germany will receive only 1050 million ECU from the EU-funds (BMELF 1995a, p. 118). Since some of the more expensive Länder programmes (Bavaria, Baden-Württemberg) had already been agreed before, the available resources for late comers, in particular for the East German Länder, were rather limited. A clear exception is Saxony with a rather well budgeted countryside protection programme.

Organic farming is considered to play a key role in German agri-environmental policies. Support for organic farming is provided not only under Regulation 2078/92 but also under the previous extensification scheme (Regulation 2328/91). Table 5.3 and Fig. 5.1 provide a general overview on the support to and the relative importance of organic farms in the different German Länder. They show that in 1994 the share of (supported) organic farms in the total agricultual area of the respective Länder differed considerably from less than 1% in Schleswig-Holstein, Lower Saxony and North Rhine-Westfalia to more than 5% in Mecklenburg-West Pomerania, Saarland and Hessia.

Examples from the Länder: Bavaria and Hessia

By providing some more specific information on the two Länder, Bavaria and Hessia, it will become even clearer that in Germany Regulation 2078/92 is implemented in very different ways.

The Bavarian Kulturlandschaftsprogramm (KULAP) is characterized by a broad approach trying to cover the entire territory and thereby implicitly having also a strong income orientation. Payments in the order of DM 400 – 1400 per farm (DM 40/ha AA) can be obtained by every farm fulfilling the following basic conditions: Maintenance of a minimum of 3 ha; no conversion of grassland to arable land; stocking rate below 2.5 LU/ha AA.

On top of these basic requirements different measures can be applied and cumulated leading to higher premia. Bavaria like other Länder has also made use of the opportunity to raise GAK premia. If for example the Bavarian situation is compared with that in Schleswig-Holstein (see Table 5.1) the differences in incentives are apparent:

Table 5.3. Organic farming in Germany in 1994. (Sources: Inquiry of responsible Länder administrations in charge of implementation, February 1995; Plankl, 1995.)

	Under Reg 2078/92 schemes				Under Reg 2328/91*		Organic farming
	Conversion and maintenance premia DM/ha AA		Farms (number)	AA (hectares)	Farms (number)	AA (hectares)	Area as per cent of total AA
Länder	Arable land	Grassland		(1)		(2)	(1) + (2)
Baden-Württemburg	260/260	260/260	495	13,700	2,030	26,300	2.7
Bavaria	400/400	300/300	3,394	26,820	3,404	72,030	2.9
Brandenburg	300/255	300/255	43	2,703	286	50,082	4.1
Berlin	300/255	300/255	0	0	1	30	1.7
Bremen	250/250	250/250	–	–	n.a.	n.a.	0
Hamburg	600/250	600/250	–	–	2	120	0.9
Hessia	350/350	450/450	188	3,400	2,300	56,000	7.5
Mecklenburg-W. Pomerania	300/300	300/300	–	–	386	75,670	5.8
Lower Saxony	300/300	300/300	n.a.	n.a.	n.a.	17,874	0.7
North Rhine-Westfalia	300/200	300/200	n.a.	n.a.	n.a.	12,300	0.8
Rhineland-Palatinate	550/450	500/450	220	3,840	890	5,650	1.3
Saarland	300/250	300/250	0	0	95	5,000	6.8
Saxony	550/450	260/260	26	799	431	17,322	2.0
Saxony-Anhalt	300/300	300/300	20	1,700	155	16,136	2.9
Schleswig-Holstein	250/0	250/0	8	327	163	9,410	0.9
Thuringia	300/300	350/350	24	1,565	73	12,790	1.8

*Organic farming under Operational Extensification Method of Regulation 2328/91.
n.a. No information by responsible ministry.
– No budgets provided under Regulation 2078/92.

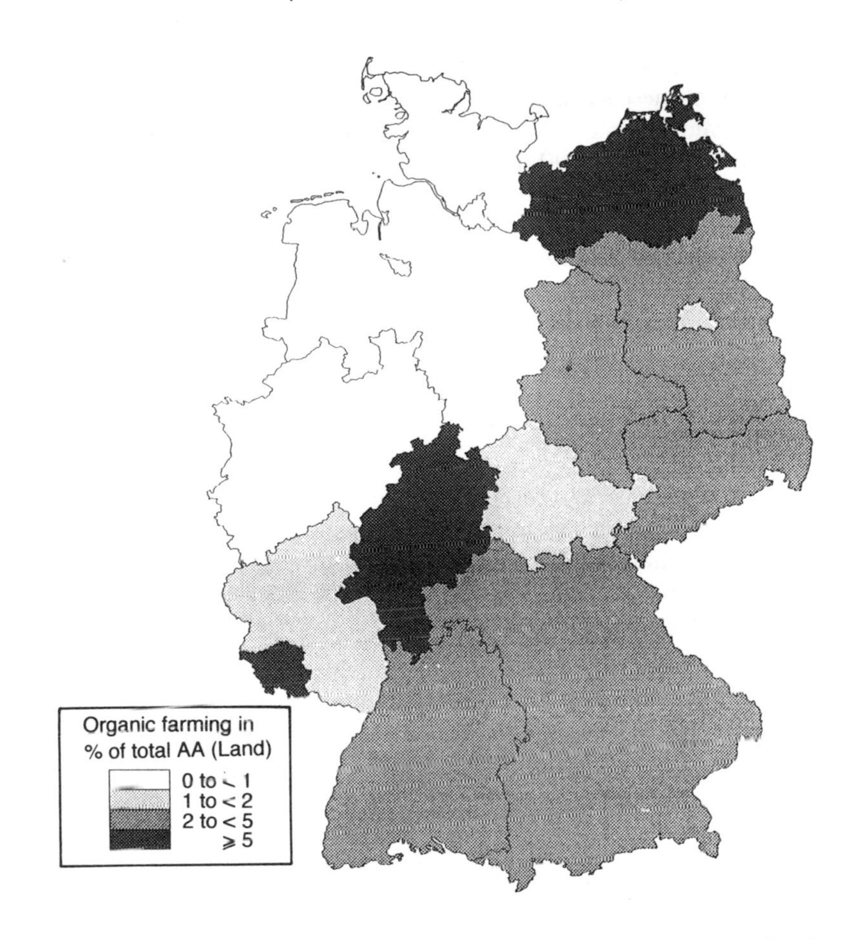

Fig. 5.1. Relative support to organic farms in the German Länder. (AA = agricultural area.)

1. whereas in Schleswig-Holstein maintenance of organic farming is not supported at all; and introduction is supported only with the payments foreseen by the GAK (i.e. DM 250 /ha arable as well as grassland);
2. in Bavaria the respective premia are DM 400 /ha both for maintenance as well as for conversion, at least up to a maximum stocking rate of 1.5 LU/ha AA; (DM 300 /ha AA up to 2.0 LU/ha AA). For grassland Bavarian premia reach DM 300/250 /ha AA.

Part 1 of the Bavarian KULAP also incorporates site-specific measures oriented towards nature conservation (e.g. old orchards, hillslope mowing, alpine grazing).

Part 2 of KULAP contains Regulation 2078/92 measures which are not co-financed by the GAK. These measures address specific kinds of land use in the peat area of Donaumoos. This is an example of a zonal programme describing specific production methods compatible with the requirements of the protection of these sites. Furthermore, Part 2 aims at long term set-aside (20 years) for environmental purposes (annual premia from DM 400–1100 /ha grassland, DM 500–1200 /ha arable land depending on yield classes).

Finally, measures under Part 3 of KULAP are not co-financed by the EU. Most of them aim at specific measures for habitat protection and maintenance of cultural landscape in cases where no support is provided under Part 1. Payments depend on the income foregone and include investment costs (such as for planting hedgerows and their management). This support scheme is available not only to farmers but also to associations and NGOs conducting landscape management.

In Hessia two main programmes address the objectives of Regulation 2078/92:

1. HEKUL (*HE*ssisches *KUL*turlandschaftsprogramm): a scheme based on GAK implementation oriented to protect the countryside as a whole; and
2. HELP (*HE*ssisches *L*andschafts*P*flegeprogramm) which provides support for site-specific landscape management for purposes of nature conservation (not co-financed by GAK).

Furthermore, schemes for training and education and long-term set-aside under Regulation 2078/92 are implemented as well. A scheme for rearing local animal breeds endangered by extinction has passed the STAR-Council in December 1994 and will be implemented from 1995 on.

In addition, Hessia has implemented under HEKUL the following schemes which are neither covered by GAK nor by Regulation 2078/92:

1. support of animal husbandry fulfilling specific requirements of animal welfare;
2. premia for beekeepers taking care of dusting old orchards;
3. marketing initiatives for foodstuffs produced under HEKUL conditions (since 1993).

Support for marketing initiatives plays an increasingly important role in Hessian agricultural policy. This approach is based on the observation that one-sided support for the supply of environment friendly production could be short-sighted without at the same time expanding the markets for such products. Furthermore, the Länder has another marketing support scheme the so-called 'product innovation and marketing alternatives' programme providing small-scale investment support for direct marketing. The Hessian marketing initiatives and support schemes are rather modest in budgetary terms but they have been very effective in developing environmentally-sensitive farming and in stabilizng small and medium sized farms (Knickel *et al.*, 1994).

Marketing initiatives are not only integral parts of Hessian agricultural policy but also considered essential for the overall rural development in

Hessia. They are a good example for the efforts undertaken in Hessia since 1991 to promote stepwise integration of agricultural, environmental and rural development policies (Jordan, 1995). The synergies resulting from such integration may explain the phenomenon that Länder like Hessia or Saarland which up to now have not spent large amounts on agri-environmental schemes although they have comparatively high rates of environmentally-sensitive farming.

Summary and Conclusions

1. In principle, the basic responsibility for implementing agri-environmental policies in Germany lies with the 16 Länder. According to the German 'Basic Act' (Grundgesetz) it is the Länder who are responsible for agricultural structures policies as well as for environmental and nature conservation policies. Since the early 1970s, however, agricultural structures policy, like regional development policy, is considered a 'Common Task' (Gemeinschaftsaufgabe) in which the Federal Government contributes to the financing (60%), and participates together with all the Länder in ensuring a common medium term planning framework (Rahmenplan).

2. Due to this constitutional situation implementation of agri-environmental schemes in Germany is characterized by a two tier approach. While a basic set of measures foreseen under EU Regulation 2078/92 is covered by the 'Common Task', other measures, in particular those more closely related to nature conservation objectives, are specifically designed and administered – if at all – by the individual Länder alone.

3. The German approach to the implementation of agri-environmental policy is characterized by the fact that the entire territory is covered, although not all measures are applied to all places. Thus, German implementation is not restricted to specific environmentally sensitive areas, as is the case of the British ESAs. To put it differently, the entire German territory is considered environmentally sensitive although to different degrees and for different purposes.

4. While total coverage could be considered positive, and should probably be maintained, a lack of proper targeting and prioritization jeopardizes the effectiveness and efficiency of the German approach. Between the various Länder, implementation of Regulation 2078/92 and of the Common Task provisions varies considerably. Within the Länder, however, there is only limited differentiation of support according to natural or sectoral conditions: Implementation of Regulation 2078/92 in Germany is not yet sufficiently adapted to zonal requirements. Preliminary data from several Länder ministries show that schemes for environmentally sensitive farming are particularly taken up in Less Favoured Areas. In particular, the Common Task framework should incorporate an obligation demanding Länder programmes to specify the requirements of environmentally sensitive farming for different regions. This could probably also enhance the effectiveness of landscape planning which is well developed in Germany.

5. The fact that only a selected number of Regulation 2078/92 measures covered by the Common Task leads to substantial distortions not only with regard to the measures available but also with respect to support intensities between different Länder. Since measures under the Common Task are co-financed by the Federal budget they tend to be more attractive for the Länder than those that have to be financed without Federal contribution. This implicitly leads to a mix of measures that is more focused on agriculture than on conservation. This is in particular true for the poorer New Länder in the East.

6. Since, contrary to initial intentions, the EU funding of Regulation 2078/92 schemes was limited to a maximum amount, the distribution of funds among the Länder turned out to be very uneven. Those Länder in the West, like Rhineland-Palatinate, Baden-Württemberg and Bavaria, that could in fact more easily afford to spend their own resources on agri-environmental schemes got their programmes agreed by the Commission in a rather early stage whereas the late-comers were already confronted with a financial ceiling from Brussels.

7. Apart from the problems of vertical coordination and cooperation between the various levels (EU, Federal, Länder) horizontal cooperation between different sectoral ministries also proved to be problematic in several instances. While agricultural ministries responsible for the measures covered by the Common Task have usually well established formal and informal links to the Federal Agricultural Ministry in charge, this is not always the case for the Länder ministries dealing with the environment and nature conservation. They are, however, often responsible for implementing those parts of the Regulation 2078/92 that are not covered by the Common Task.

8. Despite the fact that the Common Task provides a framework for consistent implementation across the entire country, the premia offered for the various options differ considerably. This would be no problem, if it reflected differences in agriculture's marginal price and cost relations. Often, however, it seems to reflect more the willingness and ability of Länder governments to support farmers. Even this would not necessarily be a problem as long as payments were explicitly targeted and addressed to reward clearly defined environmentally beneficial practices. Often, however, the designation of tasks and the measurement of results is not of great precision. Sometimes differences in the premia offered also reflect differences in the distribution of actual or presumed property rights. While in some Länder ploughing-up grassland is simply forbidden, in other Länder (e.g. Bavaria) it is subject to compensation payments. This indicates the potential conflict that often exists between Regulation 2078/92 payments and the polluter pays principle, at least in its simplest form.

9. What has become very clear from the analysis of various German Regulation 2078/92 programmes is the urgent need to set up an agri-environmental indicator scheme that would allow not only proper targeting of support to the appropriate areas and farming types but also to monitor and assess the potential and actual outcome of the policy. Indicators are indispensable tools, however, due to the great diversity in natural and structural conditions it appears impossible to rely exclusively on statistical parameters when designing, implementing and assessing agri-environmental policies.

10. In addition it appears urgent to develop institutional structures and procedures that enable rapid feedback within the system of policy administration. Such learning processes can not be organized like technocratic evaluations but have to make intelligent use of the available expertise on the spot. This implies establishing partnerships involving non-governmental actors such as farmer or nature conservation organizations as well as local administrations. Since the administrative burden caused by the present schemes is already very high, it appears reasonable to consider also changes, or at least variations, in the programme design. Rather than focusing on individual contracts with individual farmers, landscape management schemes aiming at broader agreements involving local administrations and nature conservation NGOs (also for monitoring and controlling) should be taken into consideration.

Finally, it must not be overlooked that the efficiency of agri-environmental schemes in the EU would rapidly increase if environmental aspects were better and more fully integrated in the Common Agricultural Policy. Then, agri-environmental schemes would not have to overcome competing incentives from the CAP and could be concentrated more upon targeted site-specific measures for the purpose of landscape and wildlife conservation.

Spain

Fernando Garrido and Eduardo Moyano

Introduction

This chapter analyses the introduction of the agri-environmental debate into Spanish society and shows how this subject is being incorporated within the agenda of the Spanish Government. To understand the cultural and political context in which the agri-environmental debate originates, the increasing concern for environmental problems and the emergence of the notion of sustainability are discussed initially. This is then followed by an analysis of some features of the Spanish countryside and the process of agricultural modernization to illustrate the diversity of the agricultural and social structure in Spain. Finally, the most important aspects of the agri-socio-structural policy in Spain and the current status of organized agricultural interests and environmental groups are reviewed. In particular, the implementation of Regulation 2078/92 on the agri-environmental measures at national and regional levels is assessed.

The Cultural and Political Context

There is a growing awareness of environmental problems in developed industrial societies. The concern for the deterioration of the environment and the necessity of implementing protection and conservation programmes for natural resources is becoming an important subject in both public opinion and the political sphere (Lowe, 1992). As is generally accepted, this awareness is being encouraged by several factors. First, it is encouraged by the expanded acknowledgement that the current economic development models are responsible for negative effects on the environment. This awareness arose initially from minority groups within the population (mainly from the environmental movement in the early 1970s) with only slight influence in the decision-making process. Later, it spread within the rest of public opinion until it reached the political sphere.

Second, it is encouraged by the recognition in developed countries that maintaining the force behind intensive patterns of agriculture is not a rational attitude after achieving food self-sufficiency (Mateu, 1992), since these agricultural methods produce enormous surpluses and are responsible for damage to forests and for aquatic pollution.

Initially, in the early 1970s, environmental problems were confined to the effects produced by industrial activity. These effects were identified along with the major polluting industries, so the application of a Polluter Pays Principle (PPP) was demanded for them (Baldock and Bennet, 1991; Baldock *et al.*, 1993). Until recently agriculture was respected for its responsibility as a pollution source for the following reasons.

1. Agricultural activity has been developing basically on scattered family farms. This has made it difficult, for example, to implement the PPP, since it would be implemented on sources of diffuse pollution where, in contrast with industry, it is not possible to identify the polluting agent (Baldock and Bennet, 1991).
2. For a long time, agricultural activity has been associated with nature, while farmers have been considered, from a symbolic point of view, as an example of harmony and respect to the environment (Lowe, 1992; Mormont, 1994). This idealized image has remained anchored in the consciousness of urban people – most of them coming from rural areas – in spite of the fact that environmentally harmful intensive agricultural practices are being introduced by farmers.
3. Agriculture is a strategic industry for national economies and is also the base for food supply. This is why governments of all the western countries have been interested in regulating the farming sector by implementing protectionist policies to encourage an increase of agricultural productivity. This increase has been necessary to maintain, with a reduced farming population, a high level of agricultural production capable of guaranteeing food security. Furthermore, these protectionist policies have been used by certain countries to secure the exporting potential of its agricultural sector in international trade.
4. The rural elite and farmers' unions have an important influence on the decision-making process, mainly during the stage of productive modernization (Moyano, 1988). In EU countries, farmers' unions, cooperative federations, agricultural chambers and the ministries of agriculture have together created a sort of 'agricultural front' (a particular agricultural policy community) capable of keeping agriculture and farmers out of the prevailing social and economic rationality.

This exemption, enjoyed for a long time by agriculture and farmers, has started to change rapidly in recent years. Public authorities and political forces have inserted on their agendas the necessity of reforming the current mechanisms of agricultural protection. Furthermore, they have firmly proposed the convenience of modifying the regulation patterns of agriculture, so far oriented towards the target of productivity, and of favouring the introduction of less intensive and more environmentally sound agricultural practices.

On the EU-level several factors have influenced this change in attitude of public authorities towards agriculture. Some of them have been related to rising public opinion of new non-agricultural interest groups, which demand a different regulation for agriculture, taking not only productive but also environmental factors into account. Other factors have come from the construction process of the European Union. In fact, all governments of EU countries have proposed as a priority, the control of the CAP financial budget in order to halt growing agricultural surpluses and to release funds for the development of other activities. All that is seen as a condition for advance in the process of unity and cooperation among European countries.

However, these internal factors would not have been capable alone of encouraging the CAP reforms and introducing a new discourse about agriculture and rural areas. Other factors (external factors) have accelerated the mentioned change of thinking on agricultural problems and environmentally harmful effects. Among these external factors, the following can be emphasized:

1. The recently finished GATT negotiations on the liberalization of the international trade of agricultural products. A considerable reduction in the CAP price protection system was demanded from the EU in those debates, which obliged the EU authorities to implement a new system of protection to compensate their farmers (Tió, 1991).
2. The challenge representing the consolidation of the democratic system and the market economy in Eastern European countries (Tió, 1991). This provides the challenge to the EU of the opening of agricultural markets, favouring the interchange with the less developed European economies.
3. The necessity of building on the world-level a new network of relations between developed and developing countries based on a new exchange and cooperation model.

In all these cases, agriculture plays a decisive role in the implementation of the above mentioned new exchange system because farming activity is both the base of the export market in the developing countries and is the most important productive activity. This is why the reform of the current guaranteed price policies prevailing mainly in EU countries, which were implemented 30 years ago to encourage the increase in agricultural productivity and to favour the use of intensive practices by farmers, is considered a prerequisite to building a new exchange system.

The emergence of the ideal of sustainability can be placed in that context. This notion has been incorporated into the debate on agricultural development from outside agriculture. In fact, the idea of sustainable development is one that had its origin during the 1970s, in the general discussion about the negative effects on the environment by the productive development model in industrialized countries (the two reports by the MIT to the Club of Rome in 1972 and 1974 encouraged the concern for ecological problems). Agricultural activity was excluded from that debate for reasons mentioned above and its special statute of ecologically non-harmful activity was respected.

When this statute is questioned and the appraisal of the negative effects of intensive agricultural methods begins, the damage to the environment will be incorporated into the debate on agricultural development. The notion of sustainability is initially taken as a point of reference to order the discussions within the scientific community, and then within the political sphere. Such a notion has had the virtue of putting the idea of the balance between environment, agricultural production and farmers' income into the debate, but, once that purpose is achieved, it has shown itself to be a vague and ambiguous notion with some difficulties that require understanding by the agricultural community[1].

Since the notion of sustainability originated in the framework of the general debate about development models, then it is in a strange position for farmers and their unions. They fear the notion of sustainability will reduce their freedom to choose the more profitable agricultural practices. This explains the defensive and hostile attitude held by the agricultural community (farmers' unions, rural based political groups, and agricultural public authorities).

It is argued here that the relationship between agriculture and the environment is an economic and sociocultural one. It is an economic relationship because the farmers use natural resources to produce agricultural products – mainly food – but it is also a sociocultural relationship because they use nature as a habitat and symbolic place.

It is a historically and geographically determined relationship, this is why we consider the notion of sustainability in agriculture must be defined in accordance with the particular relationship farmers have with their natural surroundings in an economic and cultural context (Ward and Munton, 1992). Nevertheless, it does not mean that each farmer has his own notion of sustainability, since in that case it would no longer be an operative and useful concept for analysing this subject. However, it is necessary to consider

[1] For a review of the notion of sustainable agriculture see for example: Terry Gips: 'What is sustainable agriculture?', in Allen and Van Dusen, 1988. *Global Perspectives on Agroecology and Sustainable Agriculture*, pp. 63–74; W. Lockeretz: 'Open questions in sustainable agriculture', in: *American Journal of Alternative Agriculture*, 1988, pp. 174–181; P. Allen *et al*.: 'Integrating social, environmental and economic issues in sustainable agriculture', in: *American Journal of Alternative Agriculture*, 1991, pp. 34–39; D.R. Keeney: 'Toward a sustainable agriculture: Need for clarification of concepts and terminology', in: *American Journal of Alternative Agriculture*, 1989, pp. 101–105.

that the relationship between agriculture and the environment varies according to the different size and productive orientation of the farm. For instance, there is not the same relationship between the environment and agriculture on an irrigated intensive farm compared with that existing in dry extensive agriculture or in greenhouses. Nor is the relationship between the environment and the small family farmer (who considers the rural area not only as a place to achieve the best economic profitable production from his farm, but also as a living and working area) the same as that existing between the environment and the large farmer (who considers the natural surroundings mainly as a factor of production).

In this chapter the notion of sustainability is approached from the particular situation of Spanish agriculture, presenting some ideas to explain the agricultural community's attitude to the subject.

Some Features of the Spanish Agricultural Countryside

Spain has a large agricultural sector, although it is not as intensive as that of other European countries, such as Denmark or the Netherlands. This is indicated for example, by the amount of nitrate fertilizer used in 1988/89, which was only 43.1 kg/ha for Spain, compared with 134.2 kg/ha in Denmark, and 215.4 kg/ha in the Netherlands (Commission of the European Communities, 1993).

The size of Spain (around 50 million hectares), and the low population density (70 inhabitants per km^2) has not created any serious competitive problems through the sharing of land between farmers and non-farmers, except in several areas near tourist resorts; the Costa Brava and the Costa del Sol. However, the population is unevenly distributed over the national territory. There is a small number of large metropolitan centres where nearly 50% of the population is concentrated (Madrid, Barcelona, Valencia or Sevilla) and there are also some vast regions affected by problems of depopulation (Castilla or Aragon).

Together with the demographic heterogeneity it must be noted that the large variation in the natural and physical environment makes Spain one of the main countries in the EU with such diverse animal and plant species. This variation is due to the conditions of the physical environment which produces a wide range of ecological niches. In effect, from the large littoral coasts, the ground gradually rises – like the South Mediterranean littoral – or becomes high in a rugged way – like the Cantabrian region – forming mountainous chains with heights ranging from 1000–2000 m. These mountain ranges have created a vast central plateau, the Castilian meseta, that has continental weather and poor soils for agriculture. Some important rivers such as the Guadalquivir and Ebro cross these mountain folds, giving rise to valleys that constitute the most fertile agricultural areas in Spain. However, the soil fertility within the valleys of several regions is limited by low annual precipitation.

Besides the topographic factors, historical factors have stressed the structural differences in Spanish territory. These factors include the resettlement and colonization process which coincided with the reconquest of the land occupied by the Muslims[2]. This process established the current structural differences in Spanish agriculture and rural society: smallholdings in areas north of the River Tagus, larger landowners in the South, together with a large agricultural working population. This agricultural and social structure has been maintained without any great changes for centuries.

Some attempts to modify the situation, essentially during the nineteenth century land alienation reforms and the agrarian reform of the Second Republic (1932), only served to reinforce the accumulation in structural problems that have not only become more difficult to solve, but have also hindered any attempt to implement new reforms. The structural reform-projects undertaken during the Franco regime (1940–1975) were mainly aimed at small farming Spanish regions. The land consolidation programmes were successful, although they had little success in modifying the social structure.

The Modernization of Spanish Agriculture

The modernization process of Spanish agriculture began in the early 1960s. However, the process was not realized within a structural reform plan as in France, where there was some specific legislation on agricultural structures[3]. The process was encouraged by factors outside agriculture and was connected with the more general industrialization process[4].

The new expanding industrial and service sectors provoked a large rural exodus from the southern regions (Andalusia and Extremadura) to the north (Catalonia and Basque Country) and to other European countries (Germany, France and the Netherlands). The exodus mainly consisted of agricultural workers and smallholding farmers, with more than one million people leaving the agricultural sector. The agricultural workforce fell by 47.6% in 1950 to 29.1% in 1970 (Tamames, 1991). Spanish agricultural production directed its market to satisfy the demands of the growing urban population. Furthermore, the agricultural sector was integrated into the agro-industrial

[2] The Muslim presence began in 711, when the Arabian army came through Tarifa, and ended in 1492, when the Catholic Kings and Queens reconquered Granada. However, since 1100, a gradual process of reconquest and resettlement of the occupied land occurred, leading to different types of settlement depending on the different historical circumstances.

[3] In France, in 1960 and 1962 the Parliament passed two important laws on agricultural structures: the loi d'orientation agricole (1960) and the loi complementaire (1962).

[4] In 1959, the Stabilization Plan was passed. This plan is considered by specialists as the end of economic autarky, opening the Spanish economy to international markets. In this way the political isolation, which it had suffered during the Franco regime, was ended.

system (Etxezarreta and Viladomiu, 1989), boosting the productive activities in the machinery sector as well as the industrial sector of fertilizers, pesticides and seeds.

However, there was no planned modernization process and policies which concentrated on production have ignored the structural side of Spanish agriculture, and have not placed it in a framework of specific guidance policy, so that their effects on Spanish agriculture varied. There were some farming regions where farmers quickly incorporated intensive agricultural methods into their farm practice, like the regions of the Mediterranean coast, the Duero and Ebro river basins, or the greenhouse areas in the Andalusian littoral.

By contrast, there were farming regions where farmers followed more extensive agricultural methods, namely, the cereal-producing areas of Castilian meseta and Guadalquivir valley, the cattle rearing zones in Extremadura, and the Andalusian olive regions.

This explains why the modernization process did not help to modify the traditional agricultural structure, whose special characteristics have been mentioned above. In fact, the process produced the coexistence of a large sector of small farms relying on a strongly protectionist price system introduced by the Franco Government, and a reduced sector of large extensive farms that was barely integrated into agri-food markets. Together with them, there was a localized (in the Mediterranean littoral) and specialized horticultural production sector with market-oriented family farms, which has become an important sector in production terms.

The Variety of Agricultural Production Systems

This approach to the modernization of Spanish agriculture and its agroecological diversity leads us to emphasize the existence of a wide variety of agricultural production systems. These systems are characterized, among other things, by having developed a specific relationship with the environment and the natural surroundings. It explains the one-sided views of Spanish farmers concerning environmental problems, that differ from one production system to another. These differences are reflected on a political level, since the Spanish state is federally organized in autonomous regional governments that have significant influence on agricultural and environmental matters.

This diversity is shown by the fact that in some Spanish regions, intensive agricultural systems producing negative effects on the environment already exist. These effects are similar to those produced in other European regions (Belgium, the Netherlands and Denmark), although they have not yet reached such serious levels (Garrido and Moyano, 1994). For instance, the excess use of fertilizers in some fruit growing areas, Valencia or Catalonia, is starting to pollute the groundwater. The massive use of plastics in greenhouse areas, Granada or Almeria in southern Spain, is creating a problem of waste disposal. With irrigation practices, environmental problems are produced through the wasteful use of water by both farmers who have had little training on hydraulics and the obsolescence of pipelines.

Finally, with intensive livestock systems, although the atmospheric pollution by gas emitted from animal waste has not yet reached emergency levels, some of these effects can already be seen in certain regions of Catalonia. In any case, the main problem of these intensive livestock systems is that they are usually based on industrial farms and located near urban centres, thus provoking water and atmospheric pollution.

Nevertheless, with extensive cereal systems, the relationship between agriculture and the environment shows the negative side in a different way, such as soil damage and exhaustion caused by deforestation and the use of intensive cultivation practices. These problems exist in the Castilian meseta and in some areas of the Guadalquivir valley. In contrast, there are in Spanish agriculture some extensive production systems where a harmonious relationship exists between agriculture and environment which should be preserved and supported. Mention should be made of the dehesa system in Extremadura, where the integration between extensive livestock and the environment has been successful (Campos Palacín, 1993), or the olive system in the Mediterranean regions.

With the brief examples showing the diversity of Spanish agricultural systems, it is clear that the relationship between agriculture and the environment in a country such as Spain, which is so heterogeneous from an ecological and structural point of view, is difficult to analyse. This situation is different from those European countries where the land is more uniform topographically, where the climate varies only slightly from one place to another and where the agricultural structure is based uniformly on family farming. In these countries, environmental problems are easily identified with the massive use by farmers of intensive practices aimed at considerably increasing the productivity of their farms.

In Spain, by contrast, there is in every production system a special relationship between agriculture and the environment, and the problems affecting the natural surroundings are not always linked to the intensification of agriculture. Sometimes the problems are caused by farms that have not been modernized (for example, they waste water in irrigation) or due to the abandonment of agricultural activities in certain areas (for example, increasing the risk of forest fires) (Vera and Romero, 1994).

Agricultural Interest Organizations and Environmental Groups

To understand the Spanish case, it is necessary to mention other factors in addition to those mentioned above on the diversity of its topography and heterogeneity of the agricultural structure. These factors include the importance and the level of articulation of environmental issues in public opinion and politics. In contrast to other EU countries, it can be said that environmental concern has relatively recently been incorporated into Spanish public debates and that it has not yet been articulated sufficiently to be considered a political and social priority in Spain (Garrido and Moyano, 1994).

The emerging Spanish environmental movement, however, is helping to widen the knowledge of environmental problems throughout the population. Over the last ten years, there has been an important increase in the number of environmental associations, which can be explained by factors related to both the advance of a new postmaterialist culture among Spanish young people and the consolidation of the democratic context in Spain (Tábara, 1995). However, the environmental movement is still poorly structured at the national level, consisting of small dispersed associations which combine the defence of the environment with other claiming issues, such as pacifism or leaving NATO. The environmental movement does not yet constitute an organized lobby in Spain and is somewhat of a social and cultural network. One of the most outstanding attempts of coordination among environmental organizations is the CODA (Coordinadora de Organizaciones de Defensa Ambiental), an umbrella organization that integrates more than 150 associations.

The lack of concern for environmental issues by the Spanish public – worried mainly about economic problems like the unemployment level and the maintenance of state welfare services – together with the weaknesses of the environmental movement, may explain to a large extent the delay in passing environmental measures by national institutions. In fact, the Spanish Government is not pressed sufficiently by civil society to implement an environmental policy that is of political priority.

The low level of articulation of the Spanish environmental groups even hinders the efforts of regional and national administration to implement a system of neo-corporatist coordination on environmental policy.

So, the creation of regional and national institutions to improve participation of environmental organized interests has met with difficulties because of the dispersion of the environmental groups and the lack of a social movement articulate enough to represent them within such institutions. It is usual that in the Spanish regions, where there is a better articulated environmental movement, more environmental measures are implemented.

In terms of politics, there is no Spanish political party for which environmental issues are a prime concern. The two most important parties in the Spanish parliament – the labour PSOE (Partido Socialista Obrero Español) and the conservative PP (Partido Popular) – are gradually introducing environmental issues into their electoral programmes, but they are doing it in a rather ambiguous way without questioning clearly the economic growth and industrial development. In this sense, more progress in terms of environmental issues has been made by the left electoral coalition called Izquierda Unida (Unit Left), which won 10% of the votes and 15 seats in the last legislative elections in June 1993. This coalition is led by the old communist party, PCE (Partido Comunista Español), and it integrates other political and social groups, such as PSOE's dissident groups and groups linked to some sectors of the environmental movement.

In particular, the environmental movement has made several unsuccessful attempts to be represented in Spanish parliament through a political party following the German model of 'The Greens'. The 'green' vote in Spain has not been very stable and has always been smaller than that won by Green parties in other European countries. It can be said that the electoral support to Spanish Green parties has varied a lot depending on the type of elections (European, national, regional or local elections) (Tábara, 1995). For example, in the last Spanish legislative elections (1993), two environmental parties put forward their candidatures but they failed to win any seats and only gained 2% of votes at the national level. Rather than changing strategies or the coalition among ecologists, it is the lack of a broad enough membership base to support environmental ideas which probably limits the environmental option in Spain. On the other hand, the difficulty of defending a purely environmental position and the spontaneous and opportunist origin of some of these groups are all factors that have not helped to increase people's trust in these parties (Tábara, 1995).

In terms of the relationship between agriculture, the environment and the introduction of sustainable methods, the public debate on the possible polluting effects of agriculture as a diffuse source of pollution did not exist before Spain entered the EU. Nevertheless, after 1986, the debate, for example, on Directive 91/676 on pollution of water by nitrates from agriculture has been rather unusual among the public at large and, in particular, within the agricultural sector. In Spanish agriculture the pollution levels in groundwater are still low, and the problem of emission and storage of manure hardly exists – except in specific areas of Catalonia – due to the low importance given to livestock in agricultural production nationally.

This explains why the public debate on environment and agriculture is more centred on aspects related to the specific context existing in each Spanish region. For instance, there is unanimity among the southern regions about the importance of considering the use of water in agriculture as the main environmental problem. In fact, currently there is a real dispute between farmers and non-farmers about the use of water and the urban population demands that farmers use water more rationally on their farms, given the drought situation. This situation has resulted in some villages having water restrictions for more than ten hours a day and an alarming decline in the quality of the water used for human consumption. Another important issue being incorporated into the environmental political agenda is that of forest fires and the role that farmers can play in their control and prevention. Likewise, the problem of plastic waste in greenhouse areas also poses doubts on the convenience of developing super-intensive agricultural methods in a limited number of regions with land expansion problems.

In short, the awareness of environmental problems and the possibility of creating an environmental national policy are gradually increasing based on the uniqueness of the Spanish situation. It makes the environmental concerns in Spain to some extent different from those of other EU Member States. To understand the real nature of the development of Spanish agri-environmental policy and the role of agricultural interest groups, it is necessary to take into account that in Spain there does not exist a close relationship between farmers' unions, agricultural cooperative federations and the Ministry of Agriculture to defend the general interests of agriculture and collaborate in the implementation of agricultural policy (Garrido and Moyano, 1994). These relations cannot be analysed as if they were an agricultural policy community.

In fact, the situation of Spanish farmers' unions is very unstable and precarious. There are three national farmers' unions – ASAJA (Asociación Agraria-Jóvenes Agricultores), COAG (Coordinadora de Organizaciones de Agricultores y Ganaderos) and UPA (Unión de Pequeños Agricultores) – with important differences from a social point of view which have hindered until now the consensus on agricultural policy (Moyano, 1993). Because of the recent Spanish political history they have not had the opportunity to participate in regular neo-corporatist relationships with the Ministry of Agriculture. The affiliation level of farmers' unions in Spain is very low in comparison with other European countries – no more than 15% of Spanish farmers are affiliated to one or more of the three farmers' unions mentioned above. They are still undergoing internal reorganization and divisions between different organizations occur continually (Moyano, 1993).

This leaves the farmers' unions as a weak policy community – nearer a network than a policy community (Frouws, 1988; Frouws and Van Tatenhove, 1993) – and they have little influence in the implementation of sociostructural policy. However, the new agri-environmental policy is seen by them as an opportunity to recover a place in rural society which has not been achieved with the market policy. This explains the initiatives taken by some farmers' unions, like COAG and UPA mainly, who represent small farmers, to coordinate their activities with environmental groups and adopt the sustainability issues in their debates. Examples of these initiatives are the Plataforma Rural (Rural Platform) adopted by COAG and some environmental associations, or the collaboration agreement between UPA and SEO (Sociedad Española de Ornitología: Spanish Association of Ornithology) for the development of joint actions capable of harmonizing agricultural development and environmental protection.

The Agri-environmental Measures: Regulation 2078/92

Although the sociostructural regulations have not had a great impact on introducing more sustainable methods in Spanish agriculture, the new accompanying measures of the last CAP reform seem to have more possibilities of success.

In particular, the afforestation programmes and the environmental protection programmes are encouraging the debate on the future role of agriculture in many poor Spanish rural areas, where agricultural activity has only been possible thanks to the existence of a price policy capable of guaranteeing the incomes of farmers. After the last reform of the CAP implying a gradual reduction of the guaranteed prices, the new accompanying measures are seen by farmers' unions and by national and regional public administrations as a good opportunity for farmers in these areas to find a complementary source of income. That is why the Spanish Ministry of Agriculture and the farmers' unions are giving so much attention to such measures. With respect to the implementation of the specific regulations, Regulation 2078/92 partly replaces Regulation 2328/91 – it replaces Article 3 concerning extensification and Articles 21 to 24 related to aids for ESAs (MAPA, 1994). In the previous period of CAP, extensification measures had not been implemented in Spain and the aids for sensitive areas were ready to be applied but were not implemented in three selected zones with specific environmental problems, namely: the aquiferous zones in western Mancha and Campo de Montiel in Castilla-La Mancha, and parts of Villafáfila and Madrigal-Peñaranda in Castilla y León. Nevertheless, those measures have had to be adapted to the new regulations set by Regulation 2078/92.

The implementation of Regulation 2078/92 in Spain has been designed through three different sets of measures. First, a set of horizontal measures to be applied to the whole national territory. Four measures have been selected, namely:

1. to encourage organic farming, decreasing the use of fertilizers and pesticides;
2. extensification in arable farming, including the use of more environmentally friendly agricultural practices;
3. the protection of breeds in danger of extinction;
4. environmental training, including activities towards the improvement of farmers' knowledge about agricultural and forestry practices suitable to the environment.

The total budget devoted to these four measures is 70,914 million Pesetas. The aids to encourage extensive production systems absorb the majority of that budget. Castilla-La Mancha and Castilla y León are the most favoured regions with respect to these horizontal measures (see Table 6.1).

These measures have already been passed at the national level as the Real

Table 6.1. Total budget for horizontal measures by region, 1994–1998 (million Pesetas). (Source: MAPA, 1994.)

Autonomous region	Organic farming	Breeds in danger of extinction	Training	Extensive cereal production	Total
Andalucía	1,014.1	461.7	510.0	7,324.5	9,310.3
Aragón	313.0	184.7	382.5	9,978.7	10,858.9
Asturias	50.0	277.1	26.8	0	353.9
Baleares	151.9	13.9	31.9	100.0	297.7
Canarias	182.5	309.4	255.0	0	746.9
Cantabria	0	138.5	17.8	0	156.3
Castilla-La Mancha	434.8	166.2	318.7	14,047.7	14,967.4
Castilla y León	424.1	461.7	510.0	12,097.8	13,493.6
Cataluña	720.9	92.3	223.1	582.0	1,618.3
Extremadura	356.0	230.9	191.2	7,095.4	7,873.5
Galicia	119.5	120.1	382.5	0	622.1
Madrid	79.0	9.2	19.1	1,732.4	1,839.7
Murcia	110.0	32.3	95.6	3,637.4	3,875.3
Navarra	70.0	18.5	12.7	955.9	1,057.1
La Rioja	144.3	69.3	95.6	300.0	609.2
Valencia	439.3	60.0	127.5	2,607.0	3,233.8
Pais Vasco	0	0	0	0	0
Total	4,609.4	2,645.8	3,200.0	60,458.8	70,914.0

Decreto (Royal Decree (RD)) 51/1995 but it is still necessary for a specific regulation to be passed by each regional government if it is to be developed and implemented in each region. The implementation will be through a specific contractual relationship between the farmer and the regional government. For Objective 1 regions, these measures will be financed as follows: FEOGA-guarantee section, 75%; national government, 12.5%; regional government, 12.5%. For remaining regions, 50% is reimbursed by FEOGA and 50% is provided by the national and regional governments in equal proportions. The projected effects along with the financial resources of these four measures can be seen in Tables 6.2, 6.3 and 6.4.

In common with the implementation of Article 19 of Regulation 797/85, one of the reasons for delay of the development and normal application of Regulation 2078/92 is the lack of coordination and the problems of administrative competence in autonomous regions. These problems occur not only between national and regional governments, but also between the different public departments within each regional government. For example, in some regions there have been new departments of environment created which have not yet defined their sphere of competence, provoking confusion among farmers and causing some departmental confrontations that are delaying the implementation of the first group of (horizontal) measures of the agri-environmental programme at the regional level. So far no regional

Table 6.2. Horizontal measures: scale of participation. (Source: MAPA, 1994.)

Measures	1994	1995	1996	1997	1998 *et seq.*
Annual number of units					
Extensification (ha)	0	808,244	1,151,780	1,435,633	2,443,187
Training (trainees)	340	4,300	4,900	4,900	4,900
Breeds in danger of extinction (UGM)	10,000	45,000	53,900	53,900	66,187
Organic farming (ha)	7,630	15,260	20,598	20,598	28,130

UGM = Livestock Unit

Table 6.3. Horizontal measures: total cost by measure per year (million Pesetas). (Source: MAPA, 1994.)

Measures	1994	1995	1996	1997	1998 *et seq.*	Total
Extensification	0	4,000	6,200	7,100	43,158.6	60,458.6
Training	50	500	640	640	1,370.0	3,200.0
Breeds in danger of extinction	100	450	539	539	1,017.8	2,645.8
Organic farming	250	500	675	675	2,509.4	4,609.4
Total	400	5,450	8,054	8,954	48,055.8	70,913.8

Table 6.4. Horizontal measures: EU co-financing by measure and per year (million ECUs). (Source: MAPA, 1994.)

Measures	1994	1995	1996	1997	1998 *et seq.*	Total
Extensification	0	17.35	26.90	30.80	187.26	262.31
Training	0.21	2.16	2.76	2.76	5.92	13.81
Breeds in danger of extinction	0.43	1.94	2.33	2.33	4.39	11.42
Organic farming	1.01	2.02	2.75	2.75	10.22	18.75
Total	1.65	23.47	34.74	38.64	207.79	306.29

government has developed these horizontal measures.

The second group of measures are for zones selected by the national government in agreement with the regional governments. These measures cover the following:

1. Areas close to National Parks.
2. Wetlands included within the RAMSAR Convention on 'Wetlands of International Importance Especially as Waterfowl Habitat'.

3. Zones of Special Protection for Birds (Zonas de Especial Protección de las Aves, ZEPAS).

There are different measures applicable to these selected areas: integrated pest management, the reduction of stocking rate and the control of erosion are some of them (see Table 6.5). The only two actions applied in 1994 have been the protection of fauna (mainly birds) in inner regions (d1) and the conservation of irrigation water (d6).

The areas with specified environmental problems addressed by these actions have been of Castilla-La Mancha and Castilla y León. The area affected by each measure, the budget devoted to them and the EU co-financing for the period 1994–1998 are indicated in Table 6.5.

The purpose of all these grants is to compensate farmers for the loss of income caused by not using some harmful agricultural techniques or carrying out some practices for bird protection. The funds for financing them are provided as follows: 75% FEOGA-guarantee section; 12.5% national government; 12.5% regional government (Objective 1 regions). Up to now only the aid programme for areas close to National Parks has been passed (April, RD 632/1995), and the other two aid programmes are waiting to be passed[5]. The area affected by this aid programme for areas close to National Parks will be 329,863 ha (see Table 6.6).

The third group of measures consists of the zonal programmes for which it is the responsibility of regional governments to define some areas of their territories as special zones from an environmental point of view. Each regional government is then in charge of developing and controlling the implementation of measures applied to those zones. For instance, in northern Spain (Galicia, Asturias or Cantabria) grazing lands and forest farms are the prevailing systems to be protected; in central Spain, there are large areas devoted to extensive agricultural production systems constituting a very beneficial habitat for some species of birds; the exotic landscape of Canary Islands can also be mentioned as a special ecosystem to be preserved. The measures to be taken in all these regions will affect 2,390,890 ha and 67,720 farmers, and the planned budget to apply them is 95,516 million Pesetas (see Table 6.7). In this case, zonal programmes are only co-financed by EU (75% from the guarantee section of FEOGA) and regional governments (25 per cent), with the national government not participating.

Generally, farmers and their organizations see the agri-environmental measures as an alternative way to support their incomes. Besides, in some environmentally sensitive areas these new programmes can serve as a feasible solu-

[5] At the time of writing this chapter, the Spanish government has also passed the RD regulating the aid programmes for Wetlands included within the RAMSAR Convention and for Zones of Special Protection for Birds (ZEPAS). This RD, however, has not been published and therefore it has not come into force yet.

Table 6.5. Measures to be applied in selected zones and in regional government zonal programmes. (Source: MAPA, 1994.) Values are per cent of total.

Measures		1994			1995			1996			1997			1998 *et seq.*		
		A	B	C	A	B	C	A	B	C	A	B	C	A	B	C
a	Integrated pest management				0.8	2.5	2.2	0.7	2.2	1.9	0.5	1.9	1.6	0.5	1.4	1.2
b	Converting arable to pasture				2.2	2.9	2.9	2.0	3.0	2.9	1.7	2.9	2.8	1.7	3.0	2.9
c	Stocking rate reductions (livestock units)				0.5	1.8	1.5	0.4	1.8	1.6	0.6	2.0	1.6	1.7	3.8	3.2
d1	Protection of fauna: inner regions	44.0	8.1	8.1	56.8	25.6	26.1	60.9	29.7	30.5	65.5	31.7	31.5	62.9	32.4	33.0
d2	Protection of fauna: wetlands				4.1	7.1	6.5	3.6	7.1	6.6	3.1	6.9	6.2	3.5	7.2	6.0
d3	Landscape conservation and fire protection				11.6	9.2	9.1	10.3	10.2	8.8	9.7	9.5	9.2	9.3	8.1	8.5
d4	Erosion control				7.3	9.3	9.3	7.3	10.1	10.2	6.1	9.4	11.2	7.3	13.8	14.2
d5	Environment: Canary Islands				1.7	5.1	5.4	1.3	4.1	4.4	1.0	3.5	3.6	1.1	3.3	3.4
d6	Conservation of irrigation water	56.0	91.9	91.9	5.9	20.4	21.5	4.4	16.4	17.4	3.3	14.0	14.4	3.2	1.2	2.8
e	Conservation of abandoned lands				6.8	9.3	9.5	7.2	9.6	9.8	6.8	11.7	11.6	6.5	11.3	11.4
f	Land set-aside				1.2	6.0	5.3	1.1	5.3	5.4	1.1	5.8	5.8	1.8	14.1	12.9
g	Land management for recreation				0.9	0.4	0.4	0.6	0.3	0.3	0.6	0.2	0.3	0.5	0.3	0.3
h	Demonstration projects				0.1	0.4	0.3	0.1	0.3	0.3	0.1	0.3	0.3	0.1	0.2	0.2
Totals		136.2	3.1	14.2	1,363.7	15.9	70.0	2,100.0	22.0	96.0	2,700.0	25.7	115.9	2,800.0	75.0	329.0

A = Surface (million ha).
B = Financial resources (million Pesetas) × 10^3.
C = EU co-financing (million ECUs).

Table 6.6. National parks, RAMSAR sites and ZEPAS. Measures, characteristics and budgets, 1994–1998. (Source: MAPA, 1994.)

Autonomous region	National Park	Measures (see Table 6.5)	Livestock (UGM)	Area (hectares)	Farmers (numbers)	Budget (million Pesetas)
Andalucía	Doñana	b,c,d2,f	3,500	32,350	4,236	5,250[1]
Aragón	Ordesa and Monte Perdido	d3,e	–	53,000	1,705	3,160
Asturias	Picos de Europa	d3,e	–	34,750	1,792	2,537
Canarias	Timanfaya	d5	–	6,483	190	1,682
	Teide	d5	–	4,190	215	556
	Garajonay	d5	–	4,990	170	610
	Caldera de Taburiente	d5	–	4,400	180	649
Cantabria	Picos de Europa	d3,e	–	9,000	528	655
Castilla-La Mancha	Tablas de Daimiel	d6	–	90,000	2,500	14,171
	Cabañeros	c,d1,d3,e,f	2,000	35,450	2,475	3,292
Castilla y León	Picos de Europa	d3,e	–	30,750	1,702	2,437
Extremadura	Monfragüe	d3,e,f	–	24,500	1,000	2,802
	Total national parks		5,500	329,863	16,693	37,801
	Zones					
	RAMSAR sites	b,d2,f		12,500	2,040	1,125
	ZEPAS	c,d1,d2,e,f,g	5,573	66,496	5,410	7,244

[1]Demonstration projects included.

Table 6.7. Zonal programmes: measures, site characteristics and budget, 1994–1998. (Source: MAPA, 1994.)

Autonomous region	Types of zone	Measures (see Table 6.5)	Livestock (UGM)	Area (thousand hectares)	Farmers (thousands)	Cost (million Pesetas)
Andalucía	Dehesas, olive groves and landscapes	b,d3,d4	–	173.0	7.3	12,250
Aragón	Monegros zone, Pyrenean Valley, Gallocanta	b,d1,e,f	–	38.6	3.1	2,948
Asturias	Grazing land in common	d3	–	147.0	10.0	3,675
Canary Islands	Three islands	d5,e	–	14.1	1.7	1,744
Castilla-La Mancha	Cereal bird habitats	d1	–	256.0	3.0	8,660
Castilla y León	Cereal bird habitats, traditional farming, forest fires	b,d1,e,f	–	1,289.9	14.9	35,138
Cataluña	Rivers, parks, lakes, livestock systems, traditional species, fruit and nuts	a,b,c,d1,d2,d3,f,d4	5,500	150.1	10.8	14,244
Extremadura	Dehesas	c,d1,g	5,400	226.2	3.3	4,058
Galicia	Forests, wetlands, vines, grazings, mountain	a,d1,d3,d4,f	–	29.3	6.3	3,674
Madrid	Traditional farming, metropolitan green area	d4,f	–	11.0	2.4	2,975
Murcia	Littoral, mountain and wetlands, grapes	a,d1,e,f	–	6.6	0.9	1,095
Navarra	West Pyrenees, other areas	b,d3,d4,e,f	–	11.7	0.8	995
La Rioja	Protected areas	d3	–	1.0	0.1	50
Levante	Natural park, valleys, hunting and other reserves	d2,d3	–	36.3	3.1	4,010
Basque country	Environment, traditional farming	d1,d2,d3,g	–	–	–	–
Total in zonal programmes			10,900	2,390.9	67.7	95,516

tion to conflicts between farmers and environmental groups. In fact, they offer a means for the environmental movement to attain its aim of conserving special natural areas while compensating farmers for the loss of income from not following some agricultural practices which could be harmful to the environment.

Conclusions

It seems too early to make a full appraisal of the implementation of the recent agri-environment policies in Spain. Only some of these policies have already been regulated and a majority of them are still waiting to be adopted and developed. In the agri-environmental field, the Spanish Government's actions have followed the steps indicated by EU and generally Spanish farmers have not shown much interest in them. Insufficient information, the lack of participation of farmers' unions in their implementation and the small amount of money devoted to those aids, can be some of the reasons for this lack of interest. However, the new environmental concern introduced within the Common Agricultural Policy and more specifically the accompanying measures passed in the CAP-reforms of 1992, have served to energize the debate within the agricultural sector, specially within the farmers' unions. This debate embraces the relationships between agriculture and environment and it has taken the notion of sustainability as a reference framework to put order on it. In the Spanish case, it becomes particularly complex due to the great heterogeneity of agricultural production systems and therefore of the different relationships between the environment and agriculture.

The implementation of agri-environmental measures in Spain has been remarkably slow compared with other European countries. The lack of economic resources, the diversity of agricultural ecosystems and the problems between national and regional governments regarding scope and level in defining their actions, are reasons for that delay.

In order to make its implementation feasible, all measures aimed at promoting the introduction of sustainable farming methods should not be considered as external actions to the agricultural community. On the contrary, the public authorities have to persuade this agricultural community (mainly farmers and farmers' unions) to identify itself with those programmes and contribute to their elaboration and implementation. In this sense, Spain has a comparative advantage in relation to other countries dominated by a powerful 'agricultural policy community'. The weakness of the agricultural lobby may well be an advantage to the introduction of new sustainable agricultural methods and the implementation of agri-environmental policy.

For Spanish farmers' unions this policy provides a very good opportunity to improve their image and social presence in rural society and to play a significant role as intermediate agents in the new relationship between the State and the farmer.

FRANCE

JEAN-MARIE BOISSON AND HENRY BULLER

Introduction

The relationship of farming to the physical environment in France is characterized by a deep-rooted and persistent agrarian tradition whose origins stretch as far back as the settlement of the French territory (Braudel, 1986). The human mastery of nature, whether it be the landscape, ecosystems or livestock is anchored in French cultural history, enshrined in the writings of Descartes, codified in the Civil Code and encouraged by a land tax system that tends to favour the exploitation over the protection of land. In many regions, the contemporary agricultural land-use structure still bears the traces of the original medieval land and social system (Dion, 1939), while the role and function of the rural environment are still largely seen in terms of agricultural use and human occupation rather than landscape protection and the maintenance of ecological diversity (Gavignaud, 1990). The history of modern France, from the *Ancien Régime* to the *Cinquième République* reveals the continual reinforcement of this exploitive, utilitarian and fundamentally anthropomorphic attitude to the natural world, one directly encouraged by post-revolutionary political and social development which initially strengthened and subsequently protected the position of a smallholding based, individualist and peasantist agricultural structure (Duby and Wallon, 1976). As late as 1950, one in four of the French working population were in agriculture, while half the national population lived in rural areas.

This historical experience is closely linked to the physical make up of France. As one of the largest European states, France is a northern and a Mediterranean nation with both maritime and continental façades. The scale of French space, the diversity of its ecosystems, climatic regimes and landscapes and the relative concentration of its urban structure have contributed to the development of this particular and original conception of the environment and of nature which has characteristically been expressed in two dominant attitudes. First, the rural environment, its spatial extent and its physical diversity, has been seen as something to conquer, occupy and exploit.

Farmers are universally known as *exploitants agricoles* and their holdings *exploitations*, with none of the negative connotations such words evoke in the English language. Second, given their heterogeneity and diversity, the French rural environment and landscape have been seen as infinitely capable of accommodating both environmental damage and nature protection without the whole ensemble being in any way threatened by either (Buller, 1992).

The French experience in agricultural–environment relations has therefore been founded upon a fundamentally different historical and cultural conception of the rural/natural environment, its role and function as well as its qualities, than that which operates in Britain. Concern for the agricultural environment initially developed out of concern, first, for the sustainability of rural communities, second, from hunting and, third, from rural resource protection; the key common feature being that these concerns are primarily inward-looking, in that they serve the agricultural population, rather than being illustrative of the increasingly dominant environmental concerns of a non-rural population. The creation of the first hunting reserves at the turn of the century (Vourc'h and Pelosse, 1985), postwar rural development policy (for example the *Rénovation rurale* programme of the State planning agency DATAR, initiated in 1967), the *parcs naturels régionaux* or PNR (created following legislation in 1967) and the designation of Less Favoured Areas can all be seen in these terms.

Farmers and the Environment in France

The agrarian tradition

Postwar agricultural modernization, though it utterly transformed French agricultural productivity to make France the largest agricultural producer in Europe and one of the principal exporting nations of the World, did not bury the hitherto dominant agrarian ideology. Its late survival and continued strength concerns us in two ways. First, it continues to influence public and political attitudes to the agricultural community and to agri-environmental relations. Although French farming has undergone a profound transformation since the end of the Second World War, the agrarian tradition which predates this transformation remains a powerful force in defining the role of farmers in the national identity and in equating agricultural health with rural well-being. Second, it underlies the basic tenet of French agriculture–environment relations which holds that farmers are the traditional creators, users and protectors of the rural landscape.

If anything the modernization process has raised the profile of French agriculture within national identity to the extent that agricultural crises are continually portrayed as national crises. Farmers remain for many the natural and defining users of French rural space (Berlan-Darqué and Kalaora, 1992; Kayser *et al.*, 1994) and continue to play a central role in the wider identification of French nationhood (Tacet, 1992). The postwar period was largely one of increasing rural mono-functionalism as the drive to agricultural modernization and agricultural self-sufficiency dominated rural policy in all but a few specifically protected areas. The period 1945–1975 which saw agricultural productivity rise also saw, however, the French rural population decline and the agricultural population plummet. Observers chronicled the decline of the peasantry and of local rural culture, documented the increasingly spatial inequalities within the French territory or revelled in the power and extent of the French agricultural transformation. The critical agricultural–environmental concern in France has therefore been agricultural land abandonment and the return of once cultivated or grazed lands to the natural state (Comolet, 1989). Such a process strikes a deep emotional cord in a nation that has traditionally prided itself on its ability to conquer nature (Terrasson, 1988). Thus the central thrust of French agri-environmental policy has been and continues to be centred essentially on retaining threshold densities of farmers in marginal areas. By contrast, only relatively recently have economists and scientists addressed the negative environmental impact of agricultural modernization (Mahé and Ranelli, 1987). The thrust of academic and policy concern was not therefore the impact of agricultural modernization on the rural environment but upon the pre-existing rural/agricultural social system of peasant farming.

The impact of rural social change

The late urbanization of France has meant that broad social demand for environmental protectionism has also been late to emerge. Whereas we can identify elite concerns for wilderness protection as early as the late nineteenth century, these remained highly marginal within a predominantly rural and agrarian society (Leynaud, 1985). The postwar demographic and social transformation of rural areas and the associated rapid growth of towns and cities have however been important elements in the emergence of concern for the environmental effects of modern agriculture.

On the one hand, socioeconomic change in rural France which, from the late 1960s onwards reversed decades of rural demographic decline, saw the growth of a non-agricultural rural population, with its attendant claims for a voice in local rural policy-making, and the demographic marginalization of the agricultural population. On the other hand, the rapid expansion of the urban population has clearly led to the emergence of new social demands for rural access and landscape protectionism (Viard, 1990).

However, we should take care not to simply read off from these new social forces inevitable challenges to traditional rural management approaches. Four caveats need to be introduced. First, although numerically and proportionally in decline, the agricultural population retains a strong position within French rural politics particularly in the more remote rural regions where a significant proportion of local political leaders continues to be drawn almost exclusively from the agricultural population. Second, the newly urbanized French population retains strong generational links with its rural predecessors to the extent that the rural experience is, for many, a familiar one. Third, the symbiotic relationship of rural to agricultural policy, born out of the agrarian ideology referred to above, remains a central feature of French countryside management (Kayser, 1994). Fourth, rural areas have not, until comparatively recently, been associated with notions of environmental decline. Pollution is still predominantly perceived as an urban issue while the physical extent and diversity of the French countryside has tended to dilute the broader impact of concern for the degradation of particular zones. As a result of these varying factors, neither the rural policy community nor civil society, in the form of pressure groups and lobbies, have been particularly vocal in their calls for change both in terms of the regulatory control over agriculture or of specific policies aimed at reconciling modern farming and environmental protection. The only real exception to this otherwise general absence of confrontational local politics is Brittany, and particularly the Breton coastal regions where environmental pressure groups, whose members are frequently drawn from the new rural population of these areas, have emerged as a major force in pressing for regulatory controls on agriculture (Bodiguel and Buller, 1989).

Agricultural policy management

A further and vital element in the agricultural–environmental equation in France has been the particular administrative structure of that nation.

It is important to locate the development of agri-environmental measures within the evolving context of national and local agricultural and environmental policy making. Despite far reaching political decentralization, which was introduced in 1982, France remains a quintessentially centralized state, not only in its political and legal structure but also in its system of specialized civil service corps. This is particularly so for the agricultural sector which is composed of a national ministry and a hierarchy of local administrative tiers, is endowed with its own secondary and higher education system, its own social security and an independent research establishment and co-manages agricultural policy with a professional farming community that, until now, has been characterized by the strength and breadth of its trade-union unity.

The Common Agricultural Policy was adopted in France on the basis of clear divisions of responsibility between, on the one hand, the inter-professional and semi-public bodies, who controlled product markets and, on the other hand, the Ministry of Agriculture, its local bureaux (DDA) and the semi-public organization CNASEA who were collectively charged with agricultural structural policy. This tightly knit policy community, linking policy makers, farmers and the food industry, has traditionally been a major player within Brussels and the CAP. Furthermore, within France, it has largely dominated rural policy making, with the broad acquiescence and even support of the French population and local political leaders. As a result, the rural environment has consistently been subsumed into the broad rural-agricultural policy domain.

It is against this backdrop that we seek in this chapter to consider the late emergence of agri-environmental measures in France. What is of fundamental importance is that the emergence of such measures, dating essentially from the 1980s has not been wholly the result of domestic concern for the environmental impacts of farming practices, or of increasing popular anxiety for rural sustainability or for potential ecological disaster. The critical force in the adoption of such measures has rather been the European Community and growing internationalization both of the environmental and the agricultural agendas. That this process of Europeanization should coincide with a more recent fundamental shift in the role and perception of rural areas within France (Lévy, 1994) indicates that the rural future as a whole rather than the specific needs to reduce agricultural production or agricultural pollution is ultimately what is at stake.

Early National Agri-environmental Measures

Paradoxically, for a nation whose traditionally holistic approach to rural management, albeit under the agricultural banner, has long been a characteristic feature, the emergence of specific concern for the protection of the French rural environment reveals the strict separation, in spatial, policy and administrative terms, of agricultural enterprise, on the one hand, and nature protection, on the other. Initial concern for the latter was limited to small, tightly regulated zones such as nature reserves and the central zones of national parks where agriculture was not generally practised. As Moreux (1994, p. 46) points out, 'Agriculture and Environment effectively divided up the national territory.' The creation in 1971 of the Ministry of the Environment perpetuated and extenuated this separation by similarly dividing up administrative responsibilities (pollution control to Environment, rural management to Agriculture) and policy styles (regulatory on the one hand, negotiative, coorporatist and contractual on the other). Beyond the defined environmental zones, farmers were considered the most suited and best placed managers of the rural/agricultural environment. This was very much the general situation up until the 1980s. In certain particular local contexts, however, agricultural and environmental objectives have developed in a mutually reinforcing way. Here, the characteristic separation of the two policy domains has been, to some extent broken down as national and local policy makers have sought to reconcile agricultural maintenance and development with species and landscape protection. These limited examples form the precursors to French agri-environmental measures.

The role of traditional farming methods in maintaining landscapes and the importance of retaining a viable agricultural population within certain national parks and their peripheries were acknowledged from the outset of French national park policy in 1960. Within the Cévennes, the only park to include an inhabited central zone, two early measures were adopted in the early 1970s, the 'Mazenot contracts' and the 'environmental plans' though similar schemes have now been extended to the Pyrénées and Ecrins parks. Faced with the possible disappearance of 64 of the 109 active farms in the central zone, the Mazenot contracts, launched in 1973, offered moneys (an average of FFr 3400) to marginal farmers in return for the undertaking of specific landscape and access improvements on their holdings. By 1988, some 262 contracts had been established which, in total, accounted for FFr 2,359,406 (Mousset, 1992).

Closer in their conception to British management agreements than to Environmentally Sensitive Area procedures, in that they seek specific undertakings rather than the maintenance of traditional agricultural practices, the Mazenot contracts have been an interesting but isolated step in the development of agri-environmental measures. However, they are limited to the Cévennes park and offer no constraint to potentially more damaging intensification, a criticism even more appropriate to the environmental plans which, despite their goal of landscape protection have tended to aid rather than hinder modernization permitting increases in stocking rate and chemical inputs.

The dualism evoked in the Cévennes example, between the need to maintain an agricultural sector and the need to fit agricultural development within broad environmental considerations, is even more apparent in the PNR policy of the 1970s. Closer to British national parks, the 26 French regional parks have a broad remit for rural protection and development. As in the Cévennes, the agricultural issue has largely been one of economic viability rather than environmental sustainability. A number of management agreement schemes have been established in a variety of parks, particularly in those that include sensitive ecosystems, such as the wetlands of the Marais Poitevin and the Marais Vernier, the pastoral systems of the Luberon and the bocage of the Brotonne Parc.

The national and regional park management agreements represent the principal thrust of voluntary agri-environmental measures in France until the entry of the European Community into this field. Although other pre-EC schemes exist, particularly in designated nature reserves, protected sites and on land held by bodies such as the *Conservatoire des Espaces Lacustres et du*

Littoral or the *Agences de Bassin*, these are, for the most part, obligatory rather than voluntary and are defined by statute and regulation.

One final 'structural' measure of considerable importance in transforming the agricultural environment in France has been the policy of voluntary holding regroupment, known as *remembrement*. Established during the 1939–45 war (though antecedent legislation dates from 1918) as a means of rationalizing the distribution of fragmented holdings, *remembrement* policy has been extended over some 43% of the French agricultural land area, particularly in the northern *départements* (Ministère de l'Environnement, 1989). In doing so, it has resulted in substantial change not only to the agricultural landscape and to ecosystems, through hedgerow and tree removal, but also to the broader agricultural environment (Brunet, 1992). Increased levels and rates of surface runoff have exacerbated the pollution of water courses by chemical entrants while also being held partly responsible for the spate of serious flooding in recent years. As a result, a number of admittedly timid and highly localized measures began to emerge in the early 1980s in an effort to protect traditional bocage landscapes (CEC, 1986). More recently, the Loi Paysage of 1993 has sought to introduce more sustainable criteria into *remembrement* policy by conserving individual hedgerows or by limiting their removal near water courses.

The Entry of the EC into Agri-environmental Policy

The conciliatory relationship between farming and environmental protection, founded upon the three alternative strategies of physical separation, limited contractual negotiation and the broad equation of agricultural practices with rural management began to break down during the 1980s. Three reasons may be advanced for this. First, a growing awareness of the impact of agricultural pollution particularly of water courses emerged in the 1970s, associated with a growing readiness on the part of the Ministry of the Environment to directly challenge the agricultural policy community on its environmental record. As a result, regulations, making certain agricultural activities such as intensive husbandry classified operations, were introduced in 1976. In addition, the ministries of Agriculture and the Environment set up in 1984 a joint committee for the reduction of nitrate and phosphate pollution of agricultural origin (CORPEN). Second, the 1980s witnessed growing popular and political concern for the excesses of *productivism* and thereby for the necessity of agriculture-related environmental change. Pressure groups began to challenge, in particular, the policy of *remembrement* and the multiplication of intensive husbandry units in ecologically sensitive areas (Bodiguel and Buller, 1989). Third, and crucially, the European Community itself began to address directly the agriculture–environment debate within the context of the CAP by proposing the creation of environmentally sensitive areas and by encouraging the extensification of farm operations.

In assessing the development of agri-environmental measures in France, it is therefore difficult to dissociate the entry of the European Community into the specific agricultural environment debate from both the impact of Community environmental policy as a whole and from the series of rural policy shifts that France has undergone since the early 1980s. On the one hand, the introduction of Community drinking water quality legislation in 1979 has had a major impact upon raising awareness, both within the agricultural sector and among the wider public, of the pollution of drinking waters by farming activities. On the other hand, the precipitous decline of the agricultural population in recent years, coupled with the centralizing tendency of the CAP (Hervieu, 1993) and, later, with imposed limitations to agricultural productivity, have undoubtedly raised the profile and the attractiveness of those measures designed not to raise farm output *per se* but to maintain sustainable forms of agriculture in areas otherwise peripheral to mainstream production and to recompense farmers for their widely acknowledged and rarely challenged role as protectors of rural resources and landscapes. The fact that the French agricultural community resisted the introduction of such measures for so long is a testament to the resilience of the agrarian and productivist ideology.

Community involvement in specific agri-environmental measures within France theoretically dates from Regulation 1872/84 relating to the Community Environmental Actions (ACE) later replaced in 1991 by the Community Actions for Nature (ACNAT) initiative. Closely linked to the implementation of the Wild Birds Directive 79/409, this measure has sought to provide financial aid (up to 50% coming from EC funds) for wild bird protection schemes in the form of grants to management bodies such as wildlife groups, regional park authorities of the *Conservatoire du littoral* who are then at liberty to establish management agreements with landholders and farmers within the defined action areas. To date, however, these actions have had only a minimal impact in France. Since 1984, some 75 operations have been established and few involve the agricultural sector (Ledru, 1994), the only major exception being the Crau region in the Vaucluse where ACE measures have been adopted to protect the natural grassland by encouraging the retention of traditional sheep grazing patterns. This site has also been designated an ESA, and we return to it below. The low take-up rate of the ACE/ACNAT measures in France is revealing. It demonstrates the general distance between the agricultural and the environmental sectors both at the national and the local level. As with domestic nature reserve policy, the zones selected have not generally been in agricultural use but are often held in private or state ownership. It serves also as a indicator for the French agricultural community's response to later agri-environmental measures contained within EC regulations 797/85, 1760/87 and 2078/92.

Regulation 797/85

The application of Article 19 of Community Regulation 797/85, the first specific agri-environmental measure at the European scale, confirmed the particularity of the French position with respect to the farming environment. In comparison with Britain and the former West Germany, France was a late participant in the application of Article 19. Unlike other Member States, France had only a very limited experience in domestic agri-environmental measures and was not a major contributor, to the elaboration of the Article 19 policy within the Community (Ledru, 1994). Indeed, French representatives to the *Comité Spécial Agricole* (CSA) were initially reluctant to agree to the linking of agricultural and environmental protection contained within the British proposition for the inclusion of agri-environmental measures in what ultimately became Article 19 of Regulation 797/85 (Boisson, 1986).

Thus, the first pilot ESAs (the Vercors, the Crau and the Marais de l'Ouest) were not identified until 1989, after EC Regulation 1760/87 had introduced the possibility of 25% Community co-financing for accepted schemes. Even then, the Ministry of Agriculture remained reluctant to give its unqualified support for a policy that, to the agricultural community, appeared to impose production limits on farmers, to belittle their role as producers while labelling them as simply gardeners of nature and to implicitly designate them as polluters and bad countryside managers (Alphandéry and Deverre, 1994).

From the outset, the application of Article 19 in France has reflected specific national concerns relating more to agricultural development issues than to landscape and species protection (Buller, 1991). To implement Article 19 policy on the ground, a specific procedural and management structure, the *OGAF-Environnement* (*Operation groupée d'aménagement foncier* or integrated land management operation) was set up. However, these procedural structures were, to a large extent, borrowed from previous OGAF models concerned with farm improvement, a common ancestry that reveals the essentially agrarian orientation of agri-environmental policy.

Interpreted principally as an agricultural protection measure rather than as part of environmental protection policy, Article 19 became in a number of areas a means of supplementing farm incomes in less favoured areas or accompanying the process of extensification. During the initial experimental phase, potential target zones were identified by the type of environmental issue to be addressed:

1. areas of intensive farming where the risks of water pollution were high;
2. areas of particular importance for rare and threatened species;
3. areas of extensive pastoral agriculture threatened by farm abandonment;
4. areas threatened by forest fires.

By 1991, some 22 ESA projects had been accepted by the National Technical Committee for Agriculture and the Environment (CTNAE), two of which concerned areas threatened by water pollution from agriculture and four, areas potentially subject to forest fires. At this point, however, only 17 were accepted by the EC for co-financing by the Commission. Later the CTNAE, ruled that the use of ESA payments to encourage farmers to reduce pollution ran counter to the polluter pays principle, while grant aiding the introduction of extensive forms of grazing to reduce the possibility of forest fires was considered contrary to the principle of reducing beef and sheep production within the Community. Since 1992, no ESAs relating to these two issues have been introduced (Leuba and Simonet, 1992).

After an initially slow start, the number of designated ESAs has grown rapidly since 1991 (Fig. 7.1). At the present time, 61 ESAs have been established in France, although to this figure one should add two additional schemes (Vosges and Isère) which are entirely financed by local government and the EC with no national Ministry funds being allocated (Barrue-Pastor, 1994). Of the 61 ESAs, some 34 have already been accepted for part-funding from Brussels (Table 7.1). Collectively, the French ESAs cover 737,000 ha of which some 205,200 ha are eligible for payments. These figures represent 2.5% and 0.7% of the agricultural area of France. The cost of current Article 19 policy stands at FFr 92.5 million per annum, an average of FFr 1.5 million per ESA.

The current orientation of approved programmes reveals strong spatial, agricultural and environmental particularities reflecting primarily the concerns of the agricultural sector (Fig. 7.2).

Programmes aimed at the protection of sensitive biotypes are concentrated in the coastal marshes of western France, many of which are affected by the implementation of the EC birds and habitats directives and have, in a

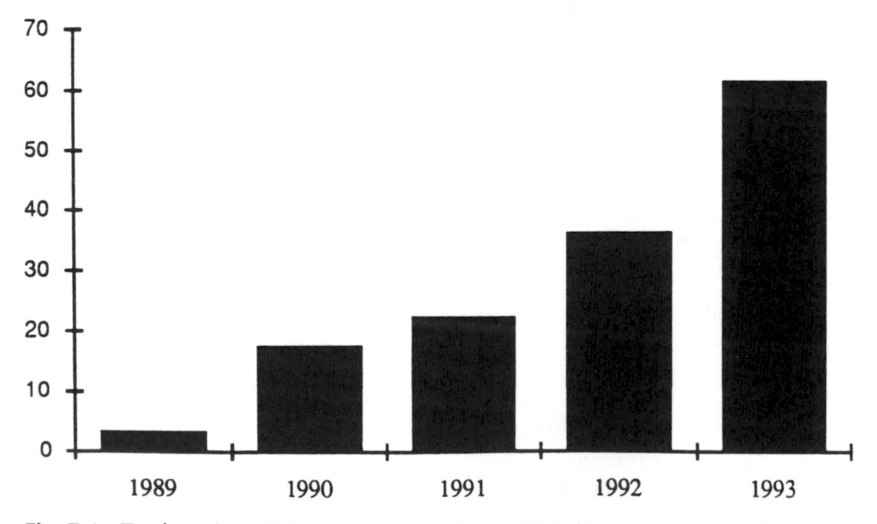

Fig. 7.1. Total number of ESAs in France, 1990 to 1993. (Source: Ministère de l'Agriculture, 1991; CNASEA, 1993.)

Table 7.1. ESAs in France. (Source: CNASEA, 1993.)

Status	No.
ESAs accepted for central government funding	61
Accepted by EC for co-funding	34
Refused for EC funding	2
Not yet submitted to EC	25
ESAs funded by local government and EC	2
Total number of ESA operations	63
Number currently operational	35

number of cases, a long history of voluntary management (for example, Billaud, 1992). In total, some 30 ESAs, covering 220,700 ha (with payments being available for 83,700 ha) specifically target sensitive biotypes (Table 7.2). For the most part, they concern areas of extensive grazing of beef or dairy cattle where the potential environmental threat comes from drainage and conversion to arable land. The ESAs aimed at combating agricultural withdrawal and the accompanying environmental consequences also display a strong spatial focus. These are the zones of marginal agricultural production in the mountain regions of the Alpes, the Jura and the Pyrénées as well as the

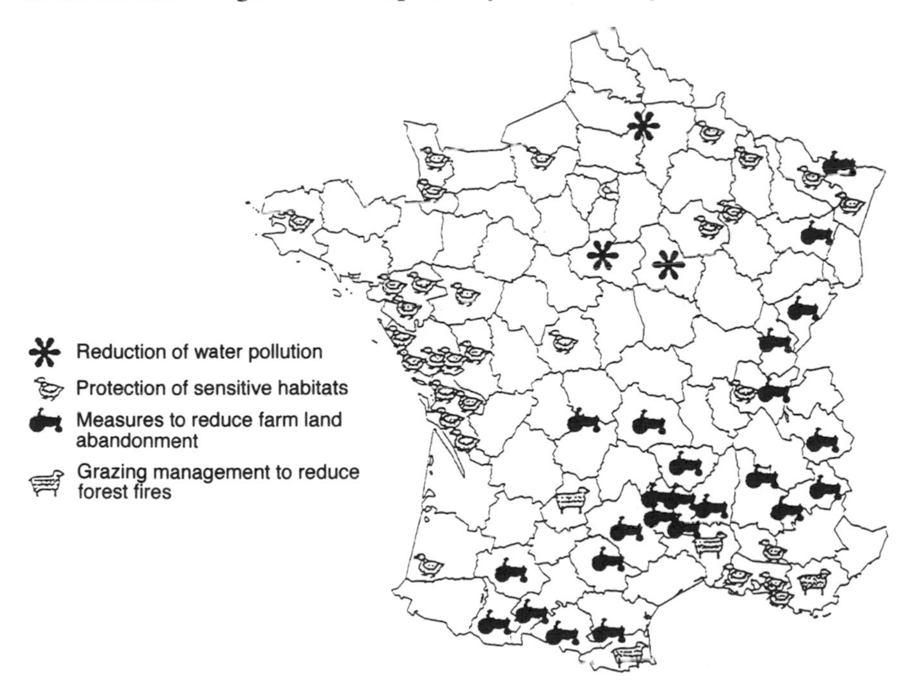

Fig. 7.2. ESAs currently designated in France, by type. (Source: CNASEA, 1993.)

Table 7.2. ESAs in France by type of zone, 1993. (Source: CNASEA, 1993.)

	Number of ESAs	Total amount of annual payments allocated (million FFr)	Surface area of ESAs (ha)	Eligible area (ha)
Water pollution	3	2.2	7,400	5,700
Sensitive ecosystems	30	52.5	220,700	83,700
Farming retreat	24	33.1	492,000	106,600
Forest fire prevention	4	4.7	17,200	9,200
Total	61	92.5	737,300	205,200

southern fringe of the Massif Central (Fig. 7.2). Here, Article 19 policy has been directly employed to ensure a minimal agricultural density, payments being made to farmers for the removal of scrub vegetation, for the establishment of rotational pastoral systems and for the use of organic rather than chemical fertilizers. In total, 24 ESAs have been established to this end, involving 492,000 ha with payments being available for 106,600 ha.

On the ground, considerable differences exist between the two principal ESA types (i.e. those seeking to counter agricultural land abandonment and those seeking to protect sensitive biotopes). The former have had a more successful take-up rate. Although the total area concerned within France is roughly half that given over to countering land abandonment (Table 7.2), the area currently under agreement is approximately double for ESAs directed at protecting sensitive ecosystems (Table 7.3). Reflecting the fundamentally different agricultural activities being undertaken in the two types of ESA, agreements established for the protection of sensitive ecosystems concern large numbers of smaller areas (2523 contractants for an average area of

Table 7.3. Characteristics of ESAs by type, 1993. (Source: CNASEA, 1993.)

	Number of agreements	Area under agreement (ha)	Average area under agreement per contractant (ha)	Total annual payment ('000 FFr)	Average payment per contractant ('000 FFr)
Water pollution	130	4,622	35.5	1,790	13.80
Sensitive ecosystems	2,523	47,265	18.7	29,125	11.50
Farming retreat	781	24,472	31.3	8,883	11.30
Forest fire prevention	85	5,086	59.8	1,955	23.00
Total	3,519	81,445	23.1	41,703	11.85

18.7 ha against 781 for an average area of 31.3 ha). Both however result in roughly the same average sums being paid to farmers, FFr 11,500 in total per year even though the average rate per hectare is substantially different, around FFr 400/ha in those areas threatened by farming retreat and FFr 600/ha in ecologically sensitive grazing areas threatened by drainage and conversion to arable (Table 7.3).

The management prescriptions for the different ESAs vary considerably in precision and in their impact upon existing agricultural practices. In sensitive ecosystems and areas of agricultural retreat, they can be relatively light even across the different tiers of constraint (Tables 7.4 and 7.5) as the threat of possible future practices rather than the impact of existing farming activities is the principal concern. By contrast, in the three experimental ESAs established to reduce farm-based pollution, where prescriptions are designed to 'complement regulatory measures' and where compensation is calculated on the basis of 'the additional costs or direct losses incurred by farmers as a result of the changes prescribed to existing practices' (Ministère de l'Agriculture, 1990), the constraints on farmers are more severe and include the establishment of fertilizer management plans, the planting of temporary cover between harvest and seeding, advancing the planting season to winter rather than spring, avoiding certain vegetable crops and reducing fertilizer and other chemical inputs. Significantly, this type of ESA has subsequently been abandoned due to the 'lack of technical and scientific bases' (CNASEA, 1993, p. 98). No new ESAs of this type have been established since 1991 though, as we show below, the reduction of agricultural pollution is emerging as a key element in the implementation of EC Regulation 2078/92.

Review

The French experience in implementing Article 19 of EC Regulation 797/85 has been limited both in the extent to which it has addressed, on a large scale, the environmental impacts of modern agriculture and in the extent to which it has seriously altered attitudes towards the relationship of farming to the environment both within the farming sector and beyond. In general, ESAs have been small in size; 12,000 ha being the average size of the broad area concerned, 3,363 ha being the average size of eligible land (Table 7.6). The highly centralized nature of the farming policy community and the administrative complexity of *OGAF-Environnement*, which seek to establish a broad consensual approach to the management of specific zones by involving not only farmers but also environmental groups, local communities and representatives of the State administration, have undoubtedly hindered the take-up of ESA agreements. Additionally, the budget allocated to the measure has been relatively small, starting in 1989, at well under FFr 10 million. Only recently has it reached levels comparable with other European states, with nearly FFr 100 million being allocated in 1993 (CNASEA, 1993).

Table 7.4. ESA: example 1. The Cotentin Wetlands Zone.

Objective:	Sensitive wetland ecosystem protection
Status:	Operational
Date:	1990
EC acceptance:	1991
Total area:	13,000 ha
Area concerned:	8,371 ha
Area available for grant aid:	3,722 ha
Finance:	FFr 1.2 million per year

Threat:

1. The abandonment of traditional grazing lands is leading to woodland and scrub encroachment and the accompanying decline in traditional species diversity.
2. Wetland drainage to permit intensification and conversion to arable and fertilized grasslands.

Contract requirements and grants:

Requirement	Grant/ha/year (FFr)
Tier 1	350
Maintain the natural wet grasslands	
Cutting after 15 July	
Stocking density of between 0.7 and 1 livestock unit/ha	
No lime or pesticide inputs	
Maximum fertilizer doses: 30 units N	
15 units P	
15 units K	
Tier 2	550
As for Tier 1 plus:	
No fertilization	
Cutting after 25 July	
Tier 3 (applicable only in a limited area)	850
Cutting after 15 August	
Daily stocking rate of between 0.7 and 1.0 livestock units/ha	
Tier 4 (applicable only in a limited area)	1100
Cutting once every 2 years on half grassland area	

The principal reason for the limited implementation of ESAs in France derives from an initial reluctance on the part of the agricultural sector, including the Ministry of Agriculture and its various associated bodies, to allow agri-environmental measures to develop in any way as a broad challenge, either to pre-existing agricultural policy or to the traditional role of the farmer in countryside management. As Deverre (1994) points out, the bulk of the agricultural trade unions at the national level have been hostile to the scheme, feeling it erodes the traditional liberty of farmers to produce in

Table 7.5. ESA: example 2. The Ariège Mountains of the Pyrénées.

Objective:	The maintenance and protection of upland pastoral landscape
Status:	Operational
Date:	1991
EC acceptance:	1991
Total area:	18,000 ha
Area concerned:	18,000 ha
Area available for grant aid:	3,690 ha
Finance:	FFr 1.2 million per year

Threat:
Agricultural retreat is threatening the sustainability of the traditional pastoral environment of an intermediate upland zone

Contract requirements and grants:

Requirement	Grant/ha/year (FFr)
Tier 1	200
Respect maximum and minimum stocking densities for cattle, sheep, goats and horses	
Maintain animals in fields during the spring and autumn	
Installation of pens and enclosures during the summer and the rotation of pastures	
Tier 2	500
As for Tier 1 plus:	
The mechanical or manual removal of encroaching scrub on pasture lands	
The maintenance of stream beds	
No pesticide or herbicide use	
Tier 3	800
The mechanical or manual removal of encroaching scrub and woodland on pasture lands	
The conversion of cleared scrub to hay meadow	
Annual grass cutting for the duration of the contract	
The maintenance of stream beds	
No pesticide or herbicide use.	

the manner they wish. At the local level, however, this initial reluctance is giving way to a qualified enthusiasm for ESAs. There are perhaps three reasons for this. First, particularly in less favoured agricultural regions, farmers have arguably been more closely involved in broader issues of rural management than in the productive cereal and husbandry areas of northern and eastern France. Thus, observers note an increasing number of ESA propositions coming directly from farmer groups within such areas often working with nature protection organizations and local hunting clubs (Alphandéry and Deverre, 1994). Second, and particularly since the

Table 7.6. Area of ESAs and Agreements, 1994. (Source: CNASEA, 1993.)

	No. of ESAs	Total designated area (ha)	Eligible land (ha)	No. of agreements	Area under agreement (ha)	Area under agreement as % of eligible land
Total	61	737,300	205,200	3,519	81,445	39.6

reorientation of the CAP in 1992, farmers in such areas are faced with a stark choice, subsidy or abandonment (Cauville, 1993). ESAs become less a specific environmental measure for individual farmers than the means to remain in business. Finally, as part of the general diversification of the rural economy, new sources of farm income are being sought. Once again, particularly in less favoured areas, farm-based tourism, often linked to environmental management, is emerging as the central theme in a growing number of rural cantons (for example, Blanc *et al.*, 1993). Agri-environmental measures, linked to *gîtes ruraux* and farm-camping become part of a new non-productive agriculture–rural vocation.

The advantages of these developments are clear, both for environmental quality and the sustainability of the agricultural population. The downside is the increasing bifurcation of French agriculture into, on the one side, an environmentally sustainable farming sector which is largely unproductive, in global market terms, heavily subsidized and oriented towards local, specialized markets and, on the other side, a highly productive, regulated farming system which, as a major contributor to the national GDP, is international in its orientation. Both models represent a major departure from the traditional relationship of farming to the rural environment in France. The former, being no longer economically viable as farming, flies in the face of the essential premise that agriculture is, above all, an economic activity.

The second, by being increasingly divorced from the rural territory in which it is located, loses its long-held claim to be the principal interest in rural management. If the 1980s brought such issues to the fore, the 1990s have, thus far, seen an increasingly wide-ranging debate develop on the ways and means of addressing them in a durable fashion. France's response to Regulation 2078/92 represents an attempt, albeit one informed as much by notions of agricultural viability as agricultural sustainability, to generalize the agri-environmental concept so as not to perpetuate the dualism identified above.

EC Regulation 2078/92

As part of the 'reform' of the Common Agricultural Policy which, both directly and indirectly, seeks to address the environmental consequences of modern farming and offer an alternative 'rationale' to the Union's more marginal farmers, the Council adopted, on 30 June 1992, Regulation 2078/92 relating to the development of agricultural production methods compatible with environmental and landscape protection. One of three accompanying measures to the CAP, this Regulation represents both an endorsement of the original ESA approach and its substantial extension into new agri-environmental domains.

From the outset, Regulation 2078/92 coincided more closely with existing French policy concerns than Article 19 of Regulation 797/85 ever did. Not only is it more relevant to the current preoccupations of a post-CAP and post-GATT agricultural community but it also offers legitimation to an extensive post-productivist, or indeed a peasant model of agricultural development, particularly in those areas where the productivist drive of the last 30 years has contributed to land abandonment and general agricultural/rural decline. Indeed, as an accompanying measure to CAP reform, the Regulation forms part of a broader rural–agricultural policy, that by encompassing surplus reduction and rural depopulation, falls fairly and squarely within dominant French preoccupations. Not surprisingly, France played a far greater role in promoting the drawing up of Regulation 2078 within the Commission than it did with respect to 797/85. Thus, compared with its predecessor, the application of 2078 in France has given rise to more numerous, larger scale, better financed and more precise measures than has previously been the case (Baillon, 1993). For the moment however, the new agri-environmental measures are still very much at the experimental stage. What we offer here is therefore more a review of current intentions and budgetary allocations than a full analysis of current practice.

The application of EC Regulation 2078/92 has been predicated upon the belief that farmers, as economic actors, should be reimbursed not only for the productive role they play in producing food and the raw materials for the food industry but also for the positive externalities, in the form of public and collective goods (the rural environment) that they provide (Bonnieux and Rainelli, 1994; Madelin, 1994; Trommetter, 1994). In this manner, the panoply of aids resulting from the Regulation's application in France are framed both in terms of payments for a public service provided as well as compensation for potentially lost income (Montgolfier, 1990).

During the final negotiations over 2078 within Brussels, the French Government independently announced in 1992 that it would be setting up a nation-wide scheme of grant-aiding livestock extensification, known as the 'grassland premium'. Subsequent to the adoption of 2078, this subsidy has become incorporated into that Regulation's application. Although the EC has agreed to co-finance the measure's application on the basis that it conforms with the intention of 2078, the French measure is tolerated rather than approved by the Commission. Thus, two levels of agri-environmental measure can be identified; those that apply generally for the entire French agricultural

Table 7.7. Agri-environmental measures adopted following EC Regulation 2078/92. (Source: Baillon, 1993; Ministère de l'Agriculture, 1992, 1994a.)

Type of programme	Objective	Measure	Annual payment
National	To maintain extensive production	Grassland premium	FFr 250/ha
Regional	Water quality protection	Reduction of inputs	
	(i) Source protection	20-year set-aside	FFr 1000–3000/ha
	(ii) River protection	Conversion to	FFr 2500–3000/ha
	(iii) Erosion protection	grassland	FFr 2500–3000/ha
	Extensification by enlargement	Reduction in stocking rate on former surface	FFr 1500/livestock unit removed
	Preservation of threatened breeds	Subsidy per animal	FFr 300/animal equivalent
	Nature protection	20-year set-aside	up to FFr 3000/ha
	Conversion to organic farming	Subsidy per hectare according to crop type	FFr 700–4700/ha
	Training		up to FFr 15,000 per person
Local	Sensitive ecosystems, land abandonment and countryside management	Subsidy in relation to constraints imposed on contractants	up to FFr 1100/ha

area and those that are specific to defined regional or local zones (Table 7.7). While both form part of the formal implementation of 2078, the former derive from a specifically domestic initiative which has sought the 'generalization' of certain agri-environmental measures. To this end, France has additionally instituted a second parallel measure, the farm-based 'sustainable farm development plans' which, though currently limited to a few experimental zones and not formally part of the application of Regulation 2078/92, are also intended as a nation-wide agri-environmental initiative.

Nation-wide agri-environmental measures

The voluntary grassland premium is designed to protect low density grazing lands and the farming systems that maintain them. It was introduced following Decree 93-738 of 29 March 1993 and is available to all active farmers operating holdings of at least 3.0 ha of usable agricultural area capable of supporting a minimum of three livestock units, providing that a stocking rate of 1.0 is not exceeded (though this can rise to between 1.0 and 1.4 if pasture lands cover at least 75% of the total agricultural area). For a

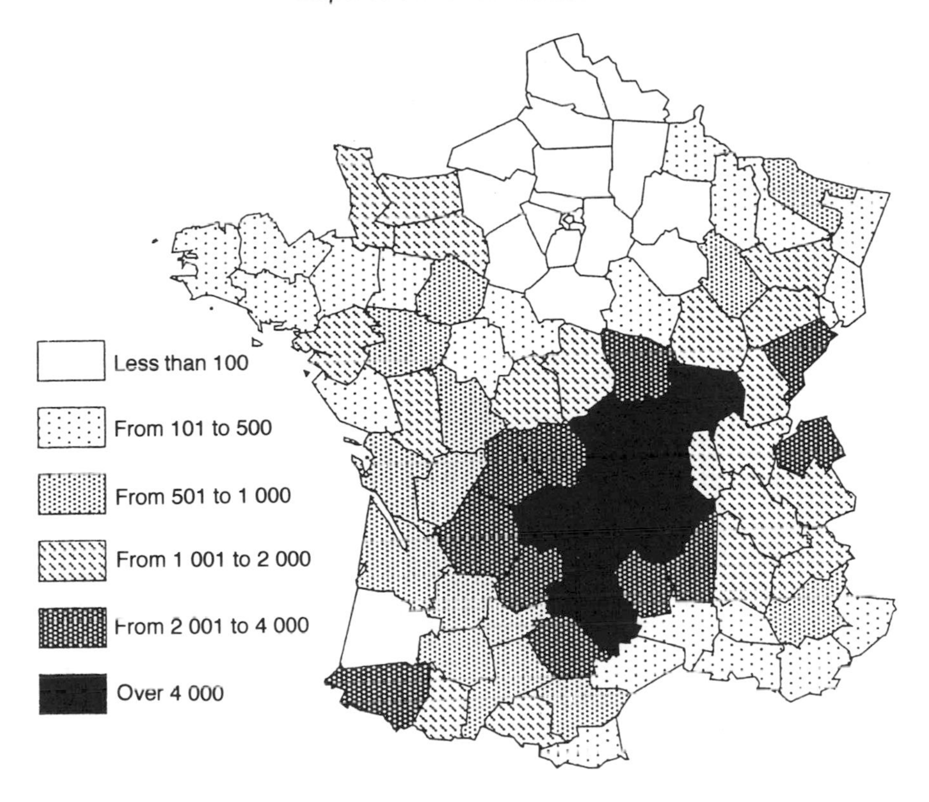

Fig. 7.3. Demands for grassland premium by Département, 1993. (Source: CNASEA, 1993.)

period of 5 years, the farmer, in exchange for a subsidy of currently up to FFr 250/ha (with a ceiling of FFr 20,000), engages to maintain existing husbandry practices, not to reduce the grassland area of his holding, to keep within set stocking rates, to maintain hedges, copses, streams and other landscape features and to undertake grass-cutting.

Since the measure's introduction, 161,989 applications have been made, of which 117,302 have been accepted (CNASEA, 1993). By March 1994, one year after the measure's introduction, some 5.7 million hectares had been contracted for a total cost of FFr 967 million. The geographical distribution of grassland premium applications reveals an unsurprising and reassuring concentration in the Massif Central (Fig. 7.3) where extensive husbandry is the norm and farm incomes among the lowest in the country (INSEE, 1993).

Although not solely an agri-environmental measure nor strictly part of the application of 2078, the sustainable farm development plans (*Plans de développement durable*: PDD) are currently poised to emerge as an important element in the French approach to improving and possibly redefining agriculture's role in countryside management. Accepted by the EC as being eligible for co-funding through FEOGA, the PDD are voluntary farm management plans, concerning the whole holding, whose goal is to promote the economic sustainability of the farm and the maintenance of the agricultural population while protecting natural resources and ensuring good countryside management practice. Thus, the PDD target individual farms rather than a broader environmentally coherent territory, ecosystem or landscape, as has been the case with ESAs. Drawn up by Ministry technicians in partnership with local groups of farmers, local policy makers and representatives of the local population, they attempt to make environmental management a key element of farm development by including detailed environmental audits and resource assessments and by proposing a series of developmental scenarios. Farmers then receive subsidies for following selected developmental trajectories for a period of five years. At the moment however, the initiative is still at the experimental phase with some 750 farms, being regrouped in 59 sites, currently engaged. The annual budget is estimated (1994) at around FFr 38 million, approximately half of which will come from the EC.

It is impossible at this stage to evaluate the success or otherwise of the PDD initiative, and some are highly sceptical of its wider application beyond fairly limited sensitive areas (Ledru, 1994). However, its significance is that it seeks to place farm management and the farm operation as a whole within a broader context of changing social demand with respect both to the quality of the rural environment and the role of the farmer in protecting that quality. The introduction of PDDs reveals a growing awareness within the agricultural policy community not only that social demand is changing but that farmers and the farming community have not, until now, changed with it. They therefore represent a considerable advance upon the predecessors, the *Plans d'amélioration materielle* (PAM) or farm development plans, whose primary role was to increase the productive capacity of farm holdings. As a recent government report maintains, the PDDs represent a new partnership between farmers, the State and the population whose aim is to 'respond to the new social aspirations' by reuniting the three key policy contexts of agriculture, rural and environment (Ledru, 1994, p. 120).

The regional and local programmes

The central thrust of Regulation 2078 as it is implemented in France is nevertheless at the 'regional' and local levels where a series of measures are currently being applied dealing with a wide range of agri-environmental concerns. The key characteristic of all these zone-based measures is that they are not generalized, like the grassland premium, but apply only within identified spatial areas defined at the regional level. Here again, two distinct levels might be distinguished, regional and local operations, though administratively all are regarded as regional measures.

Regional agri-environmental measures are, to some extent, a new departure for French policy. Coordinated at a regional level and seeking to reinforce the regional agricultural and environmental identity, the regional measures, operating within defined perimeters for a 5-year period, focus upon five environmental themes: the protection of water quality, the promotion of organic farming, the lowering of stocking rates by pasture land enlargement, the preservation of threatened breeds and nature protection (Table 7.8). In addition, two further areas of intervention have been identified by the Ministry of Agriculture, training and local agri-environmental measures. For the sake of clarity, we have chosen to deal with the latter as a separate issue, partly because the spatial context is different but partly also because the mechanisms of local operations are different from those of their regional counterparts. The regional operations comprise a set of management prescriptions, established at the national level and being universal for the type of operation concerned. There is, therefore, no variation in the prescriptions between different regions. Similarly, the payments available to farmers are equally standardized at the national level, though some variation is permitted according to the type of prescription imposed. For example, payments available to farmers seeking to convert their holdings into organic farming methods range from FFr 700/ha/year for 2 years, for those farmers converting to organic cereal growing, to as much as FFr 4700/ha/year for 3 years for those planting specialized crops (see Table 7.8).

Water protection measures include the conversion of arable land into grassland and pasture, the long term set-aside of farm-land, whatever its current use, and the reduction of chemical and other entrants in sensitive areas. Farmers voluntarily entering the scheme for a mandatory minimum 5 year period accept, against a subsidy that ranges from FFr 800–3000 according to the engagements, a number of constraints including maximum fertilizer inputs, the establishment of a formal fertilization plan, the planting of temporary covers between harvesting and sowing. The zones eligible for such subsidies are defined by the regional offices of the Ministry of Agriculture and include river catchments and drinking water sources susceptible to nitrate pollution as defined by the EC Nitrate Directive. Some 60,000 ha have been provisionally identified and a budget of FFr 241 million has been reserved. The other regional measures are summarized in Table 7.8. The initial budgetary allocations for the different measures were: FFr 241 million per year for water quality protection measures, FFr 72 million for conversion to organic farming over the 1993–1997 period, FFr 168 million over 5 years for extensification through pasture land enlargement, FFr 12 million for threatened breed preservation and FFr 32 million for nature protection within designated areas of special interest (the ZNIEFF, ZICO and ZPS). Additionally, between FFr 20 million and FFr 30 million per year has been allocated to farmer training (Table 7.9).

Specific local operations differ from the 'regional' measures with which they are usually included chiefly because the management prescriptions are

Table 7.8. Regional operations under Regulation 2078/92. (Source: Ministère de l'Agriculture, 1994b; 1994c; 1994d.)

Operation	Measures	Payment
Water protection	Long term (20 years) set-aside Conversion of arable land to pasture Reduction of entrants	FFr 3000/ha/year FFr 2500/ha/year FFr 1000/ha/year
Conversion to organic farming	Contractant agrees to follow the regulated prescriptions for organic farming methods laid down following EC Regulation 2092/91 Payment is made solely during the period of conversion to organic farming (2–3 years depending upon type of activity) All agreements are for 5 years	*Annual cultures* FFr 1000/ha/year for 2 years *Pasture* FFr 700/ha/year for 2 years *Olive groves* FFr 3000/ha/year for 3 years *Fruit and perennials* FFr 4700/ha/year for 3 years *Vines* FFr 1000/ha/year for 3 years
Extensification by enlargement	15% enlargement of pasture surface required Maintain for 5 years a maximum stocking rate of 2 livestock units/ha Maintain or increase the total grassland area	Within defined areas threatened by agricultural retreat, payment of FFr 1500 per livestock unit reduction per year will be paid
Preservation of threatened breeds	With the aim of maintaining domestic species diversity, payments are available for farmers retaining animals, above a certain threshold number, contained within a list of threatened species	FFr 300/year per livestock unit concerned
Nature protection	In order to protect flora and fauna within sensitive zones, long term (20 years) set-aside is proposed to farmers in designated areas of ecological interest (ZNIEFF), in bird habitats of community value (ZICO) and in special protection zones (ZPS). In addition to setting aside agricultural land, contractants engage to maintain the land and access to it.	Up to FFr 3000/ha/year for the period concerned

Table 7.9. Total budget for agri-environmental measures in France, 1993 to 1997. (Source: Ministère de l'Agriculture, 1994a; 1994b.)

Level	Measure	Current annual expenditure (FFr million)	Anticipated total budget 1993–1997 (FFr million)
National	PDD*	37	1,200
	Grassland premium†	967	6,800
Regional*	Water protection	241	1,205
	Organic farming	na	72
	Extensification	na	168
	Preservation of stock species	na	12
	Nature protection	na	32
	Training	30	1,500
Local*		290	1,509
Total			11,298

*Figures based upon 'Engaged credits'.
†Figures based upon amount allocated for the first year of operation.

not standardized at the national level but differ according to local (regional) agricultural and environmental contexts (Ministère de l'Agriculture, 1993). They are therefore established and managed at the regional level (Ministère de l'Agriculture, 1994a).

Local operations are essentially a continuation, under a revised structure, of the pre-existing *OGAF-Environnement* schemes established as part of the implementation of Article 19 of EC Regulation 797/85. They share with their predecessors the same twin objectives, to protect sensitive ecosystems from agricultural change and to maintain farming in areas threatened by agricultural withdrawal. However, as currently conceived, the local operations will considerably extend the ESA experience. An eligible area of 800,000 ha is anticipated (against 205,200 ha currently eligible under existing ESAs), concerning around 50,000 farms. Within the region of Languedoc Roussillon, 16 local operations have already been set up, the bulk concerned with maintaining pastoral agriculture on land threatened with abandonment.

In budgetary terms, the regional and local measures represent a significant increase in France's commitment to agri-environmental policy. For the 2-year period 1993–1994, the total budget for the regional operations (excluding local operations) amounted to around FFr 250 million. Local operations, over the same period accounted for FFr 92 million. Together, the two sets of measures represent FFr 288 million or FFr 144 million per year for the first 2 years of operation (Ministère de l'Agriculture 1994b). Since that initial start-up period, the budget has expanded. For 1995, some FFr 160 million has been allocated with the regions Midi-Pyrénées and Pays de la Loire being the largest beneficiaries (Table 7.10).

Conclusions

Agri-environmental measures have been slow to penetrate both French agricultural policy making and the agricultural mind-set of the farming community. Nevertheless, over the last 20 years, we can see a very considerable change that has, in many ways, been the result of the growing presence of the EU in this policy domain. Three broad phases emerge from the account presented above. Agri-environmental relations, until the early 1980s were characterized in France by a total separation, both physically and administratively of farming and nature/environmental protection. The gradual introduction of domestic and Community legislation led to a limited

Table 7.10. Agri-environmental measures: budget for 'regional' measures by region, 1995 (excluding animation budget). (Source: Ministère de l'Agriculture, 1994c.)

Region	FFr million
Alsace	2.2
Aquitaine	9.1
Auvergne	7.1
Basse Normandie	6.9
Bourgogne	9.9
Bretagne	10.5
Centre	11.0
Champagne Ardenne	3.6
Franche Comte	3.4
Haute Normandie	3.8
Ile de France	2.5
Languedoc Roussillon	6.3
Limousin	4.4
Lorraine	5.4
Midi-Pyrénées	12.7
Nord-Pas de Calais	4.5
Pays de la Loire	11.7
Picardie	5.7
Poitou Charentes	8.7
Provence-Alpes-Côte d'Azur	4.9
Rhone-Alpes	9.8
Corse	2.0
Guadeloupe	1.9
Martinique	1.7
Réunion	1.7
Total	151.4

coming together of the two concerns albeit within the context of negotiation and contractualization rather than regulation and penalization.

Today, a third phase is developing, one characterized by attempts at a more complete integration of sustainable objectives within agricultural management, as evinced by the nation-wide measures and the multiplication of local and regional programmes. For France, with its persistent agrarianism, this has been a major shift and one that has only really been achieved by transforming the agri-environmental agenda into an essentially domestic concern for protecting the viability of small marginal farmers. This is clear not only in the stated goals of French agri-environmental policy but also in the means to achieving those policy goals.

It is revealing, that French agri-environmental measures continue to focus upon specific targets whether they be the farm (PDD), the locality (OGAF), the protected landscape, species or water source (regional operations). They are generally referred to as operations or programmes rather than as a broad area or territory, such is conveyed by the British notion of an 'environmentally sensitive area'. Indeed, although we have employed the term ESA in this text, it needs to be noted that zones designated under Article 19 of Regulation 797/85 are rarely if ever referred to in this sense. They are known either under their operational name *OGAF Environnement* or simply as Article 19 zones. This emphasis on programmes and operations rather than territories reveals the general unwillingness of the Ministry of Agriculture to adopt a broad spatial-based policy for agri-environmental measures, preferring instead to allow farmers to select and choose from a range of different forms of aid. One of the consequences of this *à la carte* approach has been that the individual measures on their own, have not constituted a coherent approach to the spatial management of the countryside. Within the defined parameter, farmers have been at liberty to select from a wide range of different financial aid programmes some of which aim to reduce pollution and maintain landscape features but others which might have the effect of facilitating intensification. This lack of spatial and policy coherence is a central characteristic of the French experience to date, the only exceptions being those agri-environmental measures that have been introduced within the framework of broader countryside management initiatives, such as the regional parks.

Agri-environmentalism has yet to penetrate fully into agricultural policy or the agricultural endeavour at the national level. It remains essentially the concern of marginal agricultural regions. As the National Federation of Regional Parks has observed: 'Agriculture–environment relations are located at the level of the economic viability of the farm holding' (Fédération des Parcs Naturels Régionaux de France, 1994). This then is the defining feature of the French experience. Agri-environmental measures are inextricably linked to farm income diversification and to combating agricultural retreat and land abandonment. While these may not necessarily be the most pressing environmental issues *per se* within agricultural regions, they are concerns that strike at the heart of the French conception of the agricultural environment and of the relationship of human society to that environment.

Beyond the agricultural sector itself, the emergence and development of agri-environmental policy has coincided with changes within the French state. The Ministry of the Environment has, until relatively recently, had a fairly limited impact in rural policy-making other than in specific protected areas such as the central zones of the national parks and nature reserves. The 1980s however, saw major extensions to the Ministry's role in rural environmental protection, particularly in the fields of water and landscape protection. Crucially, these new roles have resulted essentially from European environmental legislation. As a result, the Ministry and its regional antennae, the *Directions de l' Environment* (DIREN), find themselves increasingly involved in rural and agricultural policy-making and implementation to the extent that the long-standing administrative association of rural and agricultural issues is appearing increasingly archaic.

The emergence of agri-environmental measures has coincided additionally with substantial political decentralization and administrative deconcentration. New powers were transferred to the communes, the *départements* and the regions and although agricultural policy remains firmly in the hands of the State, decentralization has none the less contributed to a noticeable shift towards regional policy-inputs both in the agricultural and environmental fields (Bodiguel and Buller, 1995). Within the agricultural domain, the development of agri-environmental policy has benefited at the regional level of administration and local government. The *Directions régionales de l'Agriculture et de la Fôret* (DRAF), the regional bureau of the State ministry, are playing an increasingly important role, alongside the elected regional councils, in coordinating agri-environmental policy. At the commune level, the transfer of planning powers to commune mayors has increased the possibilities for public participation in planning and has also served to raise the issue of agricultural pollution, agricultural nuisances and environmental degradation at the local level. There is, for example, considerable debate today over the extent to which local plans can be used to curb agricultural impacts on the local environment.

Taken together, these various trends point to profound changes in rural France, changes that the farming sector needs to address. Recent surveys have suggested that the French population is less happy than one might expect with paying farmers to maintain and manage the rural environment. The majority feel that farmers should be doing that anyway. Yet without subsidies and grant aid, the choice for the vast bulk of farmers currently benefiting from agri-environmental measures is not less countryside management but the total abandonment of the holding. In such areas, and there are a great many in France, it is the sociocultural role of farming as a form of land occupation within a post-productive society that is changing and not the nature of the productive activity.

Italy

Andrea Povellato

Introduction

Italian agriculture is characterized by geographical and climatic differentiations considerably affecting the adjustment of farming systems and the variety of landscape. The total land area is 35% mountains, 42% hills, whereas the plains cover only 23%. Climatic conditions vary greatly from North to South. Topographical and climatic factors are the necessary reference in describing agriculture both as productive structures and in terms of economic importance.

Since the end of the Second World War, it has not been easy to identify an autonomous development of agricultural systems, independent from public decision making. Agricultural and land-use policies have widely influenced the landscape and natural resources management in rural areas and at the same time the policies themselves were affected by regional differentiation. A correct policy evaluation of the environmental effects must consider socioeconomic development patterns, which have locally relevant differences, and the alternatives to development and conservation concern policies which differ from those in current use.

This chapter aims to highlight the relations between socioeconomic factors that seem rather distant from one another. Non-selective agricultural policy and the large number of farmers (or landowners) have somehow discouraged the urbanization process towards a vast urban concentration. In the most dynamic areas an urban planning policy paying little attention to the loss of agricultural land has encouraged the creation of isolated urban settlements in the countryside.

The uncontrolled development of industrial activities has widely contributed to the apparent welfare of several rural areas. In this situation of highly dynamic economic systems, conservation policies have proved ineffectual. Its most evident effect is landscape degradation.

In the intensive agricultural areas – frequently overlapping with those of a more dynamic economy – the non-selective agricultural policies associated with the low efficiency of central and local administrations and mandatory environmental policies has magnified the difficult implementation of the few measures aimed at limiting the production of negative agricultural externalities.

In the areas less affected by industrial development and characterized by marginal agricultural systems, an opposite process has developed, causing the abandonment of whole rural areas. It is the lack of environmental and agricultural policies promoting rational natural resource management that does not allow for the conservation of low input farming systems having great wildlife and landscape value.

Agricultural Land-use Change

The dynamics of agricultural and forestry land use provide a peculiar viewpoint for evaluating the complex relations existing between agriculture and the environment and the role of agriculture in the urban–rural context. The 1950s and 1960s saw dramatic changes in agricultural land use which led to considerable losses in area of agricultural land. The phenomenon was characterized by two different trends: on the one hand the loss of agricultural land to urban development, particularly the more fertile areas, and on the other, the abandonment of the more marginal land (Ferro, 1988).

Many authors have pointed out the different patterns of land-use change in the various Italian regions, as a result of delays and different models of socioeconomic development (Bagnasco, 1977), land availability and settlement patterns (Merlo, 1991), as well as differences in the ability of local authorities to manage territorial issues (Trigilia, 1992).

From the data in recent Agricultural Census (Table 8.1) the agricultural area, excluding woodland, decreased by 797,000 ha compared with the figures for 1982. Comparison with the previous decade, shows the decrease in agricultural area has somewhat slowed: the overall annual rate of change was –0.6% between 1982 and 1990 (100,000 ha/year) and –0.8% between 1970 and 1982 (127,000 ha/year). Land-use patterns have also changed during the past 20 years, with a gradual increase in the proportion of arable land, largely replacing permanent meadows and pastures: the crisis of livestock farming, greater profitability of arable land and abandonment of marginal meadows seem to be the main causes of this change. It should be pointed out that the most significant variations can be seen at the level of crop patterns.

Such variations are generally due to new Community measures modifying the profitability of the various crops, an example is the conversion from cereal crops to industrial crops, particularly soyabean. The introduction of

Table 8.1. Agricultural area by topographical regions. (Source: ISTAT, 1992.)

	Percentage of total agroforestry							Total agroforestry area (ha)
	Arable land	Permanent crops*	Permanent pasture and meadows	Agricultural area	Poplar wood	Woodland	Other area†	
1990								
Mountains	13.0	4.5	29.6	47.0	0.1	41.4	11.6	7,744,810
Hills	39.0	16.1	15.4	70.5	0.3	21.4	7.7	9,710,851
Plains	63.6	16.7	6.6	86.9	1.4	4.3	7.5	5,246,695
Italy	35.8	12.3	18.2	66.3	0.5	24.3	9.0	22,702,356
1982								
Mountains	13.0	4.8	30.3	48.1	0.1	41.0	10.8	8,139,467
Hills	38.8	16.2	16.4	71.4	0.3	20.7	7.5	10,091,533
Plains	62.2	17.7	7.5	87.5	1.8	3.8	6.9	5,400,533
Italy	35.3	12.6	19.1	67.0	0.6	23.9	8.5	23,631,533
1970								
Mountains	13.9	4.9	32.1	50.9	0.1	37.6	11.4	8,642,111
Hills	39.5	15.1	19.7	74.3	0.3	17.1	8.3	10,728,212
Plains	59.7	17.8	10.3	87.9	1.7	3.2	7.2	5,693,895
Italy	35.3	12.2	21.8	69.3	0.6	21.0	9.1	25,064,218

*Sweet chestnuts included.
†Set-aside included.

leguminous crops has certainly had positive environmental effects, as it has led to a significant reduction in the use of nitrogen fertilizers, as well as providing a more balanced rotation.

Differentiation according to altitude can emphasize some aspects of land use (Table 8.1). Alongside the obviously different relative weights of the various groups of crops, the rate of change in agricultural area between the plain (–0.4% annually between 1982 and 1990), hill (–0.6%) and mountain areas (–0.9%) should be pointed out. This suggests that the process of abandonment of the less fertile areas – prevailing in the hills and mountains – has a greater effect on the overall evolution of agricultural land in comparison to the demands for land for urban land uses.

A key role in land-use evolution patterns has been played by the concentration and specialization of production due to two different trends: the process of productive intensification undertaken in the most fertile areas (the plain) and the gradual abandonment and extensification of the most marginal areas (mountains and hills). The former tendency involves a greater use of inputs, which are almost always of chemical origin, with a potential negative impact on the environment. In the latter some areas are exposed to hydrogeological disorders caused by the failure to carry out adequate maintenance, while in other areas, the slow process of reforestation has been undertaken (Merlo and Boscolo, 1994).

The application of the voluntary set-aside programme in Italy in the late 1980s is to a certain extent a further confirmation of the potential abandonment of large marginal areas. Central and Southern Italy made the greatest use of this opportunity, with the percentage of arable land set-aside varying between 10 and 20%. The set-aside programme has largely involved hill regions where yields are lower and profitability is very limited.

Nature Conservation

Land use should increasingly be influenced by the widening of the protected areas, which have reached an already remarkable size and are beginning to have a relevant effect on the agricultural areas in Italy. According to the first official list of Protected Areas produced by the Ministry of the Environment, the area currently subject to specific forms of protection[6] amounts to 2,148,000 ha, equivalent to 7.1% of the total area of the country.

There is a further 713,000 ha which are officially considered protected areas under regional supervision, though subject to legislation different from the national law. According to medium term forecasts the achievement of new

[6] The criteria adopted in selecting the areas listed allow for inclusion of all those areas where the forms of protection conform with the provisions laid down in the guidelines on protected areas (Law No. 394/1991).

national and regional parks and the widening of existing ones should increase the total protected area by about 2 million hectares. As there is no official data, a survey of a sample of protected areas (Cavalli *et al.*, 1990) estimated that the area affected by farming consists of 15%, followed by 10% pasture and woodland, and 7% urbanized areas. Nearly 30% of the natural woods and forests requiring specific management should be added. Summing up the various land uses, around 60% of the total protected area calls for management of natural resources and human activity, rather than merely passive protection.

Chemical Inputs in Agriculture

The use of synthetic chemical products and the spreading of manure are the major sources of agricultural pollution in Italy. Analysis of data for the past 30 years points out drastic increases in the intensity of use of chemical products. This trend started to reverse slightly towards the late 1980s, but future trends are still uncertain. The tendency towards progressive, though slow, reduction in the use of chemical products responds to the new decoupling trends in agricultural policy. The need to reduce production costs, the evolution of agricultural practices and the growing awareness among farmers of the risk of pollution from chemical products are leading towards a general rationalization of their use. The differences seen at regional level, in terms of both intensity of use and composition of the products, and in terms of evolution in their use over time, indicate different patterns of development according to the prevailing land use and the degree of diffusion of technical innovations.

The levels of intensity of use of both pesticides and fertilizers (Table 8.2) are generally higher in the North than Central and Southern Italy (between 50 and 100% higher). Different crop patterns and constraints imposed by pedo-climatic conditions on production systems are the most evident explanations of these differences. It should be pointed out, that a clear distinction in the evolution over time between the Northern regions, where intensity of pesticide usage rose by 11% between 1982 and 1991, and the regions of Central and Southern Italy where levels of use dropped significantly (respectively by 15 and 27%). On the other hand pesticides used in the South tend to have a rather high toxicity level (19% of products in first toxicological class), as compared to lower levels seen in the Central regions (10%) and in Northern Italy (9%).

Intensive Livestock Farms

The problem of pollution due to manure from livestock farms is not connected, at least as far as the last 10 years are concerned, with the variation in the total

Table 8.2. Pesticides and fertilizers for farming use by broad geographical area (kg/ha of potentially treated area, 1991). (Source: ISTAT, 1993.)

	Fungicides	Insecticides	Herbicides	Acaricides and fumigants	Plant hormones	Total	Percentage change 1991–1982*
North	9.8	3.5	4.1	0.8	0.9	19.0	11.6
Central	5.7	1.5	1.2	1.4	0.4	10.1	−14.6
South and Island	5.4	2.4	0.7	1.5	0.6	10.6	−26.6
Italy	7.1	2.6	2.0	1.2	0.7	13.6	

	Nitrogen (N)	Phosphate (P_2O_5)	Potash (K_2O)	Total	Percentage change 1991–1982
North	86.9	63.2	56.6	206.6	−4.8
Central	61.2	46.1	13.3	120.7	−3.9
South and Island	47.0	33.3	12.6	92.8	−1.7
Italy	63.9	46.4	28.6	139.9	−4.0

*Change 1991–1982 only includes fungicides, insecticides and herbicides.

number of animals. From this point of view the 1980s saw a slight decrease in the number of head of cattle (from 9 to 8 million), while the number of pigs remained the same (9 million) and there was an increase in the number of sheep (from 9 to 10 million). The problem arises from the dramatic decrease in the number of livestock farms without a parallel increase in their agricultural area. The result is an increasing number of animals per hectare especially in the plains, and particularly in the Po Valley regions where 70% of the cattle and 75% of the pigs are concentrated. On specialized livestock farms the estimated average stocking rate is around 4–6 bovine animals per hectare at regional level, though in some areas the figure may be as high as 10 head per hectare. This intensity is directly linked to livestock rearing practices, given the high degree of specialization and economies of scale that can be realized on such farms.

Rearing of beef cattle, and to a lesser extent dairy cows, undertaken in paddocks, is based on the use of corn silage that can be purchased from nearby farms. In this way the livestock farm can be managed without a large area, and hence without sufficient areas for rational manure disposal. The type of beef cattle raised also influences the choice of stabling practice and hence the type of manure produced by the farm. A considerable proportion of the beef cattle raised in Italy are housed on slatted floors with a high production of liquid manure, which is expensive to dispose of within the code of good agricultural practices.

Institutions and Administrative Organization

The Ministry of Agriculture (MRAAF) plays a key role in managing interventions concerning the interactions between agriculture and the environment. Other ministries may be involved according to their specific competencies: among which the Ministry of Health, responsible for monitoring pollution and food safety standards, the Ministry of Public Works involved with land-use planning, the Ministry of Cultural and Environmental Goods responsible for protecting areas of outstanding natural beauty and, since 1986, the Ministry of the Environment which for the moment is mainly concerned with protected areas and environmental impact assessment. The lack of coordination between the various institutions has been seriously criticized. Recent attempts have been made to solve this problem by introducing the 'programme agreement' mechanism coordinating institutions involved in the same sector of intervention. From the experience of the past two decades the potential difficulties met at a central level seem rather understandable, in the light of the confused situation resulting from the administrative and legislative decentralization of the State to the Regions. This is certainly the most important change that has occurred in the organizational model of public intervention.

Regional legislative autonomy and administrative decentralization became operative in the 1970s, enforcing the constitutional principles (Article 117 and 118) by which the State delegates some legislative and administrative functions to the 20 Regions[7]. Public intervention in agriculture, land-use planning and the environment have largely been delegated to regional administrations and other local authorities. Under the new rules, the State essentially retains the task of outlining the overall strategies of intervention by passing guideline laws and coordinating interventions proposed at regional level. The regional administrations, with their knowledge of local economic and territorial systems, were to grade the intervention at local level. The process of decentralization took place at times in a confused, uncoordinated manner, and this is clear even in the definition of the functions to be transferred, as laid down in the legislation of the 1970s. The conflicts between central and peripheral authorities are all the more serious when it comes to sharing and assigning funds provided by national laws or EU regulations. The State's incapacity – or unwillingness – to play the role of guide and coordinator has allowed the regions freedom from control and given no incentive to plan their actions coherently. The action undertaken by the regional authorities in many cases was not an improvement on what

[7] The regional framework consists of 19 regions and two autonomous provinces of Bolzano (South Tyrol) and Trento which constitute the region of Trentino-Alto Adige. Five regions (Sicily, Sardinia, Valle d'Aosta, Trentino-Alto Adige and Friuli Venezia Giulia) are under 'special statute' and enjoy even greater legislative and administrative autonomy.

could have been done by central government.

The conflict between the State and Regions reached a climax at the time of the referendum, proposed by six northern, three central and one southern regions in 1993, to abolish the Ministry of Agriculture. The positive result of the referendum does not appear to have solved the problem. The Ministry has been re-established with more precise functions, but the difficulties of coordination and division of function still remain: the consequences can be seen in the application of the agri-environmental legislation.

Interest Groups

The farmers' unions and environmental movements are the interest groups mainly involved in agricultural and environmental issues. The increasing role of other actors should not however be overlooked. The food processing industry, together with the modern distribution system in recent years, are the most important promoters of policies aimed to increase the competitiveness and efficiency of the agricultural sector and minimizing environmental standards. On the demand side, the consumer movement requires most protective legislation as to the quality and certified local origin of food products.

The farmers' unions are in a difficult transition phase. In the past there were three organizations characterized by substantial differences in terms of their political and ideological viewpoints.

Over the past decade the emergence of powerful interest groups, especially in the agri-food sector, and the profound changes in the political geography of the country are influencing the behaviour and strategy of farmers unions. Although distinct organizational structures remain, the organizations are becoming aware of their increasingly limited influence on the decision making process of agricultural policy, the latent conflicts in various productive sectors ('corn versus horn') and the difficulty of maintaining a policy oscillating between productive and welfare objectives. Their recognition of the potential negative impact of agriculture on the environment is less evident. The organizations are beginning, however, to accept the idea that farmers should play an active role in the management of natural resources.

The role of the farmers' unions is not limited to influencing the formulation of agricultural policy, but they are also operative with their own agencies or through growers' associations. Unlike other European countries, the role of the producers' associations is still limited, even today, whereas the system of cooperatives (which are also divided on political–ideological lines) have always played an important role. The extension service, which has always been rather insufficient in Italy, is largely managed by the farmers' organizations and cooperatives. In almost all the regions, the extension service is based on a mixed model of public–private organization. Further developments in diffusing innovations should take account of this peculiar situation.

According to a recent survey carried out by the Ministry of the Environment (ISTAT, 1993) some 20 organizations in Italy are concerned in some way with environmental conservation, with a total membership of about 1.5 million. There is actually fewer than ten truly representative associations. Only some groups openly support the Green Party which has a share of about 2–4% of the vote. The organizational structure is typically based on a segmented network of contacts between small groups having different interests. As a result there are difficulties in ensuring effective collective action. Nevertheless over the past decade environmentalists have been able to influence policy formulation, both by organizing protest campaigns at national and local level and by indirectly obliging other parties to include environmental objectives in their programmes.

Since the early 1980s the environmentalist movement has concentrated increasingly on the negative impact of agriculture on the environment. Their attitude towards farmers and their organizations is therefore rather confrontational. Their campaigns culminated in the referendum proposal to dramatically reduce the use of pesticides (along with a referendum on hunting). This referendum, held in 1990, was invalidated by the fact that less than half the electorate actually voted. However, the percentage of votes in favour of imposing stricter limitations was over 80%.

Agricultural Policies

Renegotiation of the environmental problem is relatively new in Italian agricultural policy. The reasons for this delay and the difficulty of implementation met by the first programmes of the 1980s should be looked for in the peculiarity of Italian agricultural development and policies in the years following the end of the Second World War. We should however, keep in mind the dual relationship connecting – in a seemingly indirect way – the agricultural policy with other economic and land-use policies. A better comprehension of such a relationship should clarify some contradictory aspects of public decision making in agriculture during this period.

The general tendency of Italian agricultural policy concerning agricultural development and farm structures, is to identify the landowner with the farm, forging permanent links between farmer and farm, and giving priority to intervention considering small and medium family farms. Many authors have remarked on the ambivalent nature of the tendencies expressed during these years: alongside choices oriented towards increasing production (productivism), some of the policies were more inclined towards social welfare. Productivism was concentrated, especially on the more dynamic farms and areas, allowing the agricultural sector to increase its volume of production significantly. From the regional point of view, investment by the private sector was almost exclusively limited to the areas of Northern and Central Italy, while in the South investment in agriculture was generally public (Fabiani, 1986). A peculiar welfare policy originally emerged as a response to the social conflicts of the postwar years, expressed in the protective provisions of tenants and in public financing distributed without selective criteria. The main beneficiaries of these interventions were farmers employed on small farms, most of whom remained dependent on social transfer payments. Interventions on land ownership and other social transfers were hardly justifiable in terms of economic efficiency, but had a significant impact in social and political terms. The creation of a numerous petty bourgeoisie was a guarantee, for its advocates, of rather substantial returns in terms of votes.

The general approach to public intervention in agriculture has undergone no significant changes in the following decades, rather there have been merely slight changes of direction due to economic trends. The gap between production-oriented and welfare-oriented interventions has become even wider.

The objective of increasing yields, and labour productivity in particular, become a priority in the so-called 'Green Plans' of 1961 and 1966 as well as in the 'Four Leaved Clover' law of 1977. The problem of the increasingly wider gap between regions was almost entirely overlooked and most of the expenditure was devoted to the more fertile areas. The schemes did not effectively rationalize expenditure on structural incentives. Basically there were no definite selection criteria, so that small and large firms alike were beneficiaries of the allocations.

We should stress that the debate about the approval of the Agricultural Act (1977) was perhaps the last occasion when agricultural issues reached the front pages of the newspapers and agriculture became a central issue in the national economic policy. The Agricultural Act of 1977 – known as 'Four Leaved Clover' as it originally concentrated financial resources in four sectors (irrigation, afforestation, livestock and vegetables-fruit crops) – somehow attempted to reduce the wide regional gap by means of afforestation measures and interventions for hill and mountain areas. There was continuity with past policies both in the highly centralized institutional procedures and in the orientation of public expenditure particularly towards the more dynamic areas and sectors. As a result existing trends were encouraging and regional balance was not achieved. Actually this was an out-of-date approach, as by the late 1970s the problem of surpluses was becoming the major issue in the debate on the future of the Common Agricultural Policy.

Only in the mid-1980s, with the approval of the new National Agricultural Plan for the 5 years 1986–1990, were significant differences to be found. The overall objective of income support should be achieved through the improvement of productivity, rather than increasing production which had been the basis of public intervention up till that time. Defence of employment, correction of the regional gap, environmental conservation, reduction of the agro-food deficit and incentives for Southern Italy were all objectives/constraints in the plan. This was the first time that an adequate attempt had been made to provide specific intervention to deal with environmental problems.

The most apparent element emerging from this short survey of agricultural structural policies is the implicit restructuring to market forces without definite public intervention. Given the severity of institutional constraints and the lack of efficient infrastructures, the more efficient farms managed to develop – thanks to an extremely favourable price policy – while the others turn to part-time farming with the help of contractor services, to a relatively greater extent than in other European countries. It was in the 1960s and 1970s that this farm organization began to spread on an increasingly wider scale. The relative stability over time of the large number of farms (from 4.3 million in 1961 to 3 million in 1990) and the lack of development of rented land (over 40 years the agricultural area rented or under share farming declined from 50 to 15%) are both cause and effect of this phenomenon.

It is impossible to understand the evolution of farming and its policies without keeping in mind the relationship between agricultural structure and regional socioeconomic development. Even if an analysis of farm structures may highlight several similarities on a regional level, agricultural development takes specific connotations according to regional development.

The development model based on the widespread regional diffusion of industries and services found a strong ally in some farming sectors. Particularly in the areas of North-eastern and Central Italy[8] agricultural development was based on small scale production units that do not guarantee full-time employment and encourage members of farm families to seek additional sources of income in other sectors. Contractor services allow small farmers to keep their farms relatively efficient, as well as maintaining their relationship with farming and rural society in general. This integrated rural development has apparent environmental effects mainly due to the difficulty of defining rational urban planning and to the landscape degradation following uncontrolled individual cases. These areas are usually characterized by intensive farming activities causing huge environmental damage (water quality, landscape, food safety). A large number of farmers, insufficient extension services and local administrations are not ready to apply mandatory agri-environmental programmes, creating difficulties in implementing environmental policies, and structural interventions.

On the other hand, the development gap between the Centre-North and the South and between marginal and developed areas – which is also found

[8]Referring to these areas, Bagnasco (1977) defines them as the 'Third Italy', after the industrialized area of the North-west (Milan–Turin–Genoa) and the less developed areas of the South.

in some regions of Northern Italy – reinforces the need to identify a more suited agri-environmental policy for marginal areas. Unfortunately policies aimed to encourage countryside management have never been implemented on a large scale. Some examples can be found in a few mountain areas (South Tyrol, Valle d'Aosta) where tourism is favoured by the outstanding beauty of the countryside. In these areas farming is possible only if there is a relevant agricultural income support. It is not always possible to define exactly whether the approaches followed by policies are positive or negative. An example of this difficulty of assessment is the intervention of large scale land consolidation.

The decision to support small family farms has meant neglection of instruments such as land consolidation, which played a considerable role in modifying the structure of the agricultural sector in Northern European countries, especially in the 1960s. The efforts made in Italy to modernize the structures through land consolidation programmes in the 1970s and 1980s did not cover more than 50,000 hectares (as compared with 12 million hectares in France). More than 50% of the restructuring took place in one single region, Friuli-Venezia-Giulia, in the North-east. Studies undertaken in these areas (Sillani, 1987) have pointed out the loss of many semi-natural environmental features, that were also used for recreational purposes, and the creation of a simplified landscape[9]. Moreover, irrigation has contributed to increasing the intensity of use of fertilizers and weed-killers, thus aggravating pollution from agriculture. The decision – though unwilling – not to develop extended land consolidation programmes has probably had a positive indirect effect in terms of decreasing environmental impact.

The large number of small farms and the widespread use of contractor services also has other effects on the environment, which are not always clearly definable in positive or negative terms. The negative aspect concerns the temporary nature of many contracts between landowners and contractors (short-term rent contracts). This type of contract does not allow producers to pay sufficient attention to rational management of soil fertility and hardly encourages investment in land improvements. On the other hand the adoption of low environment impact practices is feasible in many cases only if radical changes are carried out in undertaking machinery management. Minimum or no tillage, fertilizer distribution and pesticide spraying require the use of new machinery and hence the farmer is faced with the choice between purchasing new equipment or using contractor services. Only in some agricultural areas are contractor services now able to supply machinery suited to new agronomic needs. Therefore the rate of adoption of environmentally compatible innovation is mainly determined by contractor services.

[9]In Italy there is a gradual landscape simplification process with the disappearance of several natural features that once characterized the countryside.

Land-use Planning Policies

Land-use planning policies form the other main set of public interventions affecting the agriculture–environment relationship. Urban planning, landscape planning, soil and water resources planning and protected area planning have significantly modified property rights concerning agricultural activities.

The main rules for land-use planning are contained in a 1942 law which deals with the allocation of agricultural land in a marginal way. The prevailing principle in urban planning until the beginning of the 1980s was to regard the countryside as an area of expansion for settlements, industrial activities and transport networks. The great exploitation of agricultural land in the 1960s and 1970s is partly due to this particular policy. Recently, the introduction of specific rules for the countryside in some regional legislation has produced a few changes. The State has not yet established new basic principles and for the moment the task of urban planning guidelines is left to the 1942 Law. The partial and haphazard application of urban legislation has largely contributed to environment and landscape degradation thanks to uncontrolled housing sprawl, new industrial activities and trade centres in the countryside. As is already stressed, in some regions the integrated rural development has provided high living standards for the rural population. Negative environmental effects are mainly derived from the almost unlimited expectation of natural resources and land uses, leading to depletion. In contrast to the inefficient urban planning in environment protection, other land-use policies have attempted to reduce the negative effect of economic development.

The Italian Constitution does not contain any clear reference to environmental conservation apart from Article 9 which lays down the principle of landscape protection[10] along with conservation of the nation's historical and artistic heritage[11]. This association with the nation's artistic and historic heritage shows that conservation was seen in a restrictive way when the Constitution was established, because the emphasis was primarily given to the cultural value of environmental goods. This aesthetic approach can be seen in Italian legislation dating back to 1939 when two laws were passed (numbers 1089 and 1497) fulfilling much of the environmental conservation policy until 1985. The laws only allowed public authorities to impose various kinds of constraints (from prohibition of building to the obligation to maintain current crop patterns) in specific sites of outstanding natural beauty. Preparation of landscape plans for specific areas was optional. At that time, the issue of conservation mainly concerned villas, gardens, areas of land

[10]In Italian, especially in the recent past, the term 'landscape' referred more to the visual, aesthetic aspects of a part of the territory, rather than its environmental and ecological features.
[11]The law also refers to Article 32 concerning protection of health and Article 44 which mentions 'rational exploitation of the land' and the 'social role of private property'.

recognized for their natural beauty and panoramas (with rights of public access), because human intervention in the landscape was still limited. The inadequacy of the legislation became evident when the spread of economic well-being brought about a growing demand for housing and industrial land.

In 1985 the field of action was drastically widened with the passing of a new law for the conservation of areas of particular natural interest (Law number 431 known as the Galasso Law). The wide range of protected areas refers to their strictly environmental value, rather than their beauty. The list of such areas includes several million hectares[12] of land where any modification in land use and buildings are prohibited. The new legislation indifferently protects certain categories of environmental goods under temporary regulations. The latter have been subsequently replaced by landscape plans, prepared by the regional administrations, which will establish a scale of constraints, limiting them to more specific areas, making them more stringent or eliminating them in other areas.

Although a specific definition of the landscape and the environment is not given, the list of the protected areas provides an idea of the explicit association between landscape and environmental values, which is far from the traditional concept of natural beauty as a visual and aesthetic value of the good to be protected (Casadei, 1991). The reference to landscape planning associated with town and country planning is also moving away from the mere imposition of constraints under the previous legislation.

Landscape planning constitutes a significant modification in property rights regarding land use, even though in the case of the agroforestry sector no authorization is required for undertaking productive activity, provided this does not involve permanent alterations of houses or productive buildings or hydrogeological constraints. There is implicit recognition of the positive role played by agriculture in landscape formation (Casadei, 1991). Some problems of interpretation arise with regard to forestry which only permits selective cutting. In the case of coppicing, this approach would not allow for rational economic management.

Constraints on farm management might constitute a remarkable problem, because the legislation does not provide compensation for farmers suffering economic loss. On the basis of constitutional law no compensation is provided in the case of constraints imposed on private use of goods characterized by 'intrinsic (original) quality of public interest'. Only in the case of a total impediment to economic use of the good – which is equivalent to compulsory acquisition for public purposes – is compensation due. It has been pointed out (Genghini and Scalzulli, 1989) that the major defect in this legislative mechanism concerns the definition of the good

[12](a) Lake shore and coastlines; (b) rivers, brooks and other water courses and their banks; (c) mountains; (d) nature parks and reserves and the protected areas around them; (e) forests and woods; (f) humid areas and (g) volcanoes.

as being characterized by an 'intrinsic (original) quality of public interest'. Environmental goods are widely recognized as a category of goods subject to continual increase in value with increasing economic development and welfare.

In general terms, this approach could lead to the gradual inclusion of new categories of environmental goods with increasingly widespread effects on agriculture. The problem becomes particularly serious in those areas, characterized by environmental constraints, where marginal farms cannot accept further constraints, lowering the adjustment process of productive systems with the consequent risk of land abandonment.

Alongside these considerations – some of which are purely academic – allowance should be made for the degree of enforcement of a law which is largely incomplete due to the continual exemptions conceded for a variety of 'urgent' reasons and the various laws passed in recent years condoning illegal building. It should also be added that there are serious delays in the preparation of the landscape plans by the regional administrations. In 1992, 7 years since the law was passed, only two regions had a general landscape plan and five other regions had set up plans for some parts of their regional territory. At present general constraints are still in force, reducing the original spirit of the law to the old concept of static constraint. Law No. 431 is an example of the risk of wishful thinking involved in 'command and control' policies. Given the urgent need to curb the changes in the environmental features, the law has underestimated the resources available in terms of control offices for the management of constraints and relative authorizations.

Protected area policy was given a sudden impetus in 1991 when the new guidelines on protected areas became law, after 20 years of preparation. The new law includes at least two positive elements, firstly concerning its general approach based on the attempt to combine conservation and enhancement of the natural heritage with development and promotion of environmentally compatible productive activities. A particular system of agreement between the State, Regions and local authorities allows for more effective protection of natural resources.

The major problem with protected areas in Italy arises from the hostility of the local population who criticize the strict constraint system imposed by the park authorities on land use and productive activities in general. There are examples of protected areas, however, where new activities have grown up in the field of tourism and recreation. Recent research has shown that the investment/employment ratio in the field of protected areas is 50–80 million lire per employee, as compared with Lit 300 million in the industrial sector (WWF, 1994). On the other hand, in some parks, such as regional ones, the relatively high population density causes continual conflicts of interest, which in some cases gives rise to spontaneous movements opposed to any proposed conservation of natural resources.

The new law foresees the establishment of management authorities, mainly representing the local population, and the approval of Park by-laws regulating permitted activities, including farming and forestry. A system of incentives is also provided to promote environmentally compatible activities and compensation for loss of income due to specific constraints. Application of the law, however, is a slow process. Bureaucratic delays, financial difficulties at national and local level, and probably the lack of political interest in environmental conservation, are causing the establishment of the new management authorities to be postponed.

In the context of land-use policy the new law on land defence was approved in 1989 (No. 183), providing new administrative and financial procedures for soil and water resource management, with particular attention paid to environmental aspects. According to the law, the national territory is divided into 11 catchment basins and other smaller regional basins. A basin authority is empowered to define the Basin Plan coordinating the activities of all the agencies supplying services connected with the water cycle.

There are numerous references to the agricultural sector, as the Plan also entails planning the use of agroforestry resources and the definition of constraints and rules concerning hydraulic works in agriculture and forestry. Apart from the positive aspects of the new law, criticism has been expressed regarding the lack of clarity about relations with the agroforestry administrations which could lead to overlapping powers or the lack of an authority on some issues. Taking account of the inevitable difficulties met in the initial phase of application, 6 years after the law was passed the Basin Authorities are still not entirely operative and only transitory planning instruments have been defined.

Policy Preventing Negative Environmental Impacts

The preventive monitoring of water resources pollution was perhaps the first really important measure in Italian environmental policy. Law No. 319 of 1976 (the so-called Merli Law) was a forerunner of much of the legislation that in the following years – and often in a rather disorderly way – aimed to protect peoples' health and natural resources from the risk of pollution. The law and its later amendments consider agriculture as a non-point source of pollution and hence overlook all those regulations regarding the use of fertilizers and pesticides and concentrate attention only on intensive livestock farms. If the livestock farm exceeds the threshold of 4 tonnes of live animal weight per hectare, then authorization is required to undertake productive activity. Authorization is given to those livestock farms that set up adequate sewage treatment plants or disposal plans lowering the amount of pollution from the discharges. Twenty years after approval of the first law, the results are not entirely positive considering that in many cases the regulation is only nominally respected. The lack of adequate monitoring is certainly the main cause of the situation which may be 'favourable' to those farmers who do not intend to face the cost of liquid sewage disposal, but risk causing considerable damage to other users such as farmers practising irrigation.

The lack of controls and the inefficiency – if not the total lack – of sewage treatment plants for urban and industrial areas explain the low quality of the water available for irrigation. There are cases in which the water is completely useless for agricultural purposes.

The various laws and regulations following one another in the 1980s, generally applying EU Directives, are gradually reducing farmers' rights in using the land and natural resources. Despite this, there is little application of even minimum measures of control on the use of potentially polluting substances. A clear example of this is the 'farm register' which the Ministry of Health decided to introduce compulsorily for farms from 1987 onwards. This register – which had been strongly encouraged by environmentalists and opposed by farmers – should record all data regarding purchase, sale and use of plant protection products by those using pesticides for agricultural and non-agricultural purposes. Postponements of this measure have continued year after year. It is currently expected to come into force from 1 January 1996.

The difficulty of solving the problem of environmental control by means of merely mandatory policies has stimulated the diffusion of programmes based mostly on information and extension services. The general target has been first of all to reduce the use of pesticides and fertilizers, making farmers aware of the cost reduction and the possibility of marketing their products at higher prices as the consumers consider them healthier. This is substantially a 'third way' besides the traditional farming and organic farming products. Among the many programmes implemented by regional administrations we should remember the 'National Plan for Integrated Pest Management', applied for 6 years (1987–1993), which was one of the heaviest agri-environmental expenditures of the 1985 Agricultural Act. Sector analysis, research and development, technical assistance, structural and organizational measures shared specific financial resources.

The actions performed have especially concerned fruits, grapes and vegetables which are particularly affected by the heavy use of plant protection products. Unfortunately, also in this case the effectiveness has varied from region to region. The main delaying cause in fulfilling the Plan measures seems to be the deficiencies of regional extension services. Among the regions mostly involved in this type of programme there are three north-east ones (Trentino Alto Adige, Veneto and Emilia Romagna). An associative structure (mostly cooperatives) aimed at acquiring technical inputs and selling products, agreements between public agencies and farmers' organizations and good market performance of high quality standard products seem to be success factors. Osti (1992) cautiously pointed out different strategies based on oligopolistic bargaining (Trentino Alto Adige) or policy networks (Emilia

Romagna and Veneto). In the former case a few large-scale actors (public agency and growers' organizations) capable of making global self-regulating agreements would prevail. In the latter case several actors having equal political power prove less effective and require more complex organizational planning.

More recently, in applying Directive 91/676 to the protection of waters from agricultural nitrate pollution, the first 'Code of Good Agricultural Practice' has been presented. The Code deals exclusively with the problem of nitrates and aims to 'contribute to the realisation of economically and environmentally sustainable agricultural models'. The Code is based on criteria of flexibility and may be the basis for preparing codes of practice suited to regional or local needs.

With regard to the part of the Nitrate Directive that calls for the preparation of compulsory action programmes for farmers operating in specific 'vulnerable areas', pilot tests have not yet been made, despite the significant effect that the regulations might have on the farm profitability. The Italian zonal plans in Regulation 2078/92 do not envisage that compensation measures should be linked to pilot tests in this specific field of action.

Countryside management policies based on financial incentives have never been widely used in Italy. Even the EC measure for Environmentally Sensitive Areas (Article 19 of Regulation number 797/85), has only been partly fulfilled by regional administrations, in 1991 some regions made the first compensatory payments to farms operating in ESAs. Only the northern regions and Tuscany in Central Italy have applied the regulation. The identification of ESAs was extended to a large proportion of the regions: generally the mountain areas were chosen indistinctly. Incentives were provided for mowing or grazing on land that had been abandoned for more than 2 years, the use of low environmental impact practices in managing pasture land, maintenance of mountain dairies *(malghe)* and paths and the conversion of arable land into meadows and pastures. There is little official data on the area and measures involved. According to local sources the initiative has been successful to a limited degree, due to the shortage of available funds, lack of information and in some cases competition from other Community support policies. Conversion of arable land into pasture land, for example, was hardly worthwhile as compared with voluntary set-aside.

The Implementation of the Agri-environmental Regulation

An evaluation of the new Agri-environmental Regulation effectiveness in Italy is premature as yet. However, a comparative analysis by examining objectives and procedures adopted in preparing the zonal programmes would help to identify the rationale of regional intervention, the priorities assigned to the various environmental problems arising and indirectly, the way in which the relations between agriculture and the environment are seen at local level. Preparation of the zonal programmes, until their approval by the Commission – generally taking more than 2 years – clearly reflects the lack of real coordination between the various national institutions involved in agri-environmental policy. Moreover, it should also be noted that this long delay was partly due to the Commission's difficulty in dealing with the work, and the novelty of this type of planning for the regional authorities. The limitations in the current State and Regional organizations can be seen from a short description of the decision-making process.

At national level the Agri-environmental Regulation is administered by the Ministry of Agriculture (MRAAF) alone and at local level by the agricultural departments of the regional governments. Consultations with the Ministry of the Environment have been very limited. Considering the possibility provided by the Regulation of formulating programmes for regional areas and the above-mentioned regional autonomy in agri-environmental policy, the Ministry of Agriculture delegated preparation of the programmes to the individual regions. The 21 regional programmes were presented to the Commission by July 1993, within the deadline laid down in the regulation, though no comparative examination had been made among them.

The zonal programmes show a list of measures directly referring to the aid scheme (Article 2 of the Regulation) in almost all the Regions. Compensations for the substantial reduction of chemical inputs or for other extensification methods ((a) and (b) measures respectively) are envisaged in all the Regions, whereas the stocking rate reduction (c) is not mentioned in seven Regions. The other environmentally compatible methods (d), the upkeep of abandoned land (e) and the 20 year set-aside (f) are envisaged in most programmes. The aids foreseen for public access land management (g) have been on the contrary unsuccessful.

Most programmes show a zoning of areas with different environmental sensitivity in order to grade the incentive payments and the priority of farmers' applications. Farmers in protected areas have the highest priority and receive the highest scheme payment along with organic farmers. ESA definition however does not exclude farms outside these areas from partaking in the programme. It is therefore a halfway situation between full segregation in well defined areas (such as ESAs) and a more flexible integration of environmental measures all over the countryside.

By the end of 1993 the Commission had only approved the programme submitted by the autonomous province of Bolzano (South Tyrol), while a painstaking process of review began between the other regions and Commission officers in order to adapt programmes to Community rules and limits on expenditure[13]. This latter aspect certainly had a drastic effect on many of the

[13] All the Regions of Southern Italy are included in the Objective 1 programme.

proposed zonal programmes. Eighteen months after the regulation was passed it has been necessary to adjust the expenditure foreseen by the individual programmes to the limitations of a national budget laid down by the Commission[14]. Overall the funds provided by the EC (with inclusion of national funds) amounted to around 40% of the expenditure originally foreseen by the regional programmes. Only one region had foreseen expenditure lower than the budget, while the estimates by the Northern regions were rather less excessive than those made by the regions of Southern and Central Italy (Table 8.3).

Examination by the Commission showed that many of the programmes were lacking in criteria to modulate payment among preferential zones; in defining minimum areas for those measures (such as measure (a)) which would make the actions environmentally significant and controllable; in presenting specific production controls; in defining abandoned land; and in specifying the advantages deriving from 20-year set-aside schemes. The programmes were also criticized for the notable differences in the methods of calculating payments under measure (d) (use of other environmentally compatible methods), lack of detail regarding training schemes and the lack of correlation between proposed new landscape features and the traditional, well established aspects of the environment.

In short, the Commission found a high degree of differentiation with respect to both planning criteria and application procedures. In some cases the Commission's calls for clarifications and amendments meant that the programmes had to be entirely rewritten. In February 1995 three of the regional programmes are still to be approved.

In broad terms, Regulation 2078 contains two objectives, corresponding to two different lines of intervention: one concerns a reduction in the negative impact of agriculture on the environment through the reduction in the use

Table 8.3. Expenditure scheduled by regions, granted by commission and spent in 1994 ('000 ECU budget). (Source: INEA, 1995.)

	Forecast (1)	Granted (2)	Percentage (2/1)	Spent 1994 (3)	Percentage (3/2)
North	765,426	488,466	63.8	20,602	4.2
Central	788,473	170,101	21.6	6,218	3.7
South and Island	1,051,073	415,836	39.6	10,381	2.5
Italy	2,604,972	1,074,403	41.2	37,201	3.5

Note: Expenditure of Zonal Programmes refers to next 4 years' implementation of programmes. All the figures refer to the total amount of funds (Community and National quotas).

[14] The considerable delay is partly due to the effects induced by the result of the referendum on the relations between the Regions and the Ministry of Agriculture.

of chemical products and the adoption of environmentally compatible practices (largely, measures (a), (b) and (c)), while the other is aimed at compensating farmers for the positive externalities connected with countryside stewardship and environmental conservation (measures (d), (e), (f) and (g)). The two strategies are extremely diversified at regional level whereas the measures related to the former objective should generally be adopted in the more fertile areas, while the actions aimed at environmental conservation mainly concern marginal areas. The allocation of funds for each measure in relation to the major ecological and land-use characteristics of the individual regions does not always appear to be followed in defining the programmes. It seems that Regulation 2078 is basically recognized as an instrument reducing the negative impact of agriculture on the environment. Countryside stewardship is recognized as an essential factor in land-use management only in some regional areas where these objectives have been pursued for a long time.

Analysing the estimates by broad geographical area, it appears that measures (a), (b) and (c) account for 65–70% of estimated expenditure (Table 8.4). Relatively high allocations are made for organic farming (a3) in Southern Italy and the reduction of livestock per hectare (c) in the Northern regions (especially Lombardy).

Other environmentally compatible methods (d1), especially replanting of hedgerows, are proposed by the Northern regions in particular, whereas the Southern regions allocate considerable funds to the conservation of agro-forestry land and 20-year set-aside. It is somewhat puzzling that the regions of Central Italy, rather than those of the Po Valley, allocate the most resources to reducing the negative impact of agriculture (almost 80% of the estimated expenditure and more than 85%of the land area).

Further analyses have been made by comparing the estimates for agricultural areas and the numbers of livestock involved with the measures, with the total agricultural area and livestock at the regional level. An overall 12% of agricultural area in Italy should be covered by the measures, ranging from 22% of the farmland in Northern Italy to 6% in the South (Table 8.5). Only 3% of the livestock would be affected by the agri-environmental measures, with little differentiation among regions. Particularly after the first 4 years of application 120,000 ha of land would be used for organic farming: a figure that is not very different from recent estimates, but much higher than the area of land currently controlled according to Community and national legislation.

Finally, a short commentary on the partial application of the regulation in some regions in 1994 is given. Definitive data on the number of accepted applications is not yet available, but according to provisional estimates it appears that the payments amount to 3.5% of the budget for the first 4 years (Table 8.3). Measure (a) accounts for the largest share of payments, particularly with regard to organic farmers who already have reliable systems of control set up by their own associations. Obviously the delays in approving

Table 8.4. Expenditure, agricultural area and livestock unit related to each measure scheduled in the zonal programmes (row percentage). (Source: INEA, 1995.)

	Measures									
	a1+a2	a3	b	c	d1	d2	d3	e	f	g
Expenditure										
North	40.6	3.4	18.5	9.3	12.7	1.8	0.0	4.1	8.8	0.9
Centre	50.0	13.9	12.1	3.3	4.7	1.7	1.1	4.4	7.4	1.4
South and Island	20.3	21.2	17.7	5.5	7.7	2.3	0.0	11.8	12.8	0.7
Italy	35.2	10.9	17.2	7.1	9.8	1.9	0.2	6.7	9.9	0.9
Agricultural area and livestock unit										
North	26.2	1.7	24.1	76.6	41.9	23.4	0.0	3.2	2.3	0.6
Centre	55.7	10.2	20.3	59.4	2.3	40.6	1.1	5.5	4.0	0.9
South and Island	24.0	18.1	20.7	67.2	11.0	32.8	0.0	18.2	7.2	0.6
Italy	29.8	6.6	22.8	72.5	29.3	27.5	0.2	6.9	3.7	0.7

Note: The division of expenditure into measures refers to next 4–5 years' implementation of programmes.
All the measures are expressed in hectares, excepting measures (c) and (d2) expressed in livestock unit.
List of measures:
(a) to reduce substantially the use of fertilizer and/or plant protection products (a1), to maintain reductions already made (a2), to introduce or continue with organic farming methods (a3);
(b) to change or to maintain more extensive forms of crop production or to convert arable land into extensive grassland;
(c) to reduce the proportion of sheep and cattle per forage area;
(d) to use other farming environmentally compatible practices, including the maintenance of the countryside and the landscape (d1), to rear animals of local breed in danger of extinction (d2), to grow crops of local breed in danger of extinction (d3);
(e) to ensure the upkeep of abandoned farmland or woodland;
(f) to set aside farmland for at least 20 years;
(g) to manage land for public access and leisure activities.

the programmes have compromised the effectiveness of the schemes, although many of the measures are unlikely to be adopted because of stricter constraints imposed by the production restrictions which are generally offset by low payment. Moreover, the question of the availability of the national share of funds should not be overlooked.

Increasing budgetary problems have forced Parliament to postpone approval of this expenditure several times. Clearly the risk of long delays in receiving scheme payments (or not receiving it at all) discourages even those farmers who are more willing to commit themselves to environmentally compatible production methods. The hypothesis that budgetary constraints may become the main impediment to applying the regulation in the future should not be underrated.

Table 8.5. Comparison between agricultural area and livestock unit scheduled by zonal programmes and data from agricultural census 1990. (Source: INEA, 1995.)

	Agricultural area ('000 ha)			Livestock unit ('000)		
	Zonal programme	Census	Per cent	Zonal programme	Census	Per cent
North	1,167.9	5,206.3	22.4	130.7	3,993.9	3.3
Centre	257.1	2,707.0	9.5	19.8	875.2	2.3
South and Island	421.6	7,132.5	5.9	53.1	2,214.0	2.4
Italy	1,846.6	15,045.8	12.3	203.6	7,083.1	2.9

Note: Agricultural area and livestock unit of zonal programme refer to 4–5 years implementation of programme.

Policy Options for the Future

The implementation of Regulation 2078/92 will certainly constitute the backbone of agri-environmental policy for the next 3–4 years. The more general framework of reference should be the two programme documents recently approved by the Government.

Late 1993 saw the approval of a document submitted by the Ministry of the Environment enforcing Agenda 21 – the protocol of intentions signed by 170 countries at the Rio Conference. The 'National Plan for Sustainable Development' contains specific indications, objectives and priorities for various economic sectors. The Ministry of the Environment has also presented plans for enforcing the Rio agreements on climatic change and biological diversity. The documents provide important guidelines, but they require the support of laws, decrees, financing and control systems that for the moment appear to be rather insufficient with respect to the objectives it is hoped to achieve.

Besides the policies outlined in the most recent government and parliamentary documents, the main issue remains the ability to respond in the future to the growing demand for nature conservation according to the resources (*sensu latu*) available at present. We can draw a few conclusions using the common distinction in environmental policy instruments, from the most strictly compulsory to the most market-oriented ones. The compulsory policies (standards, licences, planning, etc.), which were widely employed in the past, are no doubt the easiest to introduce but require an administrative capacity of control that is not available at the moment. The problem is not to impose more restrictions but to make those extant more respected or to create environmental standards that can be complied with reasonably. On the whole such policies are hardly acceptable because they might 'freeze' the land-use pattern, preventing its rational evolution.

The implementation of mechanisms of persuasion (information, advice, and extension) have given rather different results according to the organization of the agricultural sector depending on the degree of social cohesion. Their effectiveness depends on satisfying extension services and associated structures able to provide sufficient information flow. Nevertheless the experiences are concentrated in reducing the negative effect of pollution, other actions aiming to increase the environmental quality are very limited if they exist at all.

Heavy uncertainty weighs on the financial instruments (grants, incentives, tax concession) because of the great public expenditure necessary to implement these measures. The public budget deficit, maybe the heaviest legacy of the 1980s, does not encourage this type of measure. The transaction costs connected with their implementation make this strategy even less attractive. The success of the set-aside programme suggests a good farmers' availability, but active natural resources management (according to Regulation 2078) requires new technical knowledge that farmers may not always possess.

Finally the market-led measures, not strictly dependent on public expenditure (commoditization, marketing of products joined with stewardship), might become a useful tool for wide rural areas. Connections with other activities, first of all tourism, have led some regions to interesting experiences of multi-purpose countryside management.

Payments to farmers aimed at the maintenance of farming activities have a direct positive effect on the rural landscape and recreation-linked economic activities. Another example in the future might come from product licensing like appellation d'orique controlée. This scheme is a reallocation of property right: the public at large gives the farmers of a specific area the rights to produce a particular product, assigning them a potential 'monopolistic' position. The public at large requires particular obligations in terms of production standards and in the future could exact specific environmental standards.

NETHERLANDS

GERT VAN DER BIJL AND ERNST OOSTERVELD

Introduction: Developments in Dutch Agriculture

Dutch agriculture is mainly trade oriented, with more than 60% of the agricultural production exported. About 80% of this export is directed at other EU Member States. Agribusiness (primary and secondary sector) amounts to about 7.5% of the national income and employment. In 1993 the number of farms in the Netherlands was 119,000, with grazing livestock farms the most numerous, amounting to 47% of the total. Horticulture, arable, and intensive livestock farms have a share between 8 and 10% each. Between 1950 and 1990, the labour force in agriculture diminished by an average rate of 3% per year and during the same period, the use of non-factor inputs in Dutch agriculture increased by an average of 4.3% per year. The production volume of Dutch agriculture thus increased at an average rate of 3.5% per year. Development of agriculture in the Netherlands during the last four decades can therefore be summarized as follows: a decreasing labour force and less land resulting in a steady increase in production.

The increasing intensification of Dutch agriculture in recent decades has brought severe environmental problems. Agriculture plays a major role in the contamination and eutrofication of surface water and ground water and of natural ecosystems in the Netherlands. The use of non-factor inputs per hectare in Dutch agriculture is considerably higher than in other Member States of the EU, the average nitrogen production from animal manure in 1990 was 343 kg/ha, or nearly five times the average in the EU (Poppe *et al.*, 1994). The use of pesticides per hectare of plant production in the Netherlands was four times the EU-average in 1992, however, it is important to note that per unit of produce the use of inputs in the Netherlands is relatively low, indicating a high degree of physical efficiency. The return per 1000 ECU on the use of pesticides is considerably lower than the EU average: 2.1 kg/1000 ECU for the Netherlands compared with 2.6 kg/1000 ECU for the EU (Brouwer *et al.*, 1994). Nutrient use in the whole production chain per kilogramme of meat in Dutch pig farming is lower than in other important European production areas such as Denmark, Brittany and the Po Valley. For the production of pig meat, use of nitrogen is 160 g/kg compared with around 150 g/kg for other areas, and the use of phosphorus is 22 g/kg compared with 30 g/kg for other areas (Rijks Planologische Dienst, 1993, p. 50).

Interest Groups and Policy Making

The Dutch Government, political parties, agricultural organizations and agro-industry are closely linked in a corporate organizational structure, of which the leader is the so called Landbouwschap. In the Landbouwschap, farmers' organizations and trade unions for agricultural sector workers cooperate. The Landbouwschap is on the one hand a platform for communication and negotiation between the agricultural sector and the Government and on the other hand it has legislative powers and implementation roles for governmental policies in agriculture. Within this agricultural community enlargement and intensification of agricultural holdings was considered the basic aim for agricultural policy. The close relations between Landbouwschap as representative of the agricultural sector, the Ministry of Agriculture and the politicians concerned with agriculture has been an important reason for the slow implementation of environmental policies in agriculture. Consequently, a serious policy on manure surpluses has only been implemented since 1984, although the first signs of the related environmental problems were evident from the 1970s (Frouws, 1993). When signs of environmental hazards could no longer be ignored, the agricultural community mainly opted for technical solutions, such as industrial processing of manure. By the end of the 1980s, this closed agricultural community began to lose power and the influence on agricultural policy of environmental groups and the Ministry of the Environment increased.

Since the 1970s, and especially since the late 1980s, the pressure from society to reduce environmental problems has increased. A range of national and regional environmental organizations have played an important role in this process. After years of mainly protesting, the role of many environmental organizations is slowly changing towards cooperation in problem solving with the agricultural sector. A comparable development can be seen with agricultural organizations. Over the years, most farmers' organizations had a predominantly defensive attitude towards environmental issues; nowadays most regional farmers' organizations as well as local farmers' groups realize that environmental measures can be beneficial to farmers and initiate programmes for the reduction of pesticide and fertilizer pollution. A significant role in bridging the gap between environmentalists, environmental policy and farmers has been played by the young farmers organization NAJK and the Centre for Agriculture and Environment (CLM) (Termeer, 1993). The same organizations played a role in bridging the gap between farmers and nature conservationists (Janmaat *et al.*, 1995).

During the 1980s some leading conservation organizations lost their interest in the wildlife on farms and focused mainly on the creation of nature reserves and 'nature development areas'. Paramount is WWF Netherlands which only supports 'self-regulating nature', not influenced by human activities. Farmers' organizations and Landbouwschap also opted for segregation of agriculture and wildlife, because conservation was seen as conflicting with the intensification desired by the farmers' organizations. In the first half of the 1990s, the interests of farmers and some nature conservationists in wildlife on the farm was increasing. Declining incomes, fading prospects because of the CAP reforms, the outcome of the GATT Uruguay Round and new considerable land claims for the creation of nature reserves and 'new nature' has increased farmers' interests in new activities and challenged the concept of intensification. Recently the number of local initiatives by farmers taking an active part in the management of landscape and wildlife has increased significantly. In several regions farmers have started conservation cooperatives that negotiate with local governments and conservation organizations and unite farmers in their search for additional income from conservation activities (Hees *et al.*, 1994). In 1995 around 60 local farmers' cooperatives were active in wildlife and landscape conservation.

During the 1990s, the role of the Government in environmental and conservation policy has tended to change. An important reason is that faith in the capability of the government to control changes in society is decreasing. Simultaneously, there has been increased recognition that solving environmental problems demands changes in the behaviour of farmers. The Dutch Government has announced a change in environmental policy from a one-sided hierarchical method of policy formulating with regulations, commands and prohibitions towards a policy of more self-regulation and stimulation of responsibility of farmers. An important element in recent changes in government policy is the ongoing decentralization. Aspects of environmental and conservation policy and land planning, increasingly become the responsibility of provinces instead of national governments. A prerequisite for the success of this new policy of self-regulation is that the lack of openness of farmers' organizations towards general interests, changes into a more offensive attitude in which measures to reduce environmental damage or enhance wildlife values is also considered a responsibility of farmers. These changes are slowly taking place.

Farmers' organizations and environmentalists see an increasing role for market instruments to stimulate demand for environmentally sound products. The environment increasingly becomes a marketable phenomenon instrument. Fruit and vegetables sold by the largest supermarket chain in the Netherlands, Albert Heijn, have been produced under demands that meet the expected environmental policy in the year 2000. At the beginning of 1995 a committee consisting of farmers, environmentalists and traders developed criteria for a green label for potatoes, onions and wheat. These products will be marketed during 1995, with other products to follow.

Environmental Issues and Policies

In 1989 the Dutch Government agreed on an overall national environmental policy, the so-called National Environmental Policy Plan. In 1994 this was followed by the National Environmental Policy Plan 2. In these Environmental Policy Plans, the main targets for Dutch environmental policy have been laid down. The main environmental issues and policies in Dutch agriculture are:

1. manure, minerals and ammonia;
2. pesticides;
3. energy and carbon-dioxide emissions;
4. wildlife and landscape.

Manure, minerals and ammonia

The problem

Recently, the number of livestock in the Netherlands has increased tremendously. From 1950 to 1990 the number of cows doubled, the number of chickens quadrupled and the number of pigs increased sevenfold. Since 1984 the number of pigs has increased by 34% to 14.9 million, despite the introduction of environmental regulations. In contrast, since the introduction of milk quotas in 1984 the number of dairy cows has decreased by 31% to 1.7 million in 1993.

The intensive agriculture, with a high application of fertilizers and manure and the use of large quantities of imported animal feeds has caused severe negative side-effects on the environment. The emissions of the minerals nitrogen (N), phosphate (P) and potassium (K) has created a burden on the environment, causing several problems. Nitrogen in the form of ammonia is one of the causes of 'acid rain', which damages forests and ecosystems, and as a nitrate can leach into the groundwater. Phosphates accumulate in the soil, with over-application possibly resulting in saturation of the soil and subsequent leaching into ground and surface water. Leaching of nitrogen and phosphate result in eutrophication of surface water and the pollution of groundwater. Agriculture is responsible for around 32% of the deposition of acidifying elements in the Netherlands. About 400,000 ha of the sandy soils

(50%) in the Netherlands is considered saturated with phosphates. In less than 40% of the agricultural area, the nitrate content of the upper groundwater is above the 50 mg/l specified in Directive 91/676.

The policy

The aim laid down in the National Environmental Policy Plan is that by the year 2000, no more phosphate and nitrogen may enter water and soil than can be absorbed via natural processes. However, what the level of this 'balanced fertilization' should be, remains a matter for political discussion. Research has indicated that from an environmental point of view acceptable losses are 1 kg P/ha. The minimum from an agricultural point of view, however, varies from 25–70 kg P_2O_5/ha, depending on various factors, including soil type. A second aim in the plan, is to reduce the emission of ammonia by 50% compared with 1985 levels.

Initial measures to control the manure problems were taken in 1984 when prohibition of starting new intensive livestock farms was introduced in areas with a large concentration of pigs and poultry. In 1987 a maximum physical amount of manure applied to land was introduced, since 1991 this maximum has been reduced. In 1995 the amount is 150 kg of phosphate for grassland and 110 kg for arable land. Nitrogen has only been regulated in an indirect way, with chemical fertilizers not yet included in this regulation. Since 1991, the manure policy consists of a mix of regulation, research and technical measures. These technical measures were threefold:

1. adjustments in feed concentrates;
2. improvements in distribution of manure to regions where arable farming is predominant;
3. industrial processing of manure.

The industrial processing of manure has still not proved viable. To reduce ammonia pollution, there are additional regulations for covering manure storage and the reduction of emissions from livestock housing. About 15 million tonnes of animal manure, or 19% of the total animal manure production, is transported from farms with a surplus to farms that have not reached their maximum application of manure. To reduce pollution by improved utilization, the application of manure is now prohibited outside the growing season, or between 1 September and 1 February.

In solving the problems of eutrophication and acidification the system of mineral accounting plays a crucial role. In this system, developed by CLM, a farmer records his input of nutrients (for instance as a result of purchasing fertilizer, animal manure, concentrates and roughage) and his nutrient output (products supplied by the farm). The surplus which is the difference between input and output, is partly or entirely lost to the environment and may cause damage.

Mandatory mineral accounting on all livestock farms plays an important role in the most recent proposals from both government and farmers' organizations. In these proposals farmers producing surplus nitrogen and phosphorus above a certain level per hectare will be confronted with a regulatory levy. The Dutch parliament supported the introduction of a premium for farmers producing less than the fixed level of surpluses of nitrogen and phosphorus per hectare. The problem of controlling and monitoring the system after the introduction of levies will be complicated, especially if the levies are high. A system of mineral accounting with levies seems to be unsuitable in a situation where a reduction in the number of livestock is necessary. As a result, several options including a forced reduction in the number of animals are now under discussion.

The various technical measures have had limited success. The emissions of ammonia decreased by 40% between 1980 and 1992, thus approaching the final governmental policy goal of a 50% reduction by 2000. The surpluses of nitrogen and phosphate have decreased by 20 and 15% respectively between 1985 and 1992. The average nitrogen surplus – including ammonia (NH_3) – of dairy farms decreased by 20% to 391 kg/ha. Despite these results the Netherlands is not expected to meet the objectives of the 'North Sea Treaty' in which European countries agreed to reduce the emission of nitrate to surface waters with 50% between 1989 and 1995.

The whole area of the Netherlands has been put forward as a vulnerable area under the Nitrate Directive, which means that no more than 170 kg N/ha from animal manure may be applied in the future. In 1991 nitrogen production from animal manure was above this level on 63% of Dutch farms. However, the majority of these farms remove some of the manure from the farm, which means that the real problem is somewhat reduced. An important limitation of the Nitrate Directive is that it only covers nitrogen from animal manure. The Dutch policy is aimed at reducing nitrogen pollution from both animal manure and inorganic fertilizer. Moreover, Dutch policy focuses on reducing emissions rather than the production of nitrogen. These differences have resulted in extended discussions between the European Commission and the Dutch Government on both instruments and standards for manure policy.

In general, farmers in the Netherlands agree with the necessity for a policy on manure, however, the choice of instruments meets considerable resistance from farmers. Farmers feel that the existing policy of prohibitions and regulations obliges them to invest in measurements that are costly and are not always effective. Farmers who reduce their mineral losses are not rewarded and face the same prohibitions. A change to a policy that prescribes targets rather than means would be fairer and more effective. A system of nutrient accounting with regulatory taxes on losses above a maximum and premiums on reduced losses would provide farmers with the possibility of finding their own tailor-made solutions.

Pesticides

The problem

During and after the application of pesticides, a considerable amount reaches the soil, groundwater, surface water or the atmosphere. The presence of pesticides in these environmental areas may constitute a further risk to ecosystems. In several areas of the Netherlands pesticides have been found in surface water and rainwater. In 1990 the average pesticide use on arable and horticultural crops was 20.8 kg active ingredients/ha, which is considerably higher than in other European countries. Pesticide use on grassland, however, is low. One of the main reasons for this high use of pesticides is the narrow crop rotation and the resulting high use of soil fumigants. A second reason is the large area of potatoes, which have a high fungicide requirement.

The policy

In 1991 the Dutch Government introduced the Multi-Year Crop Protection Plan (MYCPP, 1991) in which three objectives have been set:

1. a reduction in the dependence on pesticides;
2. a reduction in the use of chemical pesticides of 35% by 1995 and of 50% by 2000 compared with the amount of pesticides used during the period 1984–1988;
3. reductions in the emissions of chemical pesticides into the environment, using 1984–1988 as a reference (Table 9.1)

To meet the targets, several policy instruments will be applied. Extension and education will encourage farmers to adopt integrated farming systems, they will be advised to improve crop rotation, to grow less susceptible varieties and to apply non-chemical control methods. At the same time research on integrated pest control will be intensified. Farmers will be required to obtain a certificate of competence for the application of pesticides and application equipment must also meet quality requirements. For various agricultural sectors soil disinfectants will be obtainable on prescription only and all pesticides will have to meet stricter environmental criteria. Pesticides that have an unacceptable environmental impact will be banned.

Table 9.1. Planned reductions in the emissions of chemical pesticides into the environment using 1984–1988 levels for reference.

	Reduction by 1995 (%)	Reduction by 2000 (%)
Atmosphere	38	50
Groundwater	48	75
Surface water	70	90

Initially, the MYCPP was heavily criticized by farmers' organizations and the chemical industry, mainly because of the proposed withdrawal of potentially harmful pesticides. After political pressure the agricultural sector and the Government reached an agreement in which the agricultural sector supports the objectives of the MYCPP and expresses its willingness to take responsibility for its implementation. In return, the hazardous pesticides which are essential for agriculture will remain available to farmers until alternatives have been produced.

Compared with the reference period 1984–1988 the reduction of pesticide use in 1992 was 20% and by 1993 the reduction had reached 40% (van Bruchem, 1994). The figure for 1993 has been influenced by soil fumigation which, due to heavy rainfall, was not possible during an extended period. However, the objective of a reduction in use by 35% in 1995 is likely to be met, mainly due to the reduced use of soil fumigants. The use of herbicides decreased, but the use of other pesticides has remained more or less stable. Meeting the pollution objectives will be more difficult. In several sectors a reduction in pollution of surface water with pesticides by 90% seems impossible without additional measures such as non-sprayed field margins or field margins without cultivation. In several sectors of Dutch agriculture these measures are now under discussion.

A shortcoming of the MYCPP is that it focuses primarily on reducing the amount of active ingredients in pesticides and does not consider the damage these substances can cause the environment. There is not always a clear relationship between environmental damage and the amount of pesticides used. Sometimes an increase in pesticide use can result in less of an environmental hazard. CLM has developed a so called 'environmental yardstick' for pesticides, which enables farmers to assess the environmental hazard related to the use of pesticides. This yardstick assigns 'pollution points' to each pesticide with regard to leaching into groundwater, effects on water organisms and effects on soil organisms. An analysis of pesticide use against this yardstick shows that the pollution points for leaching into groundwater and the risk to the soil environment have equally decreased with the level of pesticide usage between 1984–1988 and 1993. The risk to soil organisms has only slowly decreased (Reus, 1994)

Energy

One of the objectives in Dutch environmental policy is a national reduction of carbon dioxide emissions, which are directly related to energy use, by 3 to 5% between 1989 and 2000. The overall energy consumption of Dutch agriculture rose 19% between 1989 and 1992 to 150×10^{15} Joules (Poppe *et al.*, 1994). Horticulture under glass utilizes 84% of the energy in the agrarian sector through direct consumption. The animal husbandry sector is also an important user of energy by indirect use (e.g. energy for the production and transport of animal feeds).

In the Multi-Year Energy Agreement, the Dutch Government and the horticultural sector have agreed upon an improvement in energy efficiency (the amount of energy used per unit of product) of 50% between 1980 and 2000. Over the period 1980–1993 the efficiency has improved by 38%, this increase was almost entirely achieved between 1980 and 1985, when energy prices were relatively high. An important reason for the improvement in energy efficiency is the increase in production per hectare in horticulture under glass. Although the energy efficiency in horticulture under glass has improved over the last 15 years, the total energy use has increased considerably. The main causes for these differences are the large increase in the area under glass in recent years and the increase in production per hectare. Carbon dioxide production from horticulture under glass increased by 12% between 1989 and 1992.

Meeting the objectives of the Multi-Year Energy Agreement and especially the overall carbon dioxide objectives appears to be difficult. However, a wide range of technical possibilities for energy conservation in the agricultural sector are available. The large differences between individuals in both energy use and energy efficiency indicate many possibilities to decrease both. Several of the possible technical adjustments are not at present economically feasible (Poppe *et al.*, 1994). A tax on energy for carbon dioxide production is widely considered an efficient instrument to stimulate energy conservation, but for horticulture under glass an energy levy would have severe consequences on competitiveness and incomes. As long as the agricultural sector and especially horticulture meets the targets of the Multi-Year Energy Agreement horticulture is likely to be exempted from the intended energy levy.

Wildlife and Landscape

A recent inventory on the situation of flora and fauna in the Netherlands shows that many rare species are becoming rarer. Two characteristic Dutch habitats of international significance (nutrient poor sandy soils and wetlands) have also shown a decline in area and quality because of eutrophication, acidification, desiccation and fragmentation (Bink *et al.*, 1994). Agriculture is one of the main causes of these environmental problems. Biodiversity of flora and fauna in agriculture has decreased significantly during recent years. However, in spite of intensification, agricultural areas in the Netherlands are still of great importance to many species of plants and animals: about 40% of the vascular plant species and more than 50% of the breeding bird species in the Netherlands can be found in agricultural biotopes (Terwan, 1992).

The Relation Paper, a government paper on the relationship between agriculture, conservation and landscape, issued in 1975, marked the start of the Government's attempts at conservation in agricultural areas. The two main instruments are management agreements and the establishment of nature reserves. An area of 100,000 ha of agricultural land was designated to become nature reserve. The desired conservation measures on this land were considered to be in conflict with modern agriculture, therefore the Government aimed to buy this land and bring it under the ownership and management of professional conservation organizations. Until the time of purchase, farmers could enter into management agreements. The total area of 'Relation Paper' reserves was around 21,000 ha by the end of 1993. With another 100,000 ha eligible for management agreements, under which the farmer receives financial compensation for adjustments in his farming practice to conserve or enhance certain biotopes and/or species, such as the typical Dutch meadow birds.

Before 1988 compensation was based on income losses compared with other farmers in similar areas, but in 1988, compensation was only given for extra costs for labour and concentrate feed. Following a test year, farmers sign a contract for 5 years, in practice less than 1% of the farmers terminate the contract. In the designated areas farmers can choose between different sets of measures that have a positive influence on wildlife and landscape. These measures can vary from reducing or zero application of nutrients (animal manure and fertilizer) to not mowing during certain periods and the maintenance of field margins. Depending on the chosen adjustments in the management of the farm the compensation varies from G180 /ha for 'light' management to G1400 /ha for 'heavy' management. Although management agreements should not provide extra income, there is no doubt that especially for a number of farmers in areas with inconvenient production circumstances, they are able to continue farming because of the compensatory allowances.

The first agreement under the Relation Paper was signed in 1981. During the first year, interest was relatively low, but in recent years the number of management agreements has increased rapidly, until a total of 35,000 ha was covered in 1994 with 5000 farmers involved. This area increases by around 5000 ha/year, with a total designated area of 100,000 ha divided over almost 200 different regions (Melman and Buitink, 1994).

The effectiveness of the management agreements is monitored and evaluated thoroughly. Aims for the meadow birds have been met in most regions, and there is reason for optimism with the botanical management. The more expensive 'heavy' management is especially effective (Melman and Buitink, 1994). The cost-effectiveness of wildlife management by farmers compared with that of conservation organizations has often been debated. In 1994 a study ordered by the Dutch WWF concluded that after 40–70 years, total costs for nature reserves are less than the costs for management agreements. But in contrast, the Department of Agricultural Economics of the Wageningen Agricultural University concluded in 1994 that land withdrawal is more expensive than wildlife management through management agreements. The picture is far from clear.

In 1990 the Nature Policy Plan (NPP) was launched, defining a new nature policy. In the NPP segregation of nature and agriculture plays an important role and a central element in it was the National Ecological Network, presenting a nation-wide ecological infrastructure of natural areas. A central aim of the NPP is the withdrawal of 150,000 ha from agriculture for conservation purposes. The second phase of the Relation Paper policy, with a second area of 100,000 ha to be designated for management agreements, was incorporated into the NPP management agreements and will be mainly confined to areas in the National Ecological Network. The farmers' organizations and the Landbouwschap agreed with the concept of the National Ecological Network, because they, not very realistically, hoped that the concept would guard farmers outside the area of the National Ecological Network from constraints in their farming practice. Outside this Network farmers and other land users should preserve a so-called 'General Nature Quality'. What this Quality is and how it shall be realized has not yet been defined.

The concept of the National Ecological Network is valuable, but also has serious disadvantages. Withdrawal of large areas of agricultural land conflicts with environmental policy, for environmental reasons, extensification is necessary and requires an increase in agricultural area. The large area necessary for the realization of the Network will result in increasing pressure on the land market. As a consequence, nature policy is at right angles to environmental policy (Oosterveld, 1994). The emphasis on nature policy has been put unilaterally on the National Ecological Network. Incentives for the enhancement or preservation of biodiversity in the remaining 85% of the 2 million hectares of agricultural land hardly exists. This has led to a large number of farmers outside the National Ecological Network being interested, but not stimulated to develop conservation activities.

New policy lines

In the years after the publication of the NPP, national and provincial governments, farmers' organizations and conservation organizations have become more interested in conservation on farms outside the National Ecological Network. The increasing concern for present and future incomes and the will to find alternatives for the withdrawal of agricultural land for nature reserves, resulted in more interest among farmers in new activities. Around 60 local farmers' cooperatives started to negotiate with local governments and nature conservation organizations to find remuneration for conservation activities. Several local and national conservation organizations also became more interested in stimulating the (potential) value of nature in agricultural areas. The Ministry of Agriculture calls 'nature conservation on the farm' or 'farming with nature' one of its new priorities. Unfortunately, until now, new funds for these new priorities have been limited. In several parts of the country pilot projects have been started with remuneration to farmers for results of nature management. Practical problems to be solved are defining nature quality, controlling and pricing the nature produced: for botanical management however, the initial results are encouraging.

Regulation 2078/92

In the Netherlands the programme under the Agri-environmental Regulation was established by the Ministry of Agriculture, Nature Management and Fisheries in the summer of 1993, after consultation with representatives of different interest groups involved, such as the Landbouwschap and conservation organizations. Environmental organizations were hardly interested in participating in the consultation process, because they considered the Regulation of minor importance. Public discussion on the implementation of the Agri- environmental Regulation 2078/92 in the Netherlands was limited, especially when compared with public discussions on the desired future manure policy.

A problem for implementation in the Netherlands was that for reducing several of the environmental problems, effective programmes would require higher payments per hectare than the present maximum payments. A programme for 20-year set-aside has not been implemented, because few farmers were expected to participate. Due to the high prices of agricultural land in the Netherlands (an average of around G38,000 /ha), it was expected that few farmers would choose to participate in the programme, unless the yearly payment was considerably higher than the maximum of 700 ECU/ha. Farmers' organizations also feared that after 20 years it would be difficult for farmers to obtain permission to reverse the land use from conservation back into agriculture.

One of the main criteria in the selection of programmes for the Dutch programme was the expected ease of control and monitoring. Programmes substantially reducing the use of fertilizers and/or plant pesticides have not been implemented because they were considered too difficult to monitor. As a result, the only programmes adopted are those under section 1a of Article 2 of Regulation 2078/92, which stimulate and support organic farming. The Dutch Government has been reluctant to introduce support schemes under Regulation 2078/92 because payments for pollution control schemes were considered inconsistent with the Polluter Pays Principle or the responsibility of farmers to reduce pollution. As a result priority has been given to extension and demonstration projects, measures to improve the training of farmers with regard to farming practices compatible with the environment and to management agreements. These priorities appear clearly in Table 9.2, showing the budget for programmes under Regulation 2078/92 for the years 1994–1997. The total EU-budget for measures under Regulation 2078/92 in the Netherlands is G156.1 million for the period 1994–1997. For the three

Table 9.2. Budget in the Netherlands for measures under regulation 2078/92 (Dutch Guilders, period 1994–1997). (Source: Ministry of Agriculture, Nature Management and Fisheries, the Netherlands.)

	Total budget ('000)	EU budget ('000)	National budget ('000)
Measure			
1. management agreements (ESAs)	78.8	39.4	39.4
2. demonstration projects	122.6	63.9	58.7
3. courses	36.6	18.3	18.3
4. stimulating organic crop production	5.1	2.8	2.3
5. recreational use of land (public footpaths)	0.2	0.1	0.1
Total measures approved by EU	243.4	124.5	118.8

measures that have not yet been approved by the European Commission an EU-budget of G31.6 million remains.

Environmentally Sensitive Areas (Regulation on Management Agreements)

A programme drawn up by the Dutch Ministry of Agriculture, Nature Management and Fisheries to protect environmentally sensitive areas was approved by the European Commission on 18 October 1993. This programme 'Management Agreements 1993' is the second phase of the Regulation on Management Agreements, which has come from Article 19 of Regulation 797/91 (later 2328/91) under the EC Regulation 2078/92.

The main objectives of the programme are the management of buffer zones surrounding nature reserves (as part of the National Ecological Network), to preserve plant life and protect bird habitats. Farmers in various areas of particular ecological or scenic interest commit themselves, on a voluntary basis, to introduce and maintain farming practices which are compatible with the need to protect the environment and to implement measures to preserve wildlife and landscape. The total area of farmland under management agreements was 35,000 ha in 1994, which is equal to 1.75% of the total agricultural area.

In May 1995 a new Regulation on Management Agreements came into practice. To stimulate more drastic conservation measures, a bonus of up to 15% above the existing compensation will be paid. Management packages that are ineffective have either been altered or left out.

Demonstration and extension projects

In November 1994 the Regulation for financial contribution for demonstration projects was started. With a total budget of G122 million in 4 years this Regulation is the largest of the regulations under 2078/92. Through this Regulation, testing and demonstrating innovations at farm level and materials and activities to improve extension work can be financed. An EU contribution can be obtained for a list of eight different projects. Each year the Ministry of Agriculture defines a number of priority areas within these eight projects, for which an additional contribution from the Dutch Government can be obtained. For instance one of the eight projects is the maintenance and development of wildlife and landscape values in agricultural areas. Within this project, the following issues are eligible for additional support:

1. creation of areas of conservation value on set-aside land;
2. management of field margins (e.g. ditch sides);
3. maintenance of meadow birds.

Courses and education

Education programmes for farmers, their family members working on the farm and agricultural labourers can be supported from this scheme. Participants on courses of a minimum of 40 hours receive a payment of G250–500 each.

Stimulating organic farming

Since 1989 the area under organic farming in the Netherlands has increased by 60%, in spite of this significant growth, its area remains limited. In 1993 the number of organic farms in the Netherlands was 490, or 0.4% of the total number of farms. In 1993 the area of organic farms in the Netherlands was 10,040 ha, equivalent to 0.5% of the total agricultural area.

Regulation to stimulate change to organic arable farming
The level of support for farmers who convert to organic arable farming is based on the income losses caused by the extended 'waiting period'. This period is the time from when organic management of a certain plot or farm begins, to when the organic produce from that plot or farm can be sold. In EC Regulation 2092/81 this period is fixed at 2 years, in the Netherlands it used to be 6 months. During the 2-year waiting period, farmers apply organic management practices, but cannot sell their produce as organic products. The payment to stimulate a change to organic farming is a compensation for the income lost during these 2 years: the payment is spread over a 5-year period.

The level of payments for converting to organic farming in different sectors can be seen in Table 9.3.

Production should be in accordance with demands in EC Regulation 2092/91, with participating farmers committing themselves to organic production for at least 5 years. For most farmers interested in organic farming the cost of changing is already substantial. The contractual obligation to maintain organic farming for a minimum of 5 years additionally increases the cost of conversion to organic farming and decreases the likely impact of the measure. A second problem is that for horticulture under glass and fruit production the payment only compensates for less than one-third of the income lost during the waiting period. For arable farming and field scale vegetable farming, the payments approximately equal the expected income lost (Leferink, 1994). This means that the stimulus from the regulation in these sectors will be limited. The budget for conversion to organic arable farming was G2.2 million in 1994. Participation in the scheme was considerably less than expected, with applications totalling G0.5 million.

Area payment for existing organic farmers in arable farming

Arable farmers and horticulturists who are already practising organic farming, receive payments over 5 years (Table 9.4). The payment is equal for all sectors of agriculture, mainly because it is not a compensation for losses but a remuneration for past efforts or for pioneering. The Ministry of Agriculture expected all organic farmers to apply for an area payment, but in 1994, only half the organic arable farmers applied and it is not known why the remainder abstained.

Stimulation of recreational use of agricultural land

Farmers who make their land available for recreational use, e.g. for fishing, canoeing or public footpaths will receive a single payment of G3600 /ha. The relevant regulation will come into force in April 1995. The Ministry of Agriculture has transferred the execution of the Regulation to the Foundation for Long Distance Footpaths (in Dutch: LAW). This Foundation is a non-governmental organization that plots the course of long distance footpaths and is responsible for marking the routes and they will approach

Table 9.3. Payments for farmers converting to organic arable farming (Dutch Guilders per hectare). (Source: Leferink, 1994.)

Arable farming	500
Vegetables in the open field	1200
Horticulture under glass	1850
Fruit	1850

Table 9.4. Payments for farmers presently applying organic arable farming. (Source: Leferink, 1994.)

Year	Guilders per hectare
1	400
2	350
3	300
4	250
5	200

farmers asking them to open their land for walkers. Farmers who open their fields or farmyard for long distance walkers can receive a remuneration of G2.20 /m of footpath, for additional provisions standard payments are available.

The following regulations have not yet been approved by the European Commission and have therefore not come into power.

Extensification of beef cattle production

The only regulation for extensification that will be part of the Dutch package within Regulation 2078/92 is a regulation for (extensification of) beef cattle production. Farmers with more than 120 head of beef cattle will receive a premium of 150 ECU per head if they reduce the number of cattle in their herd. In the Netherlands this will be applicable to 700 farms. One of the main reasons for introducing this measure, is the severe income losses of beef farmers after the most recent reforms of the Common Agricultural Policy. The payment for compensation of the recent price reduction on beef has been limited to 90 head. As a result those farmers with a herd of more than 120 face serious income problems. Restructuring the sector is the primary objective of this measure, with the reduction of environmental damage as a secondary objective.

Other measures

Farmers who have certain rare livestock breeds, that are threatened with extinction will receive a yearly payment per head to promote their preservation. A support system for organic livestock farming is still in preparation.

Conclusions: The Place of Regulation 2078 in Dutch Policies on Agriculture and Environment

In environmental and conservation policy in Dutch agriculture a set of policy instruments is used including direct regulation, research, extension, education, demonstration and financial instruments. The central elements of Regulation 2078/92 are payments for accepting management restrictions to protect the environment and to maintain countryside or preserve wildlife. An important question is therefore what the role of these payments is and should be in Dutch environmental and nature policy.

The Dutch Government has accepted the Polluter Pays Principle (PPP) as one of the cornerstones in Dutch environmental policy, and as a result, payments to compensate for accepting restrictions are not a key element. Special circumstances in agriculture limit the extent to which the PPP may be applied to agriculture. Baldock and Bennett (1991) describe how the problems concerning identification of polluters, the effective enforcement of environmental controls, the allocation of control costs and the definition of the point at which an activity becomes polluting, combined with the political significance ascribed to agriculture as a strategic sector inhibit a strict application of PPP in agriculture. The result is that, however undesirable in principle, second best solutions – involving regulations and perhaps subsidies – will in practice be necessary if agricultural pollution is to be effectively controlled. Van der Weijden (1995) states that there should be exemptions from PPP to permit temporary payments so farmers can adjust to generic environmental standards and for farmers facing stricter standards. Therefore, a system as defined under Regulation 2078/92 does not necessarily contradict the application of PPP. Regulation 2078/92 provides direct compensation payments for accepting management restrictions and for producing positive externalities. Direct payments to farmers are not a central instrument in Dutch environmental policy. Demonstration and education are two of the important instruments in Dutch environmental policy. This explains why, especially in the Netherlands, these parts of Regulation 2078/92 are widely considered a useful addition to environmental policy in Dutch agriculture.

Payments for acceptance of extra restrictions inside and outside Environmentally Sensitive Areas are a central element in conservation policy in the Netherlands, especially in the policy on management agreements.

It is accepted that farmers should be remunerated if society wishes farmers to produce positive externalities such as biodiversity and landscape. Van der Weijden (1995) proposes SRP (Stewardship is Remunerated Principle) as an additional principle in European environmental policy. According to Van der Weijden SRP should be applied:

1. as a temporary exception for farmers to adjust to generic environmental standards;

2. for farmers who face stricter than the usual regulations;

3. for remuneration of producing positive externalities such as biodiversity and landscape.

In the application of Regulation 2078/92 the Dutch Government has more or less followed these three lines. In conservation policy SRP has generally been applied, but for environmental policy, payments are an exception.

An important limitation for the Dutch situation including a high value added per hectare is that the maximum payment under Regulation 2078/92 is G1850 /ha. In several situations this maximum is not sufficient for farmers to make the scheme beneficial, examples can be found in the Dutch Regulations for conversion to organic farming, where in some sectors the payment is too low to stimulate conversion. Moreover, the ongoing policy discussions concerning manure, pesticides and energy, draw most of the attention of agricultural and environmental organizations. This explains why Regulation 2078/92 has drawn relatively little attention and caused little discussion in the Netherlands. Compared with the difficulties in reducing environmental damage to an acceptable level and meeting the objectives laid down in the National Environmental Policy Plans, the measures under Regulation 2078/92 are of minor importance.

Most programmes have started recently so that a thorough evaluation of the results of the different programmes under EC Regulation is not yet possible. However, both farmers' organizations and environmental organizations in general are positive about measures in Regulation 2078/92. When the Regulation was introduced, the interest, especially of environmental organizations, was limited. The Foundation for Nature and Environment, one of the most influential environmental organizations in the Netherlands, regards Regulation 2078/92 as a convenient vehicle to encourage a shift to environmentally friendly farm-practices and proposes a broadening of the existing Regulation.

10 SWEDEN

BENGT RUNDQVIST

Background

Sweden is a land dominated by forests, approximately 63% of the total land area is covered with forests and only 8% is agricultural land. In contrast with these figures, the agricultural landscape is of great importance for biological diversity and nature conservation (Bernes, 1993). It is estimated today that of more than 400 vascular plant species that are Red Data Book listed in Sweden over two-thirds occur in the agricultural landscape. For fungi, the corresponding proportion is one-third, for bryophytes and lichens roughly a quarter, and for invertebrates around 40%. The threat to biological diversity in the agricultural landscape is due mainly to the considerable change in land use during recent years, especially the disappearance of the semi-natural grasslands and the grazing animals (Swedish Environmental Protection Agency, 1993b).

Throughout the postwar period far-reaching changes have taken place in Swedish agriculture (Bernes, 1994). Crop improvement, livestock breeding and the use of fertilizers and pesticides have brought about increasingly higher productivity. Efficiency has also been improved by merging smaller farms into larger units. Artificial fertilizers have enabled some agricultural enterprises to focus on growing crops, while others have specialized in livestock production. Agricultural policy has principally been designed to produce cheap food, however, this policy has resulted in overproduction and large areas of farmland have been taken out of production, while most of the farmland abandoned has reverted to forest.

The 1990 Agricultural Reform

The Swedish Landscape Conservation Measures were introduced in 1990 within the framework of the agricultural policy reform decided by Parliament. The general direction of the reform was of internal deregulation, abolition of export subsidies, reduction of price support coupled with an introduction of new types of direct payment for public goods, such as regional development, food security, and landscape conservation measures, but also coupled with increased education activities in the marketing of new products and nature conservation (Swedish Parliament, 1990). The aim was to find new types of direct payment for specific public services, rather than trying to achieve a number of different objectives with only one instrument, price support, as had previously been the case. The environmental objectives of agricultural policy are the protection and enhancement of biological diversity, conservation of the agricultural landscape, the natural and cultural heritage and minimizing pollution from agriculture due to nutrient leaching and pesticide use.

Whereas the general environmental effects of the agricultural policy reform were expected to be positive, i.e. through less intensive cultivation methods, there was concern that changes in land use, mainly through afforestation and less intensive use of semi-natural grasslands, would have a negative effect on the agricultural landscape. This was seen as an additional threat to the valuable natural and man-made features of the agricultural landscape. The diversity of this landscape had been decreasing continuously for decades due to long-term structural changes in agricultural land use and higher-intensity cultivation methods. It was decided that government authorities should be allowed to enter into agreements with farmers to ensure continued grazing and cultivation of the areas concerned, in exchange for direct payment for these services. The main objective for these agreements was to protect natural and cultural values of national importance and, in doing so, to maintain the overall impression of the open agricultural landscape in particular parts of the country.

The Parliamentary Decision on the Landscape Conservation Measures

In the Parliamentary decision three central government authorities, the Swedish Environmental Protection Agency, the Swedish Board of Agriculture and the Central Board of National Antiquities, were responsible for developing a proposed programme for the new Landscape Conservation Measures (Swedish Parliament, 1990). The programme was to include clear definitions of its objectives to facilitate monitoring and evaluation. The Swedish Environmental Protection Agency is responsible for monitoring and evaluating the programme on a continuous basis, in cooperation with the Board of

Agriculture and the National Board of Antiquities. The County Administration Boards (who are the regional agencies for agriculture, environment and so on) were to be responsible for the major part of the public funds available for this purpose and for concluding agreements with farmers. The Government would decide on the distribution of funds to the county boards on the basis of a joint proposal from the three central authorities concerned. The central government authorities were given additional resources for the new tasks entrusted to them.

Parliament decided to introduce the new programme during a 3-year period in order to avoid transitional problems during the initial years and to increase national budget allocations gradually from SKr 100 million to 250 million by the third year. It was difficult to estimate how much of the agricultural landscape with nationally important natural and cultural value would come under threat as a consequence of the agricultural reforms. At the end of the initial 3-year period a thorough evaluation of the programme would be undertaken.

The agreements between the County Administration Boards and the farmers were to be based on civil law and the terms adapted to the specific conditions in each case. Thus, no enabling legislation was introduced for this new measure. The payments were not to be standardized but based on the terms of the specific agreement and the type of service rendered. Within the framework of the agricultural policy reforms, a number of changes were made in the Law on the Management of Agricultural Land. Most importantly, the obligation to keep land suitable for farming under agricultural use was abolished. An amendment was made to the effect that the farmer would be obliged to notify the County Board 8 months in advance of taking agricultural land out of use. This would render it possible for the authorities to examine the area in question and decide whether or not to offer payment for the continued cultivation of that land.

Further, the Parliamentary resolution stressed that the efficient implementation of the landscape conservation measures would be dependent on the presence of economically viable farms to provide the environmental services required and this should be taken into consideration when deciding where and how to apply the measures. It was also noted that the municipalities had a responsibility for maintaining values for nature and cultural conservation of the agricultural landscape at a local level.

Implementation and Administration

The Landscape Conservation Measures are applied at the national level. The responsible central agencies for implementation, monitoring and evaluation are the Swedish Environmental Protection Agency in cooperation with the Central Board of National Antiquities and the Swedish Board of Agriculture.

At the regional level the agencies in charge are the County Administration Boards.

There are two General Guidelines for the regional implementation of the measures. These guidelines describe and elaborate on the main objectives of the programme as decided by Parliament, and also provide practical advice for the administrators of the County Administration Boards for whom the guidelines were introduced during information meetings and 1-week courses. The administrators are educated each year in evaluation and documentation of the conservation values of the agricultural landscape, formulation of aims and management demands in agreements, and in methods for following up the evaluation. Education in grazing for nature conservation purposes is also available to administrators and farmers.

The General Guidelines focus on how to produce regional programmes for estimating the values of the agricultural landscape in terms of natural, historical and cultural values. These programmes, the *Regional Programmes for Conservation of the Values of the Agricultural Landscape*, are elaborated by the County Administration Boards and are used in the counties as a basis for the work, on conservation of the agricultural landscape with a variety of means. At present the most important of these means are the Landscape Conservation Measures, which are described in this case study.

For each area described in the regional programmes, the *nature conservation* value is classified into three categories according to the following criteria:

- biological qualities representative of the region;
- high biological diversity, specifically a rich variety of habitats and areas with a mosaic structure used in a traditional way for grazing, mowing and cultivation;
- variety and density of species, particularly the ones dependent on traditional agricultural management;
- presence of semi-natural grasslands, i.e. ancient unfertilized and uncultivated grasslands used for traditional grazing (pastures) or mowing (meadows);
- presence of habitats and species that are threatened, including Red Data Book listed species.

Cultural or historical values are classified in two ways, according to the following criteria:

- representativeness of different periods of history of the region;
- structure of settlements or ground allocation typical of historical periods of the region;
- presence of landscape elements representing traditional agricultural activity such as ancient remains, fences, roads, ditches and ancient cultivated fields;
- areas with continuity in agricultural activity during long periods of time;

- presence of old agricultural landscape elements that are disappearing.

For each area the following aspects are also considered:

- the overall impression or the values of the area as a whole;
- the status of the present agricultural management;
- recreational values (aesthetic and ease of access);
- educational and scientific value.

The General Guidelines, *On the Use of the Funds for Landscape Conservation*, provide advice on:

- how to select areas and farms eligible for agreements on the basis of the most valuable areas emerging from the regional programmes (see above);
- how to reach an agreement with the farmer;
- the terms of the agreements, which values and objectives to cover, what management demands to include in the terms and the legal aspects of framing and the wording of the agreement;
- how to estimate reasonable payments;
- how to monitor the results of the agreements.

Generally, the agreements are entered into on a voluntary basis according to civil law (Swedish Environmental Protection Agency, 1994a). It is the farmer of the land, owner or tenant, who is eligible for entering the agreement and thus responsible for the management agreed upon. The selected farm should preferably be a livestock one, based on grazing and represent a viable rural enterprise so that farming is likely to prevail. The farmer should also show an interest in conservation measures. The agreements are intended to be long term, but should be reviewed every fifth year, when the agreement can either be cancelled, modified or prolonged. The agreements follow the principle of incentive payments. There is no national list of acceptable payments, but an annual payment of SKr 300–2000/ha has been recommended. The amount agreed should reflect the sum of the values of the conservation of the area and the terms of the agreement. When the requirements of the agreement increase the costs for the farmer, payments should accordingly be increased. Each agreement should be individually adapted to the particular farm, the local conditions and the demands for conservation measures. The payment is intended for specified land use, but not maintenance of buildings. The agreements should not cover nature reserves or otherwise protected areas.

Each year, the Government allocates the funds to the County Administration Boards on the basis of a proposal from the three central Government agencies. This proposal is based on statistics from the regional programmes, The National Inventory of the semi-natural grasslands, the National Agricultural Statistics and an assessment of the threat to the agricultural landscape in the various regions. As the threat to the agricultural landscape is relatively smaller in the agricultural regions of southern Sweden, this part of the country receives small amounts of the Funds for Landscape Conservation in comparison with other regions.

The agricultural reforms have so far, in contrast with expectations, shown no major negative effects on the natural and cultural values in the agricultural landscape (Swedish Environmental Protection Agency, 1994b). One reason for this is that farmers are delaying their decisions on changes in land use as a response to the Swedish application in 1991 for membership of the EC. There have also been some changes of the agricultural policy after the reforms of 1990. The concern for conservation, is the constant and long term transformation of the agricultural landscape, that occurs predominantly in the regions dominated by forests intermingled with agriculture in the southern and central parts of Sweden and in the northern regions. This transformation causes afforestation and the natural succession of semi-natural grassland and arable.

Economic and Environmental Efficiency

The three responsible agencies at the national level are continuously following up, monitoring and evaluating the results and efficiency of the programme. An annual report of the results is presented to the Government and provides proposals for improvements. This report is based primarily on statistics from the County Administration Boards but also on considerations from discussions and conferences with the relevant agencies. In the autumn and winter of 1993/94 a more thorough evaluation was carried out in eight of the 24 counties (Swedish Environmental Protection Agency, 1994a). Long term environmental effects are monitored within a monitoring programme. Results and considerations presented in this chapter have references to the annual reports and results from the evaluation mentioned above.

The Landscape Conservation Programme is now implemented throughout Sweden. For an annual cost of SKr 250 million there are roughly 15,000 agreements covering more than 376,000 ha, of which around half is arable and half semi-natural grassland (Swedish Environmental Protection Agency, 1994b).

Overall 15% of Swedish farmers are now involved in the Landscape Conservation Programme. The average payment as a mean for all agreements is SKr 700 /ha. For arable land the payment is normally between SKr 300–SKr 700 /ha and for semi-natural grassland between SKr 900–1500 /ha. The annual sum is allocated when the farmer has signed a declaration stating that he has fulfilled his obligations according to the agreement. In the rare cases when a farmer has failed to fulfil an agreement or part of it, the payment is reduced. In such a case the normal procedure is that the County Administration Board initiates a discussion with the farmer; most frequently the problems can be solved and are sometimes caused by a lack of information.

Of the total annual cost, SKr 250 million, around SKr 5 million is allocated to central authorities for information, education, administration and monitoring. Apart from the Landscape Conservation Measures these funds also cover the implementation and administration of monitoring, with the main objective of evaluating the effects on the agricultural landscape of the Swedish agricultural policy. At the county level, the total annual costs for the implementation have been calculated at 5–10% of the Landscape Conservation funds allocated to the County Administration Boards. Now, when the initial implementation period is over, the annual administration cost is calculated to be around 5%. This will cover the costs for information, administration, and monitoring.

The methods used for landscape conservation agreements are now the most important for conservation in the Swedish agricultural landscape. The areas the agreements cover are of significant value for the conservation of the flora, fauna and ancient remains. The agreements cover a considerable proportion of the priority areas for conservation according to the Regional Programmes for the conservation of Values of the Agricultural Landscape. The total area of these priority areas is estimated to be around 1.2 million ha. Not all areas require active conservation measures. The Swedish Environmental Protection Agency estimates the most valuable areas in need of conserving cover 600,000 ha (Swedish Environmental Protection Agency, 1993c). The Landscape Conservation Programme now includes half of these areas, with almost half of the semi-natural grasslands, i.e. ancient pastures and meadows, still in use included. The programme thus constitutes an essential step towards safeguarding valuable parts of the agricultural landscape.

To evaluate the efficiency of the Landscape Conservation Measures it is important to know if, or to what extent, farmers are managing these areas in a better way (in aspects relevant for conservation) than they would without the agreement. This is, however, difficult to ascertain, particularly in the short term. A comparison of areas with agreements with areas managed without agreements will be done in conjunction with the monitoring system for the Swedish agricultural landscape. However, results from the monitoring in this respect will not be available for evaluation for some years. The process of change in the management of grasslands is slow and it takes some time before the farmers are well informed about and capable of adapting to the requirements of the conservation agreements.

To obtain direct information about the farmers' response to the agreements, the recent evaluation has included questionnaires to 180 farmers with conservation agreements. The answers indicate that around half the farmers have increased their work on clearing bushes in the meadows and pastures as a result of the agreement, and a quarter have increased the number of grazing animals on the semi-natural grasslands.

On many farms this is essential because in Sweden one of the major threats to nature conservation is that of declining numbers of grazing animals. Some farmers have increased the area of mown semi-natural grassland, and most of the farmers ceased using fertilizers on semi-natural grasslands years ago. These changes in management are important and provide positive results from the conservation and environment point of view. These changes however are normally small.

Problems During the Implementation Period

The initial implementation period was characterized by work organizing the administrative system and preparing regional programmes as a basis for the conservation measures (Swedish Environmental Protection Agency, 1994a). The issues at hand are of a complex nature and have placed great demands on all the agencies concerned.

Despite the fact that the programme is being introduced gradually, the central agencies have judged the time and financial resources available for preparation and administration insufficient. The Government has however, in its yearly decision on budget allocations, decided not to depart from the initial decision to limit administration outlays to 5% of the total funds. The system of agreements adapted individually in each case puts great demands on the information and extension services to farmers to clarify the aims and terms of the agreements. This also implies that both conservation and extension officers need a sound knowledge not only of agriculture and grazing practices, but also of nature conservation, as well as conservation of historical and cultural objects in the agricultural landscape.

Results from the evaluation recently carried out in a number of counties (Swedish Environmental Protection Agency, 1994a) indicate that farmers have a positive attitude towards this new programme and that their knowledge of conservation issues has improved as a consequence. There is, however, room for improvement in the agreements, the terms of which have been too general to provide any real guidance to the farmer in managing the conservation areas. It is essential that there are clear, well defined and measurable aims and management demands, and necessary that the payment provides adequate incentives for high quality conservation management. Documentation with standardized methods concerning the status of the areas concerned is also needed for the purposes of deciding on a suitable management regime for the area and efficient monitoring of changes in the quality of natural and historical values.

Fundamental to the efficiency of the use of available funds is good information and dialogue with the farmers involved. Conservation agreements that are introduced without a visit to the farm, and with a lack of information concerning the objectives and the aims of the agreement, the actual conservation values and a dialogue on the management are not cost-effective. A necessary prerequisite for all of the conservation work is well trained administrators and close cooperation between conservationists, administrators and agricultural experts. Changes in the agricultural landscape are of a long term nature and it is difficult to judge any such changes from experience of a short implementation period. The results of the programme so far are, on the whole, encouraging and the experience is of great value for future work involving conservation of the agricultural landscape. It is important that a programme like this has a cost-effective approach for which the methods of implementation and well defined objectives are basic conditions.

Comparisons with other Conservation Measures

It is interesting to compare the landscape conservation agreements with the measures according to the Act of Nature Conservation. To attain the long term conservation objectives for the agricultural landscape different measures have to be used. For some of the most valuable areas it is necessary to use the method of protection, i.e. Nature Reserves and Nature Conservation Areas, in combination with management agreements according to an adopted management plan. The establishment of these protected areas however is administratively costly, but the advantage is that they are safeguarded in the long term. Landscape conservation agreements, however, have the advantage of being relatively easy to establish and the system is flexible, the disadvantage is that the areas concerned are not protected for more than 5 years at a time.

The concept of landscape conservation agreements was originally developed in 1986 with the programme initially called Nature Conservation Measures in the Agricultural Landscape. The main objective of this programme was to protect and enhance the flora and fauna by ensuring continued use of the semi-natural ancient grasslands in a traditional way for grazing and mowing. With an annual budget of SKr 38 million there are roughly 4000 agreements covering around 40,000 ha of mainly ancient grasslands.

The County Administration Boards use a combination of different conservation measures. In the most valuable areas they use protection with Nature Reserves and Nature Conservation Areas. In other areas they use agreements for Nature Conservation and agreements for Landscape Conservation in a way that the measures supplement each other.

The Agricultural Land Management Act since 1984 has contained legislation concerning the environment. The regulations based on this act are issued by the Swedish Board of Agriculture, and regarding nature conservation and historical preservation stipulate that certain landscape elements may not be removed or damaged and that fertilization of hitherto unfertilized grasslands is banned. The 1991 Nature Conservation Act contains legislation permitting the general protection of smaller 'biotopes', i.e. areas with habitats of special value for nature conservation, especially habitats for Red Data Book listed species. Examples of such 'biotopes' or landscape elements in the agricultural landscape are: old stone walls, small wetlands and 'field islands'. The County Administration Boards may also decide upon the protection of specific small pastures and meadows.

Acceptability and Consultation with the Target Groups

The Swedish Farmers' Association and the Swedish Nature Conservation Association are involved in the Landscape Conservation Measures as members of reference groups at the national and county level. They have a positive attitude to the whole programme which they consider valuable for conservation purposes as well as for the maintenance of agriculture and the countryside. From the inquiries made to 180 farmers previously mentioned, those involved with the Landscape Conservation Programme show the following characteristics:

1. The mean age of the farmers is around 50 years, which is comparable to the mean for Swedish farmers as a whole. The farms involved, normally concentrate on livestock husbandry with grazing animals, around 30% of the farms are dairy with the recruitment of young stock, and around 50% are beef and sheep. This means that the animal-dominated farms are more frequent than the normal Swedish farm of today. This is well in line with the Parliamentary decision concerning the Landscape Conservation Programme. One-third of those farmers involved, farm full time, half are part-time farmers and around 15% are pensioners. These figures are again comparable to all Swedish farmers.
2. Many of the farmers have changed from dairying to meat production in the last 10 years which indicates an extensification of their farming system.
3. There have been positive changes in the management on one-third of the farms due to the agreements, but most of the farmers would have managed this land in the same way without the agreements. Changes in management can not, however, be expected to occur rapidly, especially under the currently uncertain economic situation for the farmers.
4. Around half of the farmers think that the agreements will be of importance in the long run, when it comes to survival of the farm in the future and when there is a takeover from the new generation of farmers. This positive attitude is due to the economic incentives and to the new positive interest from society in the services the farmers provide for conservation.
5. Many of the farmers are now more interested in conservation and the

importance of the management of their farms for biological diversity and historical values.

6. It is of great importance to the farmers that the agreements are of a permanent nature. They desire agreements of long duration.

7. To maintain agricultural landscape conservation in the future, it is necessary to find solutions to the economic problems experienced by farmers in the regions where semi-natural grasslands still exist and are used in a traditional way. In these regions, many farmers are elderly, farm buildings are deteriorating and grazing livestock numbers are declining.

The New Swedish Environmental Programme

Sweden is now a member of the European Community and is producing a new scheme for environmental payments in accordance with the EU Regulation 2078/92. There is as yet no definitive decision in the Swedish Parliament or the EU Commission regarding this new programme but the preliminary proposals are as follows: The Swedish Agri-environmental Programme will consist of three parts, the first part concerns landscape conservation, the second, deals with Environmentally Sensitive Areas and the third part deals with ecological farming. The Programme is planned for a period of 5 years but will be evaluated after 2 years before its continuation. The total cost for the Programme is expected to be SKr 1500 million per year of which Sweden pays SKr 750 million.

Programme for Conservation of Biodiversity and Historical Values in the Agricultural Landscape and Maintenance of Agricultural Landscape in Forest Regions

The conservation of the biodiversity in semi-natural meadows has as its objective, the conservation and management of all the remaining traditionally managed meadows in Sweden, i.e. 3000 ha, in a manner that preserves and enhances the large diversity and density of species, representative flora and fauna and rare species typical of mowed unfertilized meadows. The annual payment will probably be SKr 1300–1800/ha with additional payments for scything (SKr 2500 /ha), autumn grazing (SKr 700/ ha) and for pollarding (SKr 500 /tree). However, the conservation of the biodiversity and sites of historical value in semi-natural grazing lands, i.e. 370,000 ha of traditionally managed pastures, consists of managing them in a way that preserves and enhances high diversity and density of species, representative flora and fauna and rare species typical of unfertilized grazing lands. The pastures are divided into three or four classes depending on their value for conservation. Annual payments will probably be SKr 800–1600 /ha according to the class, with the highest payments for pastures of highest conservation value.

Conservation of biologically rich habitats and valuable historic remnants, concerns the preservation of historical features in all areas of the Swedish agricultural landscape, and the flora and fauna endangered by modern agricultural practices approximately 10,000 valuable areas in 167,000 ha of agricultural land. The valuable areas must meet the criteria for very small landscape elements, landscape and buildings that reflect the history of farming. Examples of valuable landscape elements are ancient monument sites, cattle paths enclosed with stone walls or wooden fences, small waters, open ditches, hedges and avenues, banks between fields or along streams and meadow barns. The payments vary according to the intensity of management and the sum of the conservation values for the area concerned. The mean payment is estimated at SKr 600 /ha with additions for the mowing of banks.

The maintenance of an open agricultural landscape in northern Sweden and in forest regions, is aimed at promoting land use and farming practices that are compatible with the maintenance of the agricultural landscape and the countryside in specific regions. The scheme will ensure a continuation of extensive agriculture, with leys and grazing by cattle, sheep and goats. There will be three levels of annual payment, with the highest amounts for the north of Sweden, support area 1–3 (SKr 2050 /ha for pastures, other permanent grasslands and leys on arable land), lower payments in support area 4, again in the north (SKr 800 /ha for pastures and SKr 600 /ha for other permanent grasslands and leys), and the lowest payments are for the forest regions of the south (SKr 800 /ha for pastures and SKr 400 /ha for other permanent grasslands and leys).

A programme of information, education and demonstration projects will be introduced to accompany these schemes which will lead to a large demand for knowledge concerning the management of traditional grazing lands and landscape elements with historical or biological value. The information and training courses are necessary to achieve an effective execution of the whole scheme.

Programme for Environmentally Sensitive Areas

This programme involves the restoration and establishment of wetlands and ponds on arable land, the establishment of permanent grasslands, the promotion of catch crops and the conservation of local breeds threatened with extinction. The objectives and the annual payments for these various schemes consist of respectively:

1. The restoration or establishment of wetlands on 4200 ha of arable land in the south, with compensation paid over 20 years: SKr 4800 /ha during the first 5 years and then SKr 2500 /ha during the remaining 15 years.

2. To establish 13,000 ha of permanent grassland on arable land in areas sensitive to nitrogen leaching, and strips along lakes and watercourses in the south of Sweden to minimize leaching and erosion, as well as promoting the flora and fauna. Compensation will be SKr 1500 /ha for extensive leys and SKr 3300 /ha for riparian zones.

3. Establish 40,000 ha of catch crops in the southern regions to reduce nitrogen leaching, with SKr 500 /ha compensation.

4. Improve the conditions for local Swedish breeds of cows, pigs, goats and sheep, and to preserve the biodiversity of domestic animals. Compensation will amount to SKr 1000 /ha per livestock unit.

Programme for the Promotion of Organic Farming

The objective is to increase organic or ecological farming in Sweden to 10% of the arable land by the year 2000. The annual compensation will be SKr 700 /ha in the north and the forest regions of the south, and SKr 1400 /ha throughout the rest of Sweden, mainly the plain lands in the south. There will be an additional payment of SKr 300 /ha if dairy or suckler cows are included in the production system.

These preliminary proposals will extend existing policies within the European framework. What remains to be seen is how farmers will respond to the new incentives they are offered and how this will affect Sweden's natural and cultural heritage.

11 The United Kingdom

Martin Whitby

Introduction

Climate and soils ensure that agriculture in the UK is dominated by livestock production mainly from grass. The wet and poorer land to the west and north is particularly appropriate for ruminant stock rearing while arable production is concentrated in the south and east where conditions are more favourable to it.

During the past 50 years land use in the UK has changed under a number of major influences. Agricultural use reflects a combination of: the UK's international trade obligations (Commonwealth Preference up to the 1970s and the obligations associated with EC/EU membership thereafter); the shifting pattern of incentives applied through agricultural policy; the rate of change of technology available to farmers and the price of inputs they purchase; changing patterns of consumer demand and other broader developments such as the rising level of environmental awareness and concern. Transfers of land from agriculture to other uses include conversion to forestry, supported by subsidies and incentives, and urban development of all kinds as regulated through the planning system.

Land-Use Change

The dynamics of land-use change in Britain may be inferred from Table 11.1 which presents the data in matrix form, based on aerial photography, allowing a dynamic picture to emerge. The sums of the rows in the table show the 1947 totals in each use and the sums of columns relate to 1980. The lead diagonal (**in bold type**) displays the areas of land which were in each

Table 11.1. Matrix of land-use change, Great Britain. (Source: Adger *et al.*, 1991.)

Cover in: 1947/1980	(a)	(b)	(c)	(d)	(e)	(f)	(g)	(h)	Total 1947
(a) Broadleaved woodland	**6,267**	862	418	553	38	625	1,103	260	10,126
(b) Coniferous woodland	54	**1,818**	178	206	0	47	60	20	2,383
(c) Mixed woodland	105	318	**1,057**	547	0	54	80	60	2,222
(d) Upland semi-natural veg.	389	5,603	998	**65,331**	341	1,649	5,475	1,284	81,070
(e) Lowland semi-natural veg.	151	340	189	190	**529**	378	342	76	2,194
(f) Crops	202	182	49	1 042	0	**37,312**	9,118	2,232	50,136
(g) Improved grass	483	362	177	2 880	38	19,845	**38,302**	3,332	65,419
(h) Other and urban	166	47	14	349	0	659	380	**9,941**	11,557
(i) Total 1980	7,817	9,533	3,079	71,099	945	60,568	54,860	17,206	**225,106**

Values are in km^2.

category for both survey years – implicitly the land that has not changed use over this period. The off-diagonal elements show the moves between categories over time. For example the 9533 km^2 of coniferous woodland in 1980 was obtained from the 2383 km^2 in 1947 plus 862 km^2 of broadleaved woodland, 318 of mixed woodland, 5603 of semi-natural vegetation and various small areas of lowland under different uses: over the same period it lost 54 km^2 to broadleaved and 178 to mixed woodland, 206 reverted to upland semi-natural vegetation and small areas disappeared under other minor uses; meanwhile, between the two years 1818 km^2 of coniferous forest stayed under that use cover. This particular row and column thus emphasize the importance of the moves between categories of forest. The largest single category of cover that switched into forest was the upland semi-natural vegetation, which has now become recognized as a highly prized ecological resource.

A more focused source of information on land use change has recently been developed by Allanson and Moxey (1996) who have interpolated Agricultural Census data, extending it from the level of Parishes to uniform kilometre grid squares. This has allowed them to estimate the distribution of broad land uses over seven composite land type classes. It is important to note that the land classes are not equal in size: the relative size and the definitions of these classes are presented in Table 11.2. The result of this analysis is presented in Fig. 11.1 which shows the percentage distribution of uses within

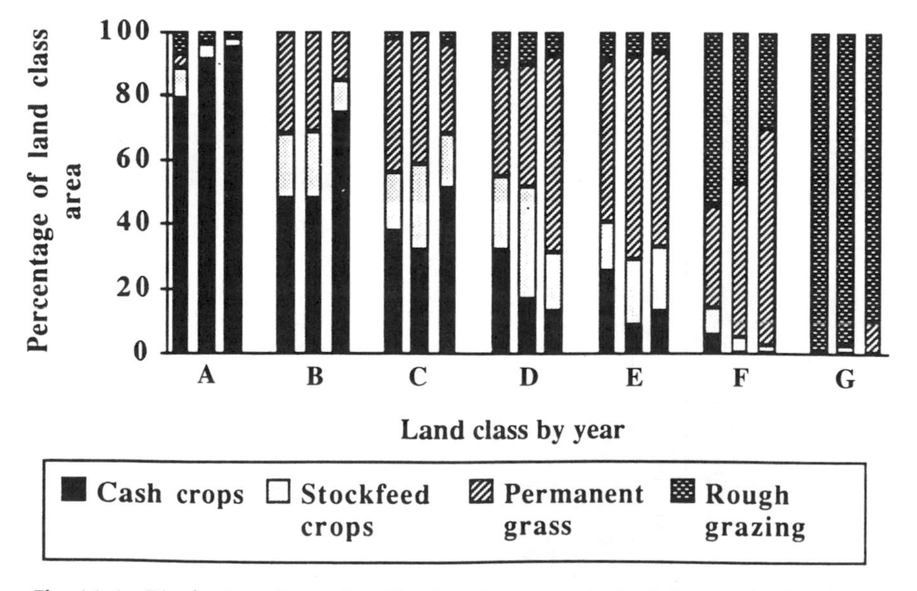

Fig. 11.1. Distribution of agricultural land use by composite land class: England and Wales, 1944, 1961 and 1992 respectively. (Source: Allanson and Moxey, 1996.)

Table 11.2. Distribution of agricultural area by composite land classes in England and Wales. (Source: Allanson and Moxey, 1996.)

Composite Land Class	Per cent of England and Wales agricultural area	Description
A	8.0	Flat fenland or floodplains in East Anglia, East Midlands and South East England
B	23.9	Flat alluvial plains and shallow river valleys in East England and Midlands
C	22.6	Flat or gently sloping lowlands, mainly in Southern England, the Midlands and NE England
D	8.8	Estaurine and coastal areas mainly in SW England, South West Wales and Anglesey
E	19.3	Undulating lowlands mainly in S, SW and Northern England and Wales
F	14.5	Mid-valleys slopes and rounded hills in SW and Northern England and Wales
G	3.0	Upper valley slopes and mountains largely in Northern England

each land class in 1944, 1961 and 1992.

The spatial distribution of agricultural activity is determined by a mixture of physical and economic forces. To the north higher altitudes, poor soils and lower temperatures tend to limit cultivation and to the west the greater rainfall encourages grass growth and livestock farming predominates. Arable production is therefore concentrated on the richer soils of the drier south and east of the country. Grazing livestock are concentrated towards the west and north. Intensive livestock systems, with their attendant environmental problems, are found in areas where proximity to markets, ports or feed growing areas offer them lower costs of production.

When agriculture has been prosperous, investment and innovation have followed: under more straightened circumstances, the pace of innovation has slowed. This follows from the fact that much of the funding of agricultural investment comes from income. Enhanced incomes encourage investment. Such investments include farm amalgamations, new buildings, the purchase of new machinery, the uptake of new technology; larger herds and flocks and higher levels of input use on farms. The broad impact of these tendencies has been that of increasing specialization and intensification at the farm level, seen mainly in the application of more inputs per hectare of land over time.

The growing intensity of land use has been accompanied by increasing concern about the environmental impact of agricultural production. These concerns have now been documented so widely that there remains little argument about their nature. In particular they grew during the 1970s, when there was an investment boom associated with the rapid tightening of world commodity markets following the entry of the USSR into the US grain market and UK accession to the EC, in January 1973. The latter is usually seen as particularly significant because of the anticipatory boom in land prices before EC entry but the relative importance of these two events remains to be established.

To sum up, past high levels of agricultural investment have been associated with recognition of increasing environmental impacts. This has coincided with rising environmental concern reflected in the political process and, to an important extent, resulting from rising levels of affluence. The changing policy scene should be analysed against these background forces.

Politics and Policies

In the 1980s the rising tide of environmental concern was reflected in domestic legislation in the UK: although previous Acts and policies had also charted the response to earlier and similar concerns related more to recreation and wildlife conservation, from the 1940s onwards. For example the National Parks and Access to the Countryside Act of 1949 provided for the designation of areas where nature conservation was to be given priority and where agreements could be negotiated with farmer and landowners for the provision of access to their land for recreation purposes. The Countryside Act of 1968 provided for a broadening of access to the countryside and led to the creation of Country Parks, of which there are now some 300. The Wildlife and Countryside Act of 1981 was the first legislative response to concerns about agricultural damage to the environment (Lowe *et al.*, 1986). Then came its revision in 1985, which imposed an environmental objective on the Forestry Authorities and the Agriculture Act of 1986 which introduced a similar constraint on Ministers of Agriculture. Other agencies with a more direct concern for conservation of landscape and wildlife have also produced policies aimed at conservation.

Commentaries and analyses of these events have been frequent. Shoard's (1980) polemic criticized the CAP as the source of problems of agricultural intensification and proposed modifications to the planning system that would constrain agricultural development and reduce its environmental impacts. Bowers and Cheshire (1983) were more implacably opposed to the CAP as such, levelling their attack at the distortions in price introduced by that system, urging a return to lower price levels to reduce the incentive for agricultural intensification. Many other academic critiques (for example Buckwell *et al.*, 1982) drew attention to the market distortions and economic inefficiencies introduced by the CAP. There was also a sustained political attack on the environmental impact of the CAP through for example, the House of Commons Environment Committee (HoC, 1985) which commented,

> 'The illogicality of one part of the Government (MAFF) offering financial inducement to someone to do something which another part of the Government (DoE and related bodies) then has to pay them not to, is clear.'

Such comments heralded the policy turnaround of the mid-1980s, when the political will to reassert the need to balance environmental management against the pressures for agricultural growth began to be formally recognized. At about the same time the UK Government proposed the now famous Article 19 in EC Regulation 797 of 1985 providing for the introduction of Environmentally Sensitive Areas (ESA). The Regulation was incorporated in the 1986 Agriculture Act and the first round of ESAs had been designated quickly: by the end of 1987 there were some 19 ESAs designated in the UK.

The precise political history of that period may show what led the Government to this position (see also, Baldock and Lowe, above): meanwhile from the outside the results can be observed in terms of policies introduced. The following sections first set out the principles involved in management agreements, summarizes the evolution of three typical types of agreeement and then compares their results in terms of cost per hectare protected by them and other measures of performance where they are available.

Management Agreements in Principle

The key aim of policies to secure provision of public goods in the countryside is that they seek to constrain the use of harmful land management practices by proprietors, or to promote the provision of public goods such as access to open countryside for those who do not own it, or to secure the availability of diverse ecosystems and landscapes. The instruments available range from simple persuasion, with or without financial incentives, through various measures of control to outright purchase (Rodgers, 1992). Over some decades a remarkable array of instruments has been developed and applied. There are too many arrangements in use to be described here and only the three most important ones are discussed.

The essence of these and most other environmental policies is the transfer of rights to use or impact on the environment between parties. Typically they will ensure that those who currently have the right, say, to cultivate land in a particular way forgo some part of that right, either for a defined period or in perpetuity, in favour of society. They may or may not be compensated for such behaviour. In the UK, practice regarding compensation varies. If land owners seek planning permission to develop their land they may be refused without compensation; if farmers are within an ESA they may not be required to farm in any externally chosen way but if they agree to do so they may be compensated by the state; by contrast, if their land falls within a Site of Special Scientific Interest (SSSI) there will be certain actions they are prohibited from undertaking without first obtaining permission. If that permission is refused they must be compensated through a management agreement but if they do not accept that option the public interest may be asserted to the extent of public purchase. The designation of a SSSI thus affects the previously unconstrained right of the farmer or landowner to undertake what are termed potentially damaging operations (PDOs) to the external sanction of the state. The transfer is perpetual, or at least as long as the SSSI is retained. The fact that compensation may be paid for this loss of right does not negate the loss.

The Polluter Pays Principle (PPP), espoused by both the British Government and the EU, is unevenly implemented (Baldock and Bennett, 1991) as can be seen from the way in which polluters are compensated for abstaining from polluting practices in Nitrate Vulnerable Zones as well as some of the provisions of ESAs. By offering such compensation the assignment to polluters of an implicit right to pollute is explicitly recognized. If they were required to pay for such social losses by a rigorous introduction of the PPP, this would reassign rights between them and society. Application of the PPP to primary land use is comparatively simple when applied to pollution; but how does it relate to the destruction of important habitat? Here producers' rights on undesignated land would appear to be paramount: if they own such land they may manage it as they will, provided they do no damage to others. Where land has been designated for particular purposes – wildlife or landscape conservation, for example – that process may allow the application of PPP. More likely, however, would be the SSSI situation where those threatening damage may be compensated through a management agreeement for abstaining from damaging the habitat. However, where pollution can be shown to have been the cause of the damage, there is a possibility that farmers or landowners may be required to bear the cost of reinstatement of the habitat. In such cases PPP would appear to be working.

The key difference between the mechanisms above is thus in the extent to which they change or attenuate the pre-existing assignment of property rights. Where they do change the rights to property, the assets of original right owners become less valuable than before and this loss provides a basis for compensation through management agreements. The general principle followed is that owners should not suffer a loss of income: compensation is therefore based on income forgone.

But the economic impact of such arrangements is broader than this. All of these policy instruments involve substantial costs in addition to compensation. They must be set up, information must be circulated about them, farmers must be persuaded to join schemes, levels of compensation must be agreed, in some cases on an individual basis, and compliance with agreements must be monitored over time. These activities tie up substantial volumes of public and private resources in what are called transactions costs. Such costs may become an important element in environmental policy because of the very problems of defining and enforcing the policy instruments available. However, transactions costs are very difficult to measure because they are not clearly identified in the same way as prices or levels of compensation may be. Consequently too little is known about them in precise terms: the information available on these arrangements is summarized at the end of this chapter.

Another important aspect of these arrangements is the sustainability of the results they produce. The public model of compensation arrangements, by concentrating on annual payments to make good income forgone, detracts attention from the alternative focus of these policies on the generation of a stock of natural capital. Insofar as that is the case, then the question of the durability of that stock is an important determinant of the future flow of environmental benefits from it. If the stock is destroyed then the flow of benefits will cease. Accordingly, a policy, which provides incentives to farmers and landowners to manage their land in such a way as to generate an accumulation of natural capital yet leaves it up to the owners of that natural capital to exploit or destroy it at will is obviously not sustainable beyond the end of the period of the agreement. To the extent that participation in such schemes is voluntary, they are mostly of short duration and can be terminated easily, which ensures that their impact can only be guaranteed by offering rates of payment that are financially attractive in the short term.

This discussion highlights several potential criteria for assessing the effectiveness of environmental policies. These may be phrased as a number of questions:

1. What would happen if the policy was not introduced?
2. Over what period will policy benefits accrue?
3. What is the compensation cost of the policy?
4. What is the transactions cost of the policy?
5. Does the policy deliver the benefits it offers?
6. How valuable are the policy benefits?
7. How certain are the costs and benefits of the policy?

In the discussion of policies we shall see that a key attribute for the lasting success of these mechanisms is whether they change the assignment of property rights to any extent or merely arrange their temporary transfer between the individual and the state. Where policies are completely unfamiliar a set of questions such as this would be difficult to answer. But experience will help in making broad estimates of outcomes in the way that officials administering policies must, if they are to make effective use of their budgets.

Management Agreements in Practice

There are too many different types of management agreement in use in the UK (see Whitby, 1996, for a more complete treatment) to be summarized in one chapter. Instead three different types of agreement are compared in terms of their uptake, costs and, where possible, benefits. The mechanisms chosen for review are those operating under Section 15 of the Countryside Act, in SSSI, those applying in ESAs and related policies and the Countryside Stewardship Scheme which applies only in England.

Management agreements on SSSI

The use of management agreements was initiated in the 1949 National Parks and Access to the Countryside Act (Section 16), which introduced them to regulate the management of nature reserves. Agreements under the 1968 Countryside Act (Section 15) are still the main type of arrangement in SSSI. The legal aspects of these arrangements have recently been reviewed by Withrington and Jones (1992) and a National Audit Office study (NAO, 1994) has examined the workings of these arrangements in England.

The present system, which also derives from the 1981 Wildlife and Countryside Act, involves compensating farmers for the net profit they forgo when prevented from carrying out operations that would damage the scientific value of a site. The process is initiated either by a farmer notifying the conservation authority that he or she proposes to carry out a potentially damaging operation (PDO) on a site. If the conservation agencies then wish to prevent the operation, they must, within 4 months, offer to enter into a management agreement providing compensation calculated in accordance with published Guidelines (DoE, 1983).

The conservation bodies have defined powers of compulsion, if the farmer is determined to carry out the operation. These are rarely used, although they are an important reserve power in conducting negotiations. Farmers and landowners may appeal against compulsion and compensation is available (on the same basis as for deferred expenditure). In the last resort the conservation body may apply to the Secretary of State for a compulsory purchase order under Section 17 of the National Parks and Access to the Countryside Act 1949.

The significant financial cost elements in this system are those of compensation, negotiation and monitoring. A study of the cost agreements operating in 1988/9 found total costs of £7.4 million, of which costs of compensation (£5.58m) were the main item. The costs of professional advisers, which arise because the NCC paid both the negotiating costs of owners as well as those it incurs itself. An estimate of the share of NCC staff 'overhead' costs brought the total of these administrative costs (in economic terms *transactions costs*) of running the system of SSSI to £1.79m. The share of transactions costs was roughly one-quarter of the total. However, the above estimates omit the substantial element of cost incurred in the designation and redesignation of SSSI. This requirement was imposed on the NCC through/after the Wildlife and Countryside Act of 1981. Whitby and Saunders (1994) estimated that the full cost of redesignation, spread over the period of the

1980s, would amount to some £3m per year, which brings the total cost to £10.37m and increases the share of cost attributed to transactions to 43% of total.

A recent NAO (1994) report adds another mechanism now used by English Nature to increase site conservation, namely the Statements of Intent it has obtained from various public bodies which 'own, occupy or influence' SSSI. It lists ten agencies including the Forestry Commission, the Association of County Councils, the Sports Council, the Ministry of Defence, National Parks and others all of whom have signed statements of intent with English Nature. In some cases statements of intent are backed up by the production of management plans.

Environmentally Sensitive Areas

The UK incorporated the EC Structural Regulation (797/85) of 1985 in the Agriculture Act of 1986 (S 17), which provides *inter alia* for the designation of Environmentally Sensitive Areas: an element which the UK had first proposed to the EC. ESAs are areas of landscape, conservation and/or archaeological interest, within which farmers are offered financial incentives to comply with a package of management practices designed to secure conservation goals. Each ESA has a separate management package which proscribes certain activities and encourages others; some are comparatively simple specifying only one level of practice to be followed, others are more complex, allowing for two or more 'tiers' of incentives for particular practices.

The 'first round' of ESAs began in 1987 and, by 1989, they covered some 0.8 million ha containing agreements with more than 4000 farmers which related to 285,000 hectares in the UK. Participation is voluntary and agreements have been taken up enthusiastically by farmers, which may be partly explained by the comparatively slight constraint they impose on the practices of many participants. Several researchers (Colman *et al.*, 1992; Hodge *et al.*, 1992; Whitby, 1994) have commented on the limited extent to which ESA management packages have obliged participants in these arrangements to farm differently from how they would otherwise have done; although such comments require the qualification that they should be based on a comparison of the behaviour of farmers *with* agreements and those on similar land but who were outside ESAs and therefore not eligible for agreements. The evidence of published monitoring studies (Whitby, 1994) confirms the difficulty of precisely identifying this counter-factual, or control, situation of participants.

The rate of uptake of ESA agreements and the area they protected grew steeply during the year following designation. Typically, by the end of the year following designation more than 90% of the final area of agreements had been negotiated. The cost data available for the early years of ESAs (MAFF, 1992) shows that the total cost of payments for the first round ESAs in England was reported as £45.3m for the 5-year period: of this, some 24% consisted of administrative costs. A proportion of the compensation paid (up to 25%) will have been reimbursed to the UK from the CAP budget.

The initial rounds of ESAs were designated for 5 years in 1987 and 1988. Consideration was given to their renewal in 1992 and they were all redesignated in that and the following year, several of them with much increased areas and virtually all with changes in their management prescriptions.

The Agri-environment Regulation (Regulation 2078/92) allows for support funding from the EU budget of up to 50% of the cost of measures introduced. For the UK the 'true' contribution from the EU budget is less than this because of the so called Fontainebleau arrangement under which the UK receives back from Brussels two-thirds of its net contribution to the budget: the impact of this on public expenditure is that the UK in effect pays up to 83% of the cost of policy instruments.

Schemes introduced under Regulation 2078/92 may include: those aimed at promoting reduced chemical use: more extensive land use: reduced stocking rates on grazing land: promotion of environmentally friendly farming and the upkeep of abandoned land; management of 20-year set-aside land; promotion of access for leisure uses of land; and provision of training in environmentally friendly farming practices. In 1993 the Ministry of Agriculture, Fisheries and Food (MAFF) announced a set of measures it proposed to introduce in compliance with this Regulation including more ESAs, the introduction of access schemes in ESAs and on other land, the introduction of more Nitrate Sensitive Areas, a moorland scheme, set-aside management schemes, two habitat improvement schemes and an organic aid scheme.

The series of ESA designations in England is divided by MAFF into four stages, depending on whether they were first designated in 1987, 1988, 1993 or 1994. It is seen in Table 11.3 that these years coincide with major increases in the area designated under these arrangements. Table 11.3 summarizes the most recent English situation with regard to agreements, by stage of designation. The other countries constituting the UK also introduced

Table 11.3. Area of ESAs and agreements by stages of designation: England 1995. (Source: MAFF, September 1995.)

	Eligible land ('000 ha)	No. of agreements	Area under agreement ('000 ha)	Land under agreement as per cent of eligible land
Stage I	147.6	2633	74.3	50.3
Stage II	156.9	637	66.4	42.3
Stage III	367.7	735	196.2	53.4
Stage IV	263.2	1000	59.1	22.5
Total	935.4	5644	396.1	42.3

Table 11.4. UK farmers' response to Regulation 2078/92. (Source: Department of Agriculture, personal communication.)

	Number of agreements	Area under agreement (ha)	Average area of agreement (ha)
Habitats	210	3776.42	17.98
Organic aids	77	3711.48	48.20
Access	80	1000.00	12.50

new ESAs and other measures under Regulation 2078/92. The total area designated ESA now approaches 3 million hectares. The area subject to agreements is some 900,000 ha on nearly 10,000 farms (Agriculture Departments, September 1995).

Other measures accompanying CAP Reform are at a much earlier stage of evolution. Thus the Nitrate Sensitive Areas scheme replaced a pilot scheme relating to an aggregate area greater than one million hectares and most of it (86.2%) was under agreement in 1993. This scheme is subject to modification as a consequence of UK compliance with the EC Nitrates Directive (91/676) and conclusions from the recent round of consultations are expected in December 1995 (Ward *et al.*, 1995). Other schemes include the provision of improved habitats, organic aid and countryside access. Their recorded initial rates of application are reported in Table 11.4 showing rather modest rates of uptake. The average areas under these schemes are predictably small with the exception of the Organic Aids Scheme which is no doubt larger because it has attracted whole rather than parts of farms. The size of areas attracting habitat and access payments would be expected to be small because the schemes are aimed at small areas.

The small number of applications for these schemes may reflect no more than their recent introduction. However, the concept of access is a difficult one for farmers and administrators alike. Although it has been possible to make access agreements with farmers in National Parks since 1949, there are very few such arrangements currently in existence. The House of Commons Environment Committee (1995) has recently reported that there are fewer than 50 agreements under that legislation currently operating in the National Parks. Farmers already find their land subject to access pressures and on much of their land the public already have some traditional rights of access. MAFF will not make agreements where such public rights already exist, confining its offer to areas where new access will be provided.

The habitat schemes offer farmers substantial payments per hectare for entering into 10- or 20-year agreements. These may relate to waterside areas, saltmarsh and former short-term set-aside land. Annual payments are between £125 and £525 /ha. The payments offered in these schemes are

Table 11.5. Development of Countryside Stewardship Scheme agreements 1991 to 1994. (Source: Countryside Commission, Cheltenham, personal communication, 1995.)

Conservation goods provided	1991	1992	1993	1994
Area of agreements (ha)	31,689	66,768	87,468	102,460
Area of access (ha)	6,591	11,654	12,919	13,803
Hedgerow (m)	151,245	312,694	460,483	1,525,916
Hedgebank (m)	6,981	17,352	24,950	58,886
Wall (m)	114,718	151,928	183,833	216,513
Linear access (m)		163,607	287,951	401,644
Agreement numbers	868	2,065	2,932	4,045
Chalk & limestone grassland (ha)	7,918	13,196	16,746	28,318
Lowland heath (ha)	2,745	7,844	11,384	12,246
Waterside (ha)	6,147	14,711	19,882	22,082
Coast (ha)	3,010	6,461	7,344	7,985
Upland (ha)	11,869	22,627	28,806	34,234
Historic landscapes (ha)	–	4,031	8,132	9,985
Old meadow and pasture (ha)	–	667	1,095	2,108
Hedgerow landscape (ha)	–	–	–	1,423
Number of old orchards	–	65	147	219

remarkable in that the highest rates would capitalize, even at 5% real interest, to two or three times the likely market price of the land. Such levels of compensation underline the comparative cost-effectiveness of public purchase of the land.

The actual and projected costs for these schemes are summarized in Table 11.5 where gross scheme costs are separated from the administrative costs reported by MAFF. The latter amount to quite a substantial addition to Scheme Costs in the current year but increase only slightly in the projected years. Slower growth in administrative costs would be expected after the initial set-up period but the relationship of total transactions costs to these estimates remains unclear. Obviously private transactions costs are excluded and it may well be that other items of cost are attributed to more general overheads in the accounts from which this material is taken. It is also unclear whether these estimates are consistent with the earlier ones reported, when the share of administrative costs for ESAs amounted to roughly 25% of reported exchequer costs.

The area of land under agreements additional to the ESA schemes introduced under this Regulaton has been slight as yet, reflecting the slow response to novel and complex schemes. The habitat schemes are for 20 years, the organic aids are for most UK farmers a radical option and the idea of making access agreements, although it has been possible legally for some decades, has not produced many agreements. The Countryside Stewardship Scheme has expanded the interest in providing access and there may be synergy between these two schemes in that sense (see below).

Countryside Stewardship Schemes

The Countryside Stewardship Scheme was introduced by the Countryside Commission in 1991 and offers agreements somewhat similar to ESAs although there is more flexibility in their operation. The flexibility is provided through the facility to pay occupiers (annually) for enhanced landscape management, restoration or re-creation; supplementary payments for landscape regeneration or access improvement; capital funds for an array of work on landscape types. The scheme does not rely on designation to define the spatial scope of agreements but is targeted at particular habitat types to which the scheme's provisions apply. The physical features that apply include the land itself, the hedgerows, hedgebanks and walls it contains; access to the agreement areas is an important feature of the scheme.

The evolution of agreements under the scheme is documented in Table 11.5 which shows a threefold increase in the area under agreement over the 4 years of the scheme. There has also been a doubling of the area subject to access provisions, which now approaches 14% of the total agreement area: there are also 400 km of linear access provided through the scheme. The relative importance of the different landscape types subject to Stewardship agreements can also be seen from Table 11.5. A total of 80% of the area is accounted for by the three most common types of agreement; chalk limestone grassland, waterside and uplands. The flexibility of the system is demonstrated by the fact that four new types of landscape have been added to the scheme since its inception and together they approach 14% of the area eligible for agreements.

The costs of Stewardship have grown to nearly £13m in 1994/5, of which less than 20% is reported to be 'overheads'. The cost has grown not only with uptake, but also with the elaboration of the scheme. For example the addition of access arrangements, both linear and spatial, has undoubtedly added to the expenditure through the scheme without necessarily increasing the total area under agreement: it has also added to the benefits achieved where access is now used. Similarly, it seems likely that the initial landscape types in the scheme will have been those which encourage the negotiation of agreements over large areas: once those agreements were made, the more recent additions would have been small areas.

This scheme has been appraised each year and the most recent report (Land Use Consultants, 1995) presents a detailed analysis of all its aspects. The report emphasizes the success of the scheme in targeting its resources to have the maximum effect:

> Overall the evidence suggests that CS has been successful in targeting its resources to landscape types and geographical areas that offer great potential for environmental improvement and public benefit and in some target areas very high uptake has been achieved.
>
> (Land Use Consultants, 1995).

While the initiation of these schemes and their administration for their first few years has been handled by the Countryside Commission and the Countryside Council for Wales (two para-statal organzations) this activity is to be taken over from them by the MAFF in 1996. The movement of a scheme from the Countryside Commission to MAFF is not a new event. This has happened with ESAs, which were piloted in the Norfolk Broads Grazing Marshes Scheme, and the Countryside Premium Scheme (to secure conservation goods from set-aside land) were also developed and tested by the Commission before handing over to MAFF. MAFF is currently involved in a public consultation process seeking views on how Stewardship should be incorporated into its existing policy set: in particular how it might be administered together with ESAs (MAFF, 1995). MAFF's initial proposals for discussion include the retention of some features of the Farm and Capital Grant Scheme (FCGS) (which is due to be terminated in 1996) as part of Stewardship. The consultation paper also suggests that Stewardship could absorb other schemes currently run as part of the UK response to Regulation 2078/92. These are the Habitat, Moorland and Countryside Access Schemes accepted under Regulation 2078/92 and currently costing some £8 million per annum. Together with a similar extension of some other elements of the FCGS this would substantially reduce the benefits of targeting which are stressed in the appraisal of Stewardship (Land Use Consultants, 1995). The paper suggests that the possible merging of Stewardship with other schemes be considered in 1998/9, when they will be reviewed.

Comparisons of Costs and Benefits

In the above three cases the pattern of property rights is modified by the policy instrument, whether by reassignment of rights, potentially with compensation (as with SSSI), or by the purchase or renting of rights, usually for a defined period (as with ESAs and Countryside Stewardship). The cost of the policy includes both the financial cost of compensation and the transactions cost of applying the mechanism. Various estimates of cost have been reported above and these are summarized in Table 11.6.

It is significant that the information on the transactions cost of most of these mechanisms is incomplete and lacks robustness. Moreover the comparisons here require so many assumptions that they can offer only the roughest guidance to policymakers. For example, averaging of costs of the areas under agreement in 12 different ESAs and the whole agricultural area of SSSI assumes that a comparable mix of habitats will be found under the two types of arrangement. Comparison of different instruments also implicitly assumes that the public goods provided by each mechanism are equally valuable.

Comparing the cost of protection in ESAs and SSSI should then be done by dividing the whole area protected into the total cost as reported above. It

Table 11.6. Comparative financial cost of conservation systems (1993 prices).

	Annual compensation cost (£m)	Annual transaction cost (£m)	Total area protected ('000 ha)	Average cost (£ /ha protected)
Management agreements on				
SSSI*	10.37	4.79	975	15.5
SSSI†	7.2	[3.33]	858	12.7
ESAs‡	8.52	2.81	113.1	105.73
Countryside Stewardship§	10.6	2.1	87.4	145.3

Sources: *Whitby *et al.*, 1990. NB Great Britain. †NAO, 1994. England only, NB Transactions cost estimated from proportions in (*). ‡MAFF, 1990, 1991. NB England only. §Countryside Commission, personal communication, 1994/5. England only.

is stressed that the data for some of these arrangements are now old and they have been indexed (using the GDP implicit deflator) to bring them to comparable prices.

The appropriate area for dividing into ESA costs is the area subject to agreements, which means that the per hectare rate of compensation payments and associated transactions costs (at some £105 /ha) will suffice as an indicator of cost. For SSSI a cost substantially less than the average per hectare level of payments is appropriate, reflecting the size of the area protected. For SSSI as a whole this would amount to £12–15 /ha, reflecting the fact that less than 10% of the agricultural area of SSSI is under agreements while the whole area enjoys some measure of protection. These estimates of £12–15 /ha reflect the lower bound of cost, because the level of protection achieved in SSSI is less than desired. Nevertheless, the NAO (1994) report records several examples of conservation improvements achieved without management agreements although it cannot be argued that all land is equally protected inside SSSI. The balance of arguments might indicate that the low estimate should be raised somewhat, but by how much must await a full study of the cost of these arrangements.

Evaluation and Prognosis

Comparison of systems in terms of cost per hectare overlooks the fact that the various systems described provide different public goods. These differences are described in discussing the schemes but there is no simple analytical means of reducing them to common terms, apart from the crude measure of cost per hectare protected in Table 11.6.

A definitive judgement as to the economic success of the schemes must await long term studies of their environental impact and rigorous studies of their effectiveness. In the case of SSSI, the level of cost per hectare protected suggests that this is a low cost mechanism. However, the conservation authorities, in particular English Nature, have been dissatisfied with the SSSI because they fail to deliver the level of management believed necessary. The critical aspect of SSSI designation is that, in contrast with other policy instruments, it does in fact change the assignment of property rights, particularly by declaring PDOs. SSSI are therefore a much more contentious instrument with farmers and landowners which would partly explain the ambivalent attitude towards them of the conservation authorities.

The effectiveness of ESAs, despite an unprecedented amount of monitoring of a 'top-down' nature remains to be seen. In particular the schemes have not been operating long enough to be appraised as mechanisms for securing conservation goods. Moreover the great difficulty of establishing what farmers would have done *had they not had a management agreement* will always cloud the issue of precisely what are the benefits from schemes such as this.

The other measures accompanying CAP Reform, apart from the extension of ESAs, cannot be evaluated yet as few agreements have yet been negotiated and none have run the full course. The long term nature of the Habitat Schemes will ensure that their evaluation will have to concentrate on the monitoring of participants' compliance with scheme requirements for the first decade or more of the agreement. How much use will be made of the Access Scheme remains to be seen.

Countryside Stewardship differs from other schemes in that its designation is more flexible, being based on landscape types, rather than on formal designated areas. The packages applying under Stewardship have varying relevance to different types of landscape and the incorporation of access into the scheme from its inception has produced some public benefits additional to those associated with conservation. The proposed merger of Stewardship with MAFFs existing (non-ESA) schemes under Regulation 2078/92, in 3 years' time may well enhance their effectiveness and their application will probably expand due to access to a wider catchment of land users and, arguably, to less constrained budgets. Whether Stewardship can retain its targeting-induced advantages as part of a much broader policy process remains to be seen. If targeting is dispensed with, perhaps following the example of the French 'prime à l'herbe', departing from the traditional attachment to land use designation as a basis for policy implementation, this may be the main radical impact of agri-enviromental policy on UK policy.

A major objection to the ESA system was raised at the beginning of this chapter regarding the low sustainability of ESA benefits of natural capital form which, because they remain in the sole ownership of private farmers and landowners, are at risk of destruction if (world) food prices swing in favour of more intensive production. The obvious offset to this risk would be to change the rights of proprietors denying them the right to destroy natural capital 'created' at public expense. Although governments are generally disinclined to introduce such swingeing reassignments of property rights, they might be persuaded to offer incentives towards such beneficial change by encouraging the extension of covenants into this area.

Covenants are legal arrangements between two people or agents whereby one promises benefits to the other. They may be used by those wishing to promote conservation by buying land and then selling or letting it with restrictive covenants attached. Naturally the severity of the restrictions imposed will be reflected in the price or rent charged. The potential utility of such arrangements for conservation purposes would appear to be mainly limited by their legal complexity. These can be avoided by the agency responsible for the covenant making sure that its provisions are met, and carrying the associated costs of monitoring.

Covenants are used widely but, because they tend to be individual legal arrangements, the extent to which they are used is not known except where a large organization such as the National Trust makes systematic use of them. The attraction of these arrangements is that they may be used for extending the life of management systems adopted under other schemes, for example by encouraging proprietors of particular types of land who have agreements to agree to extend them in this way. Hodge *et al.* (1993) make the interesting suggestion that covenants might be used to provide longer term security to the conservation benefits secured through ESAs and other arrangements. To be passed on with the sale of land covenants must meet three conditions:

1. The covenant must be negative in nature;
2. There must be land nearby which benefits from the covenant: and
3. Registration of the burden of the covenant is essential, either on the register of the title (in the case of registered land) or in the land charges registry (in the case of unregistered land).

(Hodge *et al.*, 1993)

Those holding ESA agreements could covenant to continue farming consistently with the objectives of the ESA (which themselves should have the same life as the designation, i.e. be permanent) thus lengthening the impact of such arrangements at modest cost. This idea offers a means of extending the effective life of the benefits secured through management agreements.

Covenants suffer from a problem of uncertainty as to their enforceability in law. They are complicated and their use has been limited by these uncertainties. Potentially they suffer from some of the deficiencies of tax concessions, which can only be rectified if there is full public knowledge of their existence. Such knowledge is more likely to be available if the arrangements are entered into by 'public' democratically controlled organizations which have a commitment to their members to provide access or other public goods. Several environmental groups, including the National Trust and the County Wildlife Trusts, would fall into this category.

Conclusions

The postwar process of land use change in the UK has been accompanied by increasing interest in and concern for conservation of rural habitats and landscapes in the face of increasing agricultural intensification. The chapter has reviewed major policy mechanisms that have been developed during this period.

The review directs attention to a serious problem in the interpretation of financial data on management agreements which has apparently supported important recent shifts in policy, away from individually negotiated management agreements towards the more mechanistic systems applied, for example, through ESAs. There are no doubt non-economic reasons that would support the new policies.

Progress on this important question of how to secure public availability of conservation goods requires research on the costs of the various mechanisms available. This work must focus on the individual habitats being protected if different mechanisms are to be compared.

It would be expected that, because the role of managing these systems is divided up among several departments (the Ministry of Agriculture, the Department of Inland Revenue, the Countryside Commission and the conservation agencies), there would be little evidence of coordination to be found in securing conservation. Opportunities for such coordination might be found in the legislative system or perhaps in the Treasury. Evidently a combination of both legal and financial controls is needed if a rational system is to emerge but the question of which instrument is 'best' in an economic sense has not been answered. It could be argued that the apparent specialization of the Countryside Commission in developing new policy instruments which are then taken over and run by larger government Departments has worked well in this sphere of activity. It represents an unexpected form of flexibility in government which is to be welcomed in principle.

The policies surrounding land use unavoidably pursue multiple objectives: they seek to protect farm incomes, promote forest planting, deter 'undesirable' building developments, reduce pollution and conserve heritage, wildlife and landscape features and so on. Success in pursuit of these objectives would contribute to the higher level objectives promoting economic efficiency and ensuring an acceptable distribution of the proceeds between different groups in society over time. In these circumstances, it is not surprising that a complete assessment of the desirability of particular policies to protect the environment from adverse effects of land use awaits further research.

In the midst of the plethora of pre-existing arrangements in the UK, the impact of Regulation 2078/92, still incompletely introduced, seems small indeed. This is not surprising given that the UK was a pioneer in this policy field and was the country that first pressed for EU legislation on ESAs. The main spatial effect of that Regulation so far is to be seen in the designation of a fourth round of ESAs and in the increased levels of support for those arrangements from Brussels. The UK response implies that the expansion of ESAs has not yet reduced the marginal value of their benefits to a level below their economic cost. More radical future options for this policy area would appear to lie in the broadening of applicability, perhaps in combination with the absorption by MAFF of the Countryside Stewardship Scheme, to depart from the traditional dependence on spatial designation as a basis for implementing policy. Such an innovation would have potentially important implications for the (concealed but important) transactions costs of these policies.

Evaluation

12

Environmentally Sensitive Area Schemes: Public Economics and Evidence

François Bonnieux and Robert Weaver

Introduction

ESAs represent a policy mechanism aimed at inducing changes in economic activity to increase social welfare. The value of the ESA mechanism results from the net social benefits it induces. It follows that estimation of the benefits of ESA applications must be founded on an understanding of the change in economic activity that the application hopes to induce, the changes in economic goods that result, and the sources of social value of those changes in available goods. Generally, ESAs can be viewed as a mechanism for inducing changes in current and future environmental performance. Authors have presented varied definitions of the goals of ESAs, however, the primitive concept that can be extracted from each of these definitions is that of change in environmental performance. More specifically, the environment of interest is that related to the rural land. In the UK this environment has been labelled *countryside*, while in the US literature the term *landscape* has taken on a generalized connotation.

A rather limited literature has considered the public economics of the ESA mechanism in general, or estimated the benefits associated with particular applications of ESAs. Colman (1994) presents a useful description of alternative conservation instruments and an assessment of their relative effectiveness based on a menu of criteria (e.g. capacity to protect and enhance, timeliness, targetability, monitorability, cost efficiency, political acceptability and transparency, promotion of conservation-mindedness). Bateman (1994) and Willis and Garrod (1994) present brief considerations of the micro-economics of benefit measurement and more thorough overviews of alternative approaches. However, in each case, these presentations are limited to considerations of the traditional case of benefits attributable to provision of a public good. In the case of ESAs, the linkage between a programme and public good provision is obscured by both the complexity of programmes and by numerous types of uncertainty of programme effects. Further, the intended economic effects of ESAs often go beyond provision of public goods to include management of externalities and private provision of public bads.

The objective of this chapter is to reconsider the valuation of ESAs at a conceptual level and to reassess available evidence of benefits generated by ESAs within this context. The next section reconsiders the origins of ESA mechanisms as a policy approach. This is followed by an analysis of the character of the economic goods that might result from use of the ESA mechanism, the mechanism through which the supply of these goods is induced, and the basis of the social value these goods might generate. Next, the microeconomics of valuation of these goods is reconsidered graphically and conclusions are drawn concerning what frameworks are appropriate for their general application. The fifth section presents an assessment of available evidence concerning the value of particular ESA applications and finally, conclusions are drawn.

The Origins and Nature of ESAs

The objectives of the ESA approach have been popularly understood to be the safeguarding of particular characteristics (landscape, wildlife, and historical features) of rural land through the management of agriculture's role in the landscape, Willis *et al.* (1993). The concept emerged in the 1980s in response to public concern over two types of agricultural output: (i) externalities associated with intensive agriculture and their effects on the rural landscape and (ii) perceived beneficial effects of less intensive, traditional agriculture and its role as a distinctive and valued element in rural landscapes. Traditional agriculture was perceived to contribute to the preservation of valued landscapes (e.g. meadows, grasslands and heathers); landscape functions such as wildlife habitat provision; to reduced off-site effects through hydrological stabilization; windbreaks that reduce soil displacement; and to the preservation of archaeological and architectural features. In this sense, these beneficial effects were generically classified as public goods, although economists would classify them more precisely.

The notion of an ESA was introduced in European Community law within the context of agri-environment schemes which could be adopted on a voluntary basis by Member States (Whitby and Lowe, 1994). Importantly, it emerged as a result of public recognition and valuation of the extensive variety of non-market economic goods or services produced by agricultural land use. In the UK, ESA legislation emerged under Section 18 of the Agricultural Act of 1986 and resulted in five ESAs being designated in 1987, six in 1988, six in 1993, and six in 1994. During the first rounds, farmers were offered fixed period contracts providing annual fixed payments in return for the adoption of a set of farm production practices. These practices involved requirements as well as prohibitions on daily and seasonal activities, with a focus on chemical application, stocking densities, infrastructure installation such as drainage or fencing, and management of landscape features. Often, these practices were adopted as part of an overall plan for the farm. Payments were based on the area treated by the practice or prescription. Adoption rates have been high and shown steady growth in the UK.

In France, ESA applications were initiated between 1991 and 1993. In 1993, ESA applications were brought under European Community initiative. As in the UK, the areas of application were spatially limited, however, the French applications involved more extensive negotiation with farmers of the terms of contract offered in each area. French applications focused on two goals:

1. *In situ* preservation of biodiversity or specific biotopes (30 ESAs by 1993 encompassing about 220,700 ha with about 83,700 ha enrolled).
2. Preservation of extensive agriculture and the reduction of land abandonment in areas of traditional, yet highly dispersed agriculture (24 ESAs by 1993 encompassing about 492,000 ha of which only about 106,600 ha was eligible for financing).

The second goal was intended to protect natural areas from abandonment or encroachment of more intensive uses. In total these two goals involved about 700,000 ha of which only 205,000 ha were eligible for participation (i.e. farmed land). The budgetary commitment was FFr 92.5 million. A small number of ESAs were initiated, though later abandoned, which were focused on water pollution and fire protection in forested pasture. Implementation led to a substantially smaller number of ESAs due to administrative delay, or under enrolment and closure. Despite farm representation in the contract specifications, participation was low.

Most recently, the Agri-environmental Regulation established a package of measures that are mandatory for all states to implement through zonal programmes in their territories. Rather than limiting these schemes to particular areas, their zonal focus would allow them to be available to all areas within the territory with such zonal characteristics. Each of these programmes offers standard payments to farmers to take actions that produce public goods and reduce externalities. These actions are intended simultaneously to maintain agricultural land use while shifting agricultural practices toward those that benefit the environment. Under the Regulation, Member States presented proposals for such zonal programmes. The details of these new applications have been reviewed by other chapters in this volume.

Throughout the historical evolution of ESAs, political and social debate has obscured the underlying public economics of ESAs. We next review these foundations to clarify the origins and characteristics of changes in social value they might induce. Accepting the focus of ESAs on altering the performance of the rural environment or countryside, we begin by reconsidering the nature of the performance with which we are concerned. Most typically cited as the target of ESAs are changes in externalities and environmental outputs produced by agriculture (Whitby and Lowe 1994). This definition can be extended to include changes in the production of negative public goods (public bads), such as the generation of carbon dioxide and ozone destruction or habitat destruction and displacement. Within a temporal sense, this list of performance goals appears to focus on contemporaneous effects. However, it is clear from most discussions about ESAs that their intent is to alter the intertemporal effects of agriculture on the environment. In this sense, the rural environment is viewed as a dynamic system and the objective of public management is to divert that system from its current or future paths associated with purely private decisions to a path which is viewed as having greater social value. It is within this sense that ESAs encompass goals such as improvement of habitat, stimulation of wild species, or the elimination of invasive species.

The focus of ESAs is on agriculture. Within the rural setting, man's impacts occur through agriculture, forestry, recreation, resource extraction, and infrastructure installation and use (e.g. highways and dams). Agriculture is the predominant use of rural land and, therefore, dominant as an economic activity through which man affects nature. Hence, it is natural that agriculture was targeted as a basis for affecting the value of nature. Further, agriculture involves active management of the rural landscape. Numerous analyses of ESAs present commentary on the role that the post war period of intensification (chemical use, varietal homogenization, more frequent cropping) and expansion (drainage, grassland conversion) of agricultural activities has played in altering the non-market goods supplied by such activity (Whitby, 1994).

The ESAs as a Policy Approach

We next consider the mechanism through which ESAs operate. Here, we depart from the literature to focus on the economically operative elements of the mechanisms. We define the ESA mechanism as involving standardized contracts to purchase changes in agricultural production practices (by requirement or prohibition) affecting land through payment systems. They are offered to farmers within a targeted subset of the population. Until recently, targeting was accomplished on a geographical basis imposing spatial limits based on specific environmentally sensitive areas allowing targeting of the contracted group of providers, the ecological and other physical characteristics of the environment affected, and the beneficiaries of the public good or reduction of the externality. In the CAP reform, the basis for targeting was redefined to include land or ecological characteristics, or zone types.

Most generally conceived, an ESA represents a social contracting scheme that purchases changes in production practices which are hoped to induce private provision of public goods, reduction of public bads and the reduction in externalities from private owners of rural land. Weaver (1996) presents an economic theory *of environmental effort* by farmers which clarifies the need for such social contracting. In brief, he notes that where farmers hold non-hedonistic preferences, a contracting approach can encourage private voluntary contribution in a way which is not possible using traditional Pigouvian instruments such as regulatory standards or incentives such as penalties or taxes. Despite the ESA mechanism being a contractual purchase arrangement it is centrally directed and typical applications have offered homogenous standards and incentives to heterogeneous populations of farmers. In all past applications, the mechanism has operated indirectly in the sense that contracts are for private good production practices which are predicted to result in production of desired public goods or reduction in externalities or public bads. Often the contracted production practices may produce joint multiple outputs or effects. The contracts have involved restriction of production practices either by prohibiting use of particular activities which are perceived to result in public bads or externalities or by requiring use of activities that are perceived to result in public goods production.

From this perspective, Crabtree (1992) and others have questioned whether the ESA mechanism can be interpreted as a property rights transfer mechanism. For this chapter, it is of interest to ask whether such an interpretation provides a useful basis for valuation of ESA applications. Crabtree (1992) properly noted that the ESA mechanism involves a restriction of the production possibilities available to farmers. Whitby (1989) adds the additional presumption that farmers hold a property right to choose production practices freely. Based on this presumption, ESA mechanisms which impose restrictions on production practices can be viewed as involving a transfer of property rights from the farmer to society in return for a cash payment. The cash payment can then be interpreted as equal to or greater than the farmer's valuation of that property right. This perspective would seem appropriate for contracts which prevent the farmer from using practices or taking specific actions that produce public bads, or externalities. With the case of contracts for the production of public goods, the property right transfer interpretation is less useful. In this case, the contract represents a purchase of private provision of a public good.

Where contracting is for direct provision of a public good, the contract can be viewed as providing direct incentives for a particular output and no property rights transfer is implicit. In contrast, where public goods are only indirectly contracted, e.g. by contracting only for the use of particular production practices which are perceived to result in production of the desired public good, a property right transfer interpretation would appear to remain useful. As a starting point for valuing ESA applications, the property right transfer interpretation provides a basis by which the ESA value could be viewed as founded on social valuation of ownership of the right. That is, social value would indicate society's valuation of the farmer not having the right to pursue particular production activities. However, from a practical standpoint, because this right is at best only indirectly linked to the reduced production of an externality or public bad, or increased production of a public good, this type of property right provides a very limited basis for establishing the value of an ESA application.

The ESA mechanism applications have also allowed for more direct contracting for the provision of public goods. The European Community legislation clearly recognizes farmers as having the potential to produce environmental goods through their direct management of the countryside for example, farmers can privately invest to produce various landscape elements such as woodlands or wetlands, or landscape characteristics such as wildlife through the introduction of species. In these cases, property rights transfers are not involved, and instead fairly traditional production economics suffice to analyse the supply of these goods through the acceptance of contracts. From the perspective of valuation, provision of these public goods will typically involve marginal changes in the aggregate supply allowing marginal valuation. In many cases, however, valuation will be complicated because the goods are not pure public goods, but may involve some aspects of limited access, at least in a spatial sense.

Given that ESA mechanisms can be interpreted as contracting mechanisms, it is useful to clarify the extent to which they are interpretable as management agreements. Colman (1994) defines management agreements based on their historical application in the UK, as individually negotiated contracts between a public agency and a land owner to restrict or initiate land management activities. They classify ESA mechanisms as standard payment systems in which contracted practices and payments are the same across participating farmers. Here, our intent is not to confuse the literature, but to distinguish among the operative elements of the ESA mechanisms. Clearly, they involve contracting for private management either restricting or requiring particular practices.

To summarize, several features of the ESA approach are important to note:

1. It may involve either indirect or direct provision of public goods, reduction in public bads and externalities, often these outputs are multiple and jointly produced (along with other public as well as with private outputs) by particular environmental efforts of the farmer.
2. It allows targeting in a variety of dimensions.
3. It relies on voluntary participation by farmers.
4. It involves contracts with standard payments as incentives and standard management agreements as regulatory constraints across participants.

The specific characteristics of particular ESA applications can now be placed within the context of this more general set of characteristics. Specific applications of ESAs have been presented by numerous authors (Dubgaard *et al.*, 1994). Each of these applications involves multiple standard management agreements and standard payments, which farmers choose from a menu of available options. Thus, while the general mechanism is uniform across applications, the composition of the menu offered in terms of contracted or restricted production activities varies widely as do the perceived public goods, bads, or externalities that are the objective of the contract. In practice, the objectives of each ESA and the expected effects or outputs necessarily vary depending on the character of the demand for and potential supply of its public characteristics. Applications reported in Whitby (1994) illustrate this variation as well as the variation of participation and choice of menu elements across the heterogeneous farm population.

A review of the possible non-market goods induced by ESAs highlights the variety of types of public outputs expected from these programmes. Outputs include pure public goods such as preservation of habitat diversity to encourage speciation to congestible, spatially limited quasi-public goods such as landscape use in hunting, viewing, or exercising. Importantly, the outputs are often multiple and jointly produced by the farmer's environmental effort, for example, actions taken by a farmer to enhance the visual amenities of the farm site will typically alter habitat inducing, long term benefits related to changes in biodiversity and shorter term benefits related to wildlife use activities (e.g. hunting, viewing). Outputs also vary in the extent to which they generate utilitarian versus non-utilitarian benefits, the length of the horizon over which these benefits will be generated, the uncertainty with which they will be generated, the dynamic path of supply associated with the goods or their characteristics, as well as their jointness in supply and/or demand or benefit characteristics. Finally, the interaction of these goods or outputs with private goods or services also varies substantially.

The Social Value of ESA Applications

The origin of the social value that ESA mechanisms hope to create lies ultimately on the value that humankind places on nature. ESA mechanisms hope to create a change in the environmental performance of the rural landscape. In this sense, they attempt to affect marginal changes in the current and future state of nature. Operatively, ESA mechanisms affect nature through standard management agreements. While the ESA designation of an area may have some impact independently of any effect produced through changes in management, the basis of such changes is unclear. Here, we consider only changes induced through standard management agreements. At the most general level, the notion that nature has value is complex

(Weaver and Kim, 1994). It is clear, however, that across cultures and time humans have placed value on nature. Within this context, nature has been understood to mean all things not created by man. The value of nature has been extensively considered within a number of disciplines. Economists have focused on utilitarian value, while other disciplines have recognized non-utilitarian values associated with nature (Norton, 1994). Utilitarian value can be decomposed into direct consumption value, option value, and existence value. Non-utilitarian value systems necessarily go beyond homocentric perspectives that view nature as an entity that serves humankind (Kim and Weaver, 1994). Within this taxonomy it is clear, from the list of environmental effects or outputs expected from past ESA applications, that the social value of these effects or outputs may involve all elements of this taxonomy.

Based on the characteristics of the ESA mechanism as well as the diversity of economic goods its application is hoped to induce, the challenge of valuation of a particular ESA application is striking. While the challenges of valuing landscapes have often been noted (Kim and Weaver, 1994), their consideration has often been limited to the problem of valuing a particular known change. In the context of ESA applications, numerous types of environmental efforts are implemented, which are hoped to induce a diversity of public goods, bads, and changes in externalities. Thus, the valuation problem must necessarily consider a vector of changes. While the jointness in production of these outputs has been noted, their consumption may also be joint. Focusing on single output effects, Bateman (1994), among others has presented reviews of valuation approaches that may be useful. However, these reviews consider only valuation of known changes in a particular public good. As was argued in the previous section, the problem of valuation of a particular ESA application is considerably more complicated. To simplify, while highlighting this complexity, suppose an ESA application produces a single public good, however, suppose we allow it to have characteristics the supply of which are dependent on:

1. uncertain participation by farmers,
2. uncertain efficiency of farmers production,
3. uncertain technological feasibility;

and where there may be said to be characteristics:

1. with utilitarian value,
2. with non-utilitarian value,
3. with uncertain current value,
4. with uncertain future value,
5. with pure public good aspects,
6. with spatially limited access aspects,
7. which substitute with some private goods, and

8. which complement other private or public goods.

The problem of valuation must necessarily involve:

1. valuation of different types of goods,
2. valuation over different horizons,
3. valuation under different extents of uncertainty,
4. valuation based on different approaches, and
5. aggregation of these heterogeneous sources of value, possibly recognizing complementarity in demand.

Given this complexity, the necessity of treating each output of an ESA application is clear Willis *et al.* (1993) provide illustrations of such individualized treatment.

At this point, we are in a position to consider an obvious question: what are the implications of this complexity for comparative assessment of value across applications? To suppose comparative assessment is of interest requires that the same change is made across applications. The complexity of the ESA mechanism and its variation across applications implies that such a common change does not exist, in general. Even where the same practice is restricted in a set of applications, the output goals and effects may be different. Where the same goal or output effect can be found, comparative value assessment might be feasible. However, such feasibility would rely on independence of that output from others produced by the applications, unless those other outputs are also identical across applications. Where such independence does not exist, comparison would be difficult. Consider an example, where in two applications the public good output of improved visual value of the landscape is sought and attempted by field perimeter plantings of poplar trees. Suppose in both applications, intensive field cropping occurs, however, assume in application A cold winters imply that substantial wind induced soil erosion occurs while in application B, such erosion does not occur. To compare the visual value of the treatment across the two applications would clearly require identification of the value of the erosion protection and air quality services in application B. While this example is sufficiently simple to suggest that such evaluation would be possible, typical ESA applications involve a large number of outputs and interacting practices.

Valuation when Public and Private Outputs are Joint

Before moving to consider and interpret available evidence of the value of ESA applications, we present an economic framework which clarifies many of the complications already discussed. In particular, the problem of evaluating the public good is extended to consider the case where it is jointly produced with private goods, as is the case of ESA applications. Consider the simple case where the ESA application results in a vector of public goods (e.g. ecological and amenity) and a vector of private goods (agricultural outputs). We would like to determine the value of public goods, i.e. the social rate of exchange or trade between the public and private benefits generated. Define social welfare U as a function of a vector of public goods Q and a vector of private agricultural commodities X:

$$U = U(X, Q) \tag{12.1}$$

Define the product transformation function G that bounds the set of all technologically feasible combinations (X, Q):

$$G(X, Q; Z, \theta) = 0 \tag{12.2}$$

where Z is a vector of environmental inputs and θ is a vector of quasi-fixed, private inputs.

The social choice problem is to find a feasible combination (X^*, Q^*) which maximizes social welfare (Equation 12.1) subject to available technology described by (Equation 12.2). Graphically, the solution of the problem is illustrated in Fig. 12.1 by A^* which lies at the tangency of the isosocial welfare curve U^* and the production possibility frontier G. At the tangency, the exchange rate defined in terms of social welfare (marginal rate of social welfare substitution, MRSWS) between the goods is set equal to the exchange rate between the goods defined by technology (marginal rate of technical transformation, MRTT) between the goods. Stated differently, the rate at which society is willing to trade Q for X is set equal to the rate at which society can actually trade Q for X as defined by technology.

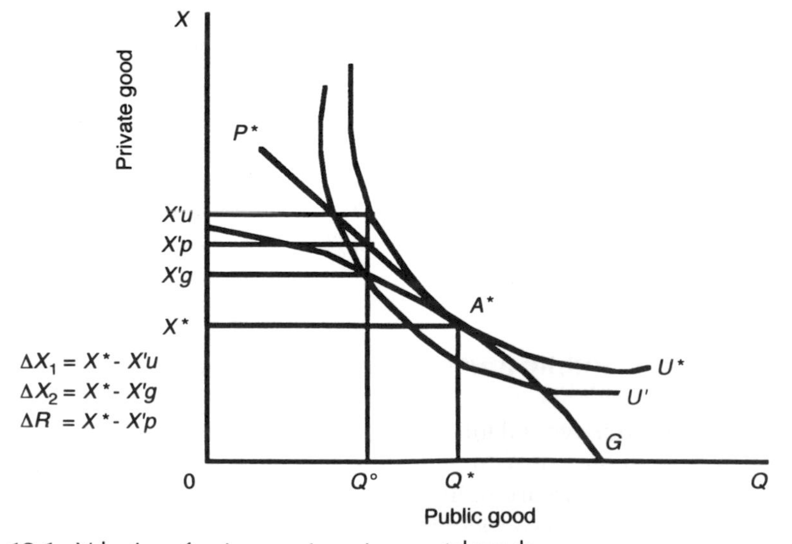

Fig. 12.1. Valuation of an increase in environmental goods.

The problem of valuation is clear from Fig. 12.1 and the character of this tangency. For market goods, prices serve as signals to which consumption and production adjust to establish the conditions at point A^*. Where prices do not exist, we seek alternative means of evaluating the exchange rates of Q for X defined at point A^*. The first conclusion that is apparent from this simple graphic and the logic of economic optimization is that when prices are not available, it is more difficult to determine the position of A^*. What we need in terms of valuation is not simply a value at A^*: instead we seek values at a suboptimal point such as Q° in Fig. 12.1. At that point, we are interested in values as a basis for information to help us define a point that will allow society to attain the allocation at A^*. In terms of the graphics of Fig. 12.1, what we need are estimates of how the rates of exchange defined by MRSWS and MRTT change as Q and X are changed. Suppose at point Q° an increase in public goods was offered to consumers. Of interest is whether such an increase would be valued more highly by consumers than by the producers of the increase in the public good. That is, to achieve the increase, a reduction in private goods must be exchanged for the increase in the public good. Given their limited budgets, consumers must give up X to finance an increase in Q. Producers likewise must use their limited resources such that Q is increased, and this can only be accomplished with a decrease in X.

The value of an increase in Q at point Q° could be established by standard methods of valuation. One approach would be to measure the willingness of society to exchange X for Q as defined by MRSWS. At point Q°, society can only achieve a level of utility indicated by the social indifference curve noted U' in Fig. 12.1. At point Q°, producers must give up X to produce more Q at a rate defined by the slope of the production possibility frontier G. At Q° and along U', MRSWS$'$ > MRTT$'$ signalling that an increase in Q and a decrease in X would increase social welfare. In Fig. 12.1, this increased welfare is indicated by the indifference curve U^*.

This same story can be told from another perspective. Consider a change in the environment from Q° to Q^*, i.e. such that $\Delta Q > 0$. In Fig. 12.1, ΔX_1 measures the amount of X society would be willing to pay (WTP) (in units of X) for the increase in Q. We measure this WTP at the utility level U^* achieved after the change in Q. Alternatively, ΔX_2 indicates how much X would have to be paid by producers to produce Q^*. Under standard curvature properties shown in Fig. 12.1, $\Delta X_1 > \Delta X_2$ which implies society would be willing to pay more than producers would have to pay to increase Q from Q° to Q^*. Finally, in Fig. 12.1, ΔR defines the amount of X that a market, if one were operative in both goods, would demand for an increase from Q° to Q^*. As is clear from Fig. 12.1, such a market based value could lie between consumer and producer valuations, i.e. $\Delta X_2 < \Delta R < \Delta X_1$ as $Q^\circ < Q^*$.

Finally, the same logic can be applied to when Q denotes a public bad as described in Fig. 12.2. Here, free disposal is not assumed for the public bad.

To consider hedonic price estimation approaches, Figs 12.1 and 12.2 are

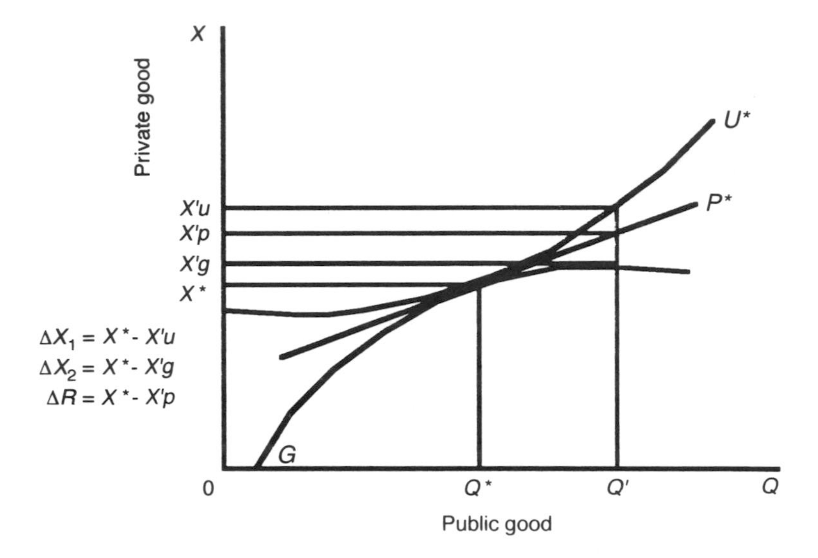

Fig. 12.2. Valuation of a decrease in environmental bads.

also useful. For example, in Fig. 12.1, the slope of line P^* indicates the relative prices of X and Q, i.e. P^*_Q/P^*_X where the asterisk indicate prices at the social welfare optimum A^*. Hedonic approaches attempt to estimate P_Q of the observed set of goods. However, where no markets are available to force P to P^* and goods to A^*, there is no basis for determining the relative prices and hedonic approaches are infeasible. In Fig. 12.1, when the line P^* is not defined relative to the slopes of G and U^* at A^*, no basis exists for fixing its position and its slope is indeterminate. Where an implicit market exists for Q, hedonic methods can be used to estimate the implicit price P_Q. For example, the use of a rural setting could be estimated from prices of leisure homes in the setting (Willis and Garrod, 1993).

Valuation of Rural Landscape Amenities

Given the conclusion that the value of ESA applications is best based on the valuation of each of multiple and possibly joint public effects which are the goals of the application, we now provide a review of available evidence concerning the value of one of the most typically expected products of ESA applications: rural landscape amenities. Several studies have been conducted in which the value of agricultural landscape was estimated by means of contingent valuation. In comparison with contingent valuation (CV) studies in the field of recreation, application of the contingent valuation method (CVM) to rural amenities is more subject to misspecification of the product. In particular, geographical part–whole bias and policy-package part–whole bias are likely to result from inadequate specification of the amenity.

Geographical part–whole bias occurs when a respondent values a landscape whose spatial attributes are larger or smaller than the spatial attributes of the researcher's intended landscape. There exists empirical evidence that people are able to discriminate between a local landscape and a national landscape. Thus, specification of the geographical scope of the amenity is essential. Consider a respondent who is asked how much he is willing to pay for maintaining today's landscape in an ESA. If the respondent is unable to isolate that area in his mind from a larger area covering the specific ESA, the respondent may in fact value a larger area than intended by the researcher. Policy-package part–whole bias occurs where a respondent values a broader or a narrower policy package than the one intended by the researcher. This issue is of particular importance within the context of ESAs. Given that agricultural activities and particular ESA applications can produce a wide variety of public effects which may be associated with both current and future time periods, the precise nature of the public effect to be valued by respondents must be specified. Where this has not been accomplished, response cannot be associated with particular public effects and in the extreme, may be based on a vague and poorly specified notion of the entire menu of public goods supplied by agriculture. In the context of the Common Agricultural Policy reform, people may treat the ESA scheme as symbolic of larger policy goal (e.g. preservation of farming as a way of life) and assign to the scheme some of the values they have for this more general goal.

Table 12.1 gives a sample of studies whose objective is to value the preservation of today's landscape. Studies covering very limited portions of land such as specific habitats or Sites of Special Scientific Interest (SSSI) are not included. First, the challenge of comparison of results from various ESA applications is clear. Not only does methodology differ across studies, but studies rarely provide sufficient information to allow even the calculation of WTP in a common currency or measure.

As already noted, the particular policy package involved in each study differs and it is unclear how estimated values can be related to particular policy elements, or management practices. Drake (1992) considers WTP to maintain an open landscape in the whole of Sweden whereas Le Goffe and Gerber (1994) and Dillman and Bergstrom (1991) address the landscape issue at a local level. Taking into account the average size of a Swedish household, Drake's figure is the highest and appears consistent with our expectations. However, no objective basis exists for validation of these estimates. Hackl and Pruckner (1994) estimated roughly a similar level of WTP for the provision of agricultural landscape cultivating services in Austria. Nonetheless, the context of their research is different since the countryside is considered an important input for the Austrian tourism service. The large difference in WTP estimates between the two other studies are attributable to the difference in local context. It is not easy to compare WTP

Table 12.1. Willingness to pay (WTP) for landscape and countryside according to geographical level.

Author	Geographical level	Type	WTP (1992 ECU)
Drake (1992)	National		92/person/year
		Resident	(16–74 years old)
Bateman *et al.* (1992)	Regional	Visitor	95.3–280.6/household/year
		Non-user	5.1/household/year
Willis *et al.* (1993)	Regional	Resident	22.0–34.5/household/year
		Visitor	14.8–24.4/household/year
		Non-user	2.5–3.1/household/year
Riera (1994)	Regional	User & existence value	6.5/visit
Rebolledo and Perez y Perez (1994)	Regional	User & existence value	9.6/visit
Le Goffe and Gerber (1994)	Local	Resident	60.6/household/year
Dillman and Bergstrom (1991)	Local	Resident	6.7–10.5/household/year

across the other studies since some refer to the household and others to visits. Riera (1994) considers the benefits of preserving a large area of high landscape quality in the Catalan Pyrénées, Rebolledo and Perez y Perez (1994) consider a broader good which is the maintenance of a natural park close to the Aragon Pyrénées.

While estimates from these studies appear consistent, a review of the literature shows a diversity of estimates, for example, Merlo and Della Puppa (1994) have reviewed Italian estimates of the public benefit values of forestry and farming areas and found a high degree of occurrence. This difference is due both to differences between sites and to differences in the way in which methods have been applied. The majority of benefit values range between 3 and 10 ECU per person per visit. Bateman *et al.* (1992) refer to the Norfolk Broads which are under threat from flooding, whereas Willis *et al.* (1993) consider ESAs based on the South Downs and the Somerset Levels and Moors. So differences in WTP could be attributable to differences in goods which are valued.

In our opinion a more important point has to be made: the value of a specific ESA is conditional on the continued protection and preservation of the other ESAs. So it can be assumed that the marginal benefits of an ESA decline as the number of ESAs increases. This implies that aggregate total benefit curves increase at a decreasing rate with the protection of additional areas. This issue was made clear in the context of the preservation of wild and scenic rivers in Colorado (Walsh *et al.*, 1985) and has been taken into account by Willis *et al.* during their investigations.

Table 12.2 provides aggregate values for several areas, with the range of WTP per hectare being large. The dispersion cannot be attributed to

Table 12.2. Comparison of nature conservation estimates.

Author	Area	Good being valued	Aggregate WTP ('000 ECU)	Average WTP (1992 ECU /ha)
Bateman *et al.* (1992)	The Norfolk Broads (UK) 29,980 ha	Preservation of the landscape from increased risk of flooding	User values 8,620	288
			Non-use benefits 137,613	4,590
Willis *et al.* (1993)	The South Downs (UK) 26,738 ha	Maintaining the ESA scheme in an environmentally sensitive area	User values 60,985	2,281
			Non-use benefits 3,951	148
	The Somerset Levels and Moors (UK) 27,170 ha	Environmentally sensitive area	User values 13,477	496
			Non-use benefits 52,462	1,931
Le Goffe and Gerber (1994)	Pacé and Saint-Gilles (F) 5,580 ha	Preservation of today's landscape	Resident 803	144
Rebolledo and Perez y Perez (1994)	Parque Natural de la dehesa del Moncayo (SP) 1,389 ha	Maintaining a natural park	User and existence values 1,334	960

differences in unitary values, it mainly comes from the definition of the population concerned with the good. For the resident component, the situation is quite clear, but it is more difficult to delineate the visitor population especially for an unbounded good such as the landscape. Moreover, we are confused with regards to how to define the non-user category.

There is a limited number of hedonic price studies whose objective is to value landscape and countryside amenities. Two specific studies referring to the UK (Garrod and Willis, 1992) and France (Jacques and Blouët, 1994) need to be mentioned. The first study encompasses some 4800 km^2 of central England and the Welsh borders and is concerned with the market of houses in rural areas (the data base includes 2000 observations). It is constrained by empirical data since the environmental data are neighbourhood-specific rather than house-specific, relating to one square kilometre in which the house is located. It was not possible to estimate the demand for particular countryside characteristics. The second study encompasses all of France and there are 3300 observations, with the dependent variable equalling the rent for a vacation house ('gîte rural'). This model suffers the same limitations regarding environmental data as defined at the 'Département' level. The variables include specific components of land use, e.g. arable or forest land and it is possible to infer the impact of various changes in land use on rents using the hedonic function. For example in a specific 'Département' ('Manche') the increase in total rent would equal FFr 45 /ha in response to the conversion of 1000 ha of arable land to permanent pasture.

Very little work has been accomplished concerning the willingness of farmers to accept (WTA) changes in farming practices to benefit the environment. Weaver (1996) presents a microeconomic theory of such behaviour and calls it *prosocial behaviour*. The theory illustrates how producer preferences over the environmental outcomes of their production activities can result in private provision of environmental goods even in the absence of pecuniary reward. Empirical evidence of this type of behaviour is reported in Weaver and Abrahams (1994) and a recent study conducted in Belgium, Semaille (1994) also finds evidence of such behaviour. As Weaver (1996) notes the presence of this type of behaviour complicates the estimation of WTA and the valuation related benefits.

Conclusions

The objective of this chapter was to consider the public economics of the ESA mechanism and the valuation of the effects of ESA applications. At the core of the ESA mechanism lies a presumption that manipulation of private good production activities will affect environmental processes to generate a change in associated public good effects. This logic relies on the existence and nature of a relationship between private and public good production processes. Our first conclusion is drawn from our consideration of this relationship. As we noted, the technology of producing public effects through agricultural activities is one that involves three important characteristics; multiple private and public effect outputs which are often joint, uncertainty due to randomness of environmental inputs and site specificity. With respect to the latter characteristic, scale, scope, biophysical, and manager characteristics are likely to be important. The immediate implication of these characteristics is that even if the ESA mechanism involved only one instrument, multiple interacting effects would be expected, implying that valuation must span the set of generated effects. Unfortunately, the ESA mechanism involves multiple instruments rendering the problem even more complicated, it is unlikely that the effects of an individual instrument will be identifiable because of the jointness of the underlying technology and the multiple effects are likely to be interactive. The conclusion can, therefore, be drawn that the valuation of ESA mechanisms must simultaneously consider the application of the entire package of instruments involved in the ESA and value the package of effects generated, including both private and public effects. Going one step deeper into a potential mire, it is interesting to note that since the application of the package of instruments induces changes in production practices, the cost of an application is not temporally separable from the benefit stream as in many cases of public investment. It follows that valuation of effects (e.g. *ex post* analysis) must be broadened beyond benefit analysis to include cost analysis as well.

A second conclusion can be drawn from the theoretical consideration of the ESA mechanism. In many cases, ESA applications will generate both local public goods which verge toward club goods for which exclusions are feasible, and, more classic public goods. Consider the case of actions directed towards biodiversity. Maintenance or redevelopment of hedges can be expected to generate local public benefits in the form of aesthetics or use (e.g. hunting) while at the same time generating a classic public good of biodiversity preservation. While the local–national dichotomy has often motivated concern for geographical part–whole bias, we note that the local versus classic public good nature of effects, will also involve what we label conceptual part–whole bias. That is, if respondents hold a conceptual basis for valuation that is not identified in the survey process, their response may follow from either a local geographical or a local conceptual base. While geographical part–whole bias may be addressed through off-site surveys, the conceptual part–whole bias must also be addressed through specification of the effects of the ESA. Clearly, where the conceptual basis is unknown to the researcher, aggregation of individual benefit estimates will be problematic.

With respect to methods, it is clear from our analysis that contingent valuation approaches, though limited by the realm of the complications we have discussed, remain preferable to hedonic or travel cost approaches. From the perspective of part–whole biases, it is clear that these latter approaches offer limited opportunities for consideration of such biases. None the less, our review provides a solid basis for the conclusion that the utility of CVM studies could be expanded by the adoption of a common basis for their implementation. Through such harmonization in approach and method, the development of an expanding data base would be ensured as well as would an increased level of comparability across studies. For this reason, we recommend that effort be placed toward such harmonization in both the valuation and in the design of ESA applications. Such harmonization, in our view, should be targeted at ensuring an adequate base is developed to provide scientific comment on the following policy questions:

1. What is the level of net benefits generated by ESA applications?
2. To what extent does ESA apply?
3. To what extent are ESA instruments neutral in their effects on production decisions?
4. What are implications of past ESA applications for future applications?

To provide a basis for comment on such questions, we see it as essential that guidelines be established to harmonize data collection and valuation approaches. In both cases, we do not support rigid standards, instead we propose the adoption of guidelines that ensure room for innovation while retaining a sufficient basis for comparative analysis. We propose that the bases for such harmonization be developed through a multinational effort that ensures the eventual convergence of approaches and confidence in results.

13

The Prospect for Agri-environmental Policies within a Reformed CAP

Martin Whitby

Introduction

The major forces determining the evolution of the European agri-environment include the attempt to reform the CAP as part of the continuing debate resulting from various policy failures, including the advent of excessive output, low farm incomes and budgetary pressure, as well as the increasingly recognized impact of agriculture on the environment. It is thus essential to keep in sight the original motivation for recent reforms and to consider whether the measures introduced meet the objectives originally espoused and, to the extent that they do not, whether this has implications for the European environment. This chapter therefore seeks to draw together common threads from the preceding chapters and attempts a tentative prognosis of what they imply for the European environment. This chapter seeks to draw out major themes and to identify outstanding questions, rather than provide an alternative to reading the book.

The predecessor of Regulation 2078/92 was Regulation 797/85 which, under Article 19 had provided for the establishment of Environmentally Sensitive Areas. That had limited impact and was only taken up by northern Member States. Regulation 2078/92, with its greater resources, its range of conservation options, and its obligatory requirements on Member States, is having a much wider impact. In consequence, some states are introducing agri-environment policies for the first time; others have been using such policies for a number of years and have expanded and adapted them to the present Regulation. It does provide for the first time a common European framework for national policies in the agri-environment field. The Regulation also sets certain precedents for agricultural policy which may have long term consequences. It established the principle that farmers, for both environmental and production control benefits, should be paid to de-intensify production and to manage the countryside. There remain grounds for concern over whether support for de-intensification may clash with the polluter pays principle in some instances.

The Nature and Scope of Agri-environmental Resources

The scope of this book is notable in that, although its individual chapters deal with only seven of the current EU of 15, the countries covered include the five largest – Spain, Germany, France, Italy and the UK. They cover some 80% of the utilized agricultural area (UAA) of the EU and the aggregate land-use pattern they present is closely similar to that for the whole Union. However, in environmental terms it is not appropriate to focus on aggregates: here heterogeneity is the desideratum and diversity the norm. Hence a fuller appreciation of the mixture of habitats within national boundaries may be gained from surveys such as Bignal and McCracken (Chapter 3). By concentrating on features of particular ecological and landscape significance, and using detailed surveys, they can begin to measure more accurately the richness which agri-environmental policies seek to conserve. They highlight the importance of low intensity farming systems which are particularly evident in Southern Europe, with Spain, Portugal and Greece having 60% of their UAA under such systems. In general such systems have long since disappeared in Northern Member States and it would be a serious misfortune if other EU policies for agriculture promoted the appearance of similar levels of intensity in these as yet undeveloped areas.

The country chapters have emphasized the extreme diversity of Members States' responses to Regulation 2078/92. The variety of national and regional responses has already thrown up some anomalies. For example, the Bavarian extensification programme pays farmers up to a stocking density of 2.5 LU/ha; this compares with lower ceilings elsewhere, including other German Länder. Certain variations of this kind might be desirable in relation to different local environmental conditions, but the Bavarian ceiling might be considered unacceptably high in relation to any environmental benefits it might possibly achieve. The French extensification approach has been to adopt a single national ceiling of 2.0 LU/ha, but the problem here may be a lack of sensitivity towards localized conservation needs; for example, the ceiling is too low to support the management of important Alpine meadows in the French Jura. Other anomalies raise questions of agricultural competitiveness. For example, organic farmers are to receive very different levels and types of support in different regions.

Although the Regulation includes a diversity of conservation options, it adopts a common measure to achieve them, namely payments to farmers and their voluntary involvement in agri-environment schemes. Implicitly or explicitly therefore it promotes a particular model for resolving the tensions between agriculture and the environment and of the property rights that should regulate the matter. There are already signs that this nascent European model is challenging other approaches, including those traditionally pursued at the national level. We have already indicated that, in certain circumstances, it may undermine the polluter pays principle. It also challenges national systems where, in the past, the maintenance of low-intensity farming systems was ensured by restrictions on property rights (for example, through zoning or land use planning restrictions). In addition, it may challenge the rationality of traditional approaches to the resolution of agriculture–environment tensions pursued through the sharp geographical segregation of agricultural and environmental functions (farm land versus natural areas).

The diversity of response has also been picked up as well by other authors, for example Delpeuch (1994) comments on some of the difficulties of interpreting Regulation 2078/92. He particularly cites the difficulty of interpreting the regulation's requirement to 'reduce substantially' the use of fertilizers and/or pesticides. The word 'substantial' is open to a wide range of interpretations. Similarly the idea that the Regulation should reduce stocking levels is one which he discusses briefly, very effectively demonstrating the complexity of this concept. Further there are well known difficulties in basing policy criteria on average stocking rates, wherever large areas are grazed extensively. As the areas managed become larger the average stocking rate becomes increasingly irrelevant, being replaced by the local intensity of stocking in particular parts of such areas. Problems of soil erosion and destruction of semi-natural grazings arise where stock congregate. It is difficult to design policy incentives to apply at the farm level which will avoid such concentrations of stock. Delpeuch also discusses the interesting aspect of the Regulation concerned with the eligibility of rare breeds for support. This eligibility is defined in terms of the total member state population of particular breeds and the instrument is designed to encourage the maintenance of these breeds through the support of the farming systems that sustain them. This is a notably different approach to the rather 'living museum' mechanisms that work in northern European states where some livestock societies have tended to promote their individual breeds rather as fossilized curiosities. Perhaps the nearest analogue of this approach is found in the LFAs of Northern Europe where policies effectively support extensive farming systems. By working through farming systems the regulation may have less impact on the purity of genetic lines but will undoubtedly serve to inject money into very specific types of agricultural system.

CAP Reform

The enduring problems of budgetary escalation, surplus disposal and the more recently recognized pollution problem resulting from agricultural production peaked in the late 1980s and from the ensuing debates a commitment to major reform of the CAP emerged. The final set of policy decisions ratified the MacSharry proposals which had been well diluted from their original form. The essence of the reform was partially to decouple income support from product prices, thus allowing reduction of farm-gate prices towards the level of world prices. This would remove the surplus commodity problem and, eventually ameliorate the budget problem too. Having decoupled income support from farm-gate prices it was then re-introduced using arable area payments and livestock headage payments. These payments require compliance with the Arable Area Payments Scheme (set-aside), under which farmers are compensated for setting aside 15% of their arable area, and for livestock producers Headage Payments are subject to quotas of livestock per farm. Saunders (1995) points out that, once introduced, such schemes will have the effect of freezing the cropped area and livestock numbers on farms. Switching between crops and grazing animals would now carry a penalty of lost entitlement to either area payments or headage payments. Ways may be found round such constraints, to the extent that entitlements may be bought and sold, but that is unlikely to be sufficient to prevent the reformed support schemes presenting a substantial source of friction in farm development. It also remains to be seen whether the policy makers will be able to deliver their commitments to reducing prices at the rates originally envisaged.

Following the logic of decoupling it is now being asked whether recoupling area and headage payments to the production of desired public goods is appropriate. Given that a major share of the present cost of the CAP will be these new decoupled compensation payments, the possibility of transforming them into incentives to provide public goods may seem tempting. However, the evidence presented in this volume so far would not give full support to such an outcome. So far many new schemes have been introduced, often in great haste, which could constitute sufficient incentives to farmers to produce the desired countryside goods. Yet the experience from Regulation 797/85 can not yet be taken as encouraging. Even where that Regulation has been extended to the full it remains to be seen whether countryside goods will be delivered as a result. This is unavoidable simply because of the time scale of ecological processes at work. The same effect also brings the risk to these policies that the finite duration of agreements offers no protection for the goods created beyond the end of contracts, leaving these environmental goods vulnerable to swings in market forces favouring a return to intensive production. Such a result could lead to the destruction of natural capital accumulated at public expense over a number of years. Such an out-turn could be defended in terms of the supremacy of market forces but its political acceptability must await the event.

Policy Costs and Benefits

The financial cost of Regulation 2078/92

The budgetary cost of this Regulation is met from the Guarantee Section of EAGGF: a much larger budgetary heading than the Guidance Section from which Regulation 797/85 was financed. The funds allocated to the Guarantee Section through various stages of the budgetary process are displayed in Table 13.1. The Table shows a small decline in the share of total budget devoted to agriculture over the 2 years while the budget for accompanying measures has made a massive leap from 0.39% to 1.85% in this short space of time. This partly reflects the slow initial uptake of these measures, first announced in 1992, but its fivefold rate of increase between 1994 and 1995 indicates the gathering level of enthusiasm for these measures among Member States.

Perhaps of greater interest is the way in which individual states have responded to these incentives. Preliminary data are presented in Table 13.2, which also reports the share of Member States UAA entered, reflecting the level of enthusiasm for these measures as well as the mechanisms used in individual states. For example, Denmark and the UK have followed the Netherlands (as confirmed in the chapters above) in applying measures based on spatially defined areas giving a small share of UAA involved but, with the exception of the UK, expect to apply comparatively high payments per hectare. The lower payment for the UK shows that although using defined areas, some of these are the poorest land on which the income per hectare would be very low. In the case of France, by contrast, a comparatively non-discriminating use of the Regulation (through the prime à l'herbe) yields a combination of

Table 13.1. EU budgets, 1994 and 1995. (Source: Commission of the European Communities, 1994.)

	Budget 1994		Budget 1995	
Item	billion ECU	per cent of total	billion ECU	per cent of total
Market measures	34.520	50.50	35.559	46.47
Accompanying measures	0.267	0.39	1.417	1.85
Total guarantee section	34.787	50.89	37.926	49.56
Total guidance section	3.343	4.89	3.316	4.33
Total EU budget	68.355	100.00	76.527	100.00

Table 13.2. Estimated Member States' budgets for Regulation 2078/92. (Source: Derived from de Putter, 1995.)

State	Per cent of UAA entered	Total cost estimate: million ECU per year	Total cost estimate: per cent of total	Expected EAGGF contribution per hectare participating
Belgium	4.6	7.75	0.49	4.38
Denmark	7.5	18.58	1.17	88.5
Germany	25	426.00	26.85	142
Spain	15	139.65	8.80	30.6
France	21	325.5	20.52	51.3
Ireland	18	69.00	4.35	66.6
Italy	8.4	110.40	6.96	74.4
Luxembourg	12	2.63	0.17	160.5
Netherlands	3.3	9.75	0.61	144.7
Austria	91	335.3	21.14	105
Portugal	19	47.5	2.99	83.2
UK	16.0*	94.4	5.95	31.5*

The annual amounts budgeted have been derived from available 3-year estimates.
*Editor's estimate for the UK: 3.0 million ha of land entered.

a high share of UAA in the scheme, close to its (25%) share of UAA under grass and meadow. France will thus have more relevant experience if there is a move to recouple payments to environmental benefit provision in contrast with other northern states who have based their systems on precise designation of policy-relevant areas.

Net policy benefits

The potential benefits from policies such as this include adjustments to the financial cost of other policy instruments, more efficient resource allocation and improvements in the distribution of income. The first steps towards agri-environmental policies were taken when budgets were seriously under pressure from growing surpluses. The CAP was absorbing up to 70% of the EC budget, engendering serious frustration in other parts of the Commission interested in developing other types of policy. That pressure is now less intense with the CAP accounting for 54% of the total budget (1995), nevertheless critics of the CAP will point to cogent reasons why this does not yet provide grounds for complacency. These will focus partly on the continuing substantial economic and budgetary cost of the CAP and partly on its evident potential to resume rapid growth as a result of enlargement, good harvests, world trade developments and so on. Indeed, the question of what would be an appropriate budget share for agriculture in the EU has yet to be considered:

so far the concern has been to bring down the budget share assigned to agriculture but in terms of the sector's relative economic importance it would be difficult to justify half the present budget.

The mechanisms of the CAP and the high levels of expenditure it demands both confirm the economic inefficiency of the system. Diverting budgetary resources towards other more desired objectives would improve the economic efficiency of the system and it can be argued that that is what agri-environmental policies do. The problem comes when we try to establish first, how strongly those ends are desired and, second, the volume of the resources devoted to them. Bonnieux and Weaver (Chapter 12) who have presented a careful dissection of the elements of economic benefit that might be expected to arise from these policies mention the very few studies that attempt to measure the extent of such benefits in quantitative terms. This may be because of the inherent difficulty of such measurement, which Bonnieux and Weaver discuss in some detail. Nevertheless such methods are now available and are producing answers to the question of value which may be defended minimally as better than the alternatives. Correctly, Bonnieux and Weaver also give a clear impression of the reservations that remain after such work has been done. Serious problems of measuring the benefits of environmental policies concerned with land use include that of the long lead time needed before such policies pay off and the problem of specifying a counter-factual (policy-off) situation against which policy (-on) outcomes may be compared. The lead time before such policies can meaningfully be evaluated must either be at least one policy cycle – in the case of most of the instruments discussed here that means 10 or 20 years – or the projected effects must be evaluated. Indeed the great difficulty of monitoring these polices may eventually lead to their withdrawal as they become more expensive to fund and the results are more difficult to identify and to evaluate.

The cost of resources devoted to these policies is also an important variable which must be less than the benefits if the policy is to be justified. Assessment of resource costs would require enumeration of the social opportunity costs of all the inputs into the agricultural system, with and without the operation of this policy. These costs are calculable from the changes in output attributable to the policy. For example they would include savings due to reduced fertilizer application on farms in an ESA and changes in output subsidies needed as well. The changes in cost should in principle be followed right through the system so that the effect of other farms expanding output elsewhere to take up any slack in markets would also add to policy costs and benefits. However, such detailed changes are usually ignored as being too difficult to trace. The effect of such cost calculations is generally to reduce the recorded costs of policies such as ESAs. For example Willis and Garrod (1994) report that the net estimated costs to the public exchequer, in the case of the South Downs ESA (in the UK) were less than half the exchequer cost of payments to farmers.

An interesting question also hangs over the policy benefits of agri-environmental arrangements associated with their durability. The problem was mentioned in the UK chapter, but is a general one in that it relates to the extent to which policy benefits may continue to flow over time. It might be expected that benefits would continue to be available during the lifetime of a policy and they probably will be, given compliance with policy prescriptions. However, they might also continue to flow after policy payments have ceased. Indeed, some groups within society might expect them to continue after the end of the policy payment period. Thus environmentalists would be expected to see the revived ecosystems and improved landscapes resulting from these policies as a form of natural capital. They might reasonably argue that society has paid for the investment needed to bring into existence the natural capital from which these benefits arise: their disposal after that is then not a matter to be determined by farmers alone. On the other hand farmers might, especially if the market incentives encouraged them, decide to reinstate their agricultural productivity to levels which had existed before they signed their contracts with governments. Both sides in this debate would claim the support of moral principle but the existing law would probably support the farmers in most states. This argument raises a quite general problem about the durability or sustainability of the benefits of environmental policies which will require resolution at the political level.

The discussion so far has concentrated on the question of efficiency benefits of policy: arguably more important are the distributional aspects of policy and the changing institutional basis for policy. Distribution is both generally important in agricultural policy which, because it is traditionally based on support distributed in relation to the size of farm output (or some related factor such as land area) and specifically important in the case of Regulation 2078/92. The latter arises from the fact that these measures accompanying CAP Reform are aimed to make good the loss of income resulting from the reform. There are thus two questions to consider. First, does the policy contribute in any way to the notoriously regressive distribution of policy benefits through the CAP? Second, are the redistributional effects of the policy likely to contribute effectively to a genuine restoration of the situation before CAP Reform?

The first question requires only brief consideration. It has been pointed out elsewhere (Whitby, 1994) that the impact of ESA-based policies on incomes may be interesting as an example of the decoupling of support from the incentive to produce. To the extent that the policy achieves this it will make a significant contribution to welfare. However it will be difficult to judge whether this has in fact occurred and it is likely to take some time before it can be demonstrated not least because of the time-lags in producing farm account data. Several of the socioeconomic evaluations in the UK presented some evidence of decoupling but, although little of it was based on actual recorded impacts on incomes, it did prompt the conclusions that those with

ESA agreements had made significant gains in income. It may be that such benefits are distributed less regressively than farm policy benefits based on production supports because the farmers entering into contracts may form somewhat smaller areas than those gaining most of the benefits of conventional policies.

These preliminary results offer some support for the capacity of Regulation 2078/92 to restore farm income reduced by CAP Reform. It is also notable that the French use of the Regulation, through 'prime à l'herbe', while seeming a potentially profligate application of the measure is likely to have its main impact within the French LFA (Boisson and Buller, Chapter 7). Even more, the Austrian commitment of more than 90% of its UAA to accompanying measures schemes indicates either substantial transactions costs are to be incurred or that compliance monitoring will be slight. Evidently, scope remains for careful studies of the impact of these schemes on farm incomes and the extent to which genuine decoupling of incentives is achieved.

But whilst land use diversity indicates the scale of environmental resources worthy of conservation, many other factors determine the scope and structure of a viable European agri-environmental policy. Whatever policy is implemented must contend with the notable diversity of legal and administrative systems within Member States which will determine both the effectiveness of the policy and the cost of implementing it. Thus, although Britain has made most use of the access provisions of Regulation 2078/92, this has to be seen in the context of the contrasting legal basis of access to rural land and the contribution of different national values between it and other states must be recognized. This British emphasis may well be explained by differences between states in the structure of property rights and values regarding access. The interest in the provisions for rare breeds is probably more evident in southern Europe than in the North and probably arises from a mixture of current values and the present structure of agriculture: it is nevertheless an important possible use of these funds.

The levels of policy cost will also vary widely from state to state. Although the principle of compensation used throughout the Regulation is to reimburse farmers for income foregone in providing public goods, that is only part of the cost of these measures. In addition to that, such polices require the existence of substantial administrative structures in any member state and may also make heavy demands on their bureaucratic and legal systems both within the states and in Brussels. Part of these costs will be paid through the public sector out of taxation but a part is absorbed by private individuals seeking to benefit from the policies. The famous case of the Integrated Administration and Control Systems (IACS) forms introduced 2 years ago is only one example of the transactions cost imposed on private individuals seeking to comply with a policy. In that case some farmers found themselves obliged to incur substantial costs in order to obtain what had previously been available in the form of income support through the CAP.

Transactions costs are a critical element in determining the rational scale of environmental policy. For the time being, Member States are apparently prepared to shoulder these costs but for those who are net contributors to the EU budget it would be irrational to sustain such costs, unless they have other reasons for wishing to establish a substantial bureaucracy. Moreover, the poorer members of the EU may be unable to mobilize the resources to fund the transactions costs of these policies and their compliance, which confers benefits on all EU members, may thus be constrained. Such potentialities would support the case for the transactions cost of policies, or at least their public transactions costs, to be supported from central EU funds. Environmental policies involving a detailed knowledge of the circumstances of individual pieces of land and control of the way in which such areas are managed, bring much greater transactions costs than the broad price support mechanisms which have hitherto been the main instruments of the CAP. Recognition of transactions costs as a part of the price of environmental policies is a precondition for rational policy design. Before that can happen Member States and the EU's central bureaucracy will have to collect reliable data on the incidence of such costs.

At least part of the case for demanding central funding of transactions costs will arise from the wide divergence of legal structures in Member States. Thus the numerical predominance of small-scale peasant producers in most Member States who have historically been supported by the agricultural policies pursued, will influence the rate of response to environmental policy. Larger scale producers may well be more opportunistic in responding to such policies and better equipped to benefit from them because of their broader financial management skills. Similarly the predominance of owner-occupied farming in most Member States would indicate the likely success of incentive systems such as Regulation 2078/92. However, its impact is likely to be very different between owner-occupied farms and rented farms. The inheritance laws within Member States are also notably different. The British tendency towards primogeniture is well known and is regarded by many other states, where the Napoleonic code prevails, as profoundly unfair to siblings of the eldest son. However, a more important structural effect of such legal habits is in the size distribution of farms; the UK is well known for the importance of its substantially larger farms than in much of the rest of Europe. Such factors are important in determining the motivation of farmers and hence influencing their response to all forms of policy.

Future Dynamics

Further CAP reforms

The main thrust of this book has been to explain and describe the processes of reform of the CAP with respect to the environment, during the past decade. The agri-environment debate can be seen as a part of the continuing process of reform of the CAP and to have been initiated mainly during the budgetary crises of the 1980s. These imperatives have continued in the last few years under the pressures for reform as part of the round of GATT discussions initiated in Uruguay in 1986. But other pressures on the EU's agricultural sector will be felt as the debate about further enlargement of the Union develops. The question of enlargement and GATT are discussed below: first we should consider the forces for reform that remain to be accommodated.

It would be optimistic indeed to claim that the present round of reform is any more than a faltering step in the right direction. The novelty of agri-environmental policies at the EU level, the noted diversity of Member States' agricultural systems as determined by their resources, administrative structures and environmental traditions are a long way from full integration 10 years after the introduction of Regulation 797/85. As revealed in the individual country chapters above, the implementation of that directive was mainly confined to Northern European Member States while the Mediterranean countries, for various reasons, were slow to take advantage of that opportunity. The same gap in implementation can be seen with Mediterranean members taking less than a third of the budget for the accompanying measures and the majority going to France and Germany (de Putter, 1995, see Table 13.2). This may be simply a reflection of the time needed, in the case of Spain and Portugal, to adjust to membership of the EU and the full complexities of the CAP before beginning to participate in environmental policies. As Baldock and Lowe point out in Chapter 2, Regulation 797/85 was not translated into Portuguese, despite its binding nature. Indeed it must be recognized that that Regulation was very much a Northern European concept, unfamiliar to several Member States and it is therefore not surprising that it has been slow in implementation.

Regulation 2078/92 differs from its predecessor in drawing on the much larger Guarantee Fund and in offering twice the rate of reimbursement for participants. Moreover, it appeared as the slower adopters were beginning to develop their own response to the earlier Regulation, and were therefore better placed to take advantage of it.

EU enlargement: budgetary pressures and the resurgence of productivism

Following the latest round of EU enlargement, the countries now seeking entry are mostly from East Europe. These countries are undergoing modernization processes which will include raising standards of technical performance and productivity in agriculture and it seems most likely that this phase will have to be passed through before they will begin to show serious interest in and concern for the environment. Moreover, the generally low levels of income in such countries and their heavy dependence on agriculture as a productive sector will mean that their entry to the Union will add further to the budgetary pressures within it. This debate is summarized in *Agra-Europe* (18.7.95) where it is pointed out that the Agriculture Commissioner appears to be taking a complacent position regarding enlargement because he is arguing that Central and Eastern European Countries (CEECs) are unlikely to become significant exporters for the first two decades of the next century because of the structural macroeconomic problems they must first overcome. However, *Agra-Europe* argues against this that only a 1.5% per annum increase is needed to take them above their highest output years during 1980–95 and a 2% increase would take them well above that level. At the same time it praises the UK Ministry of Agriculture for its report claiming that:

> The prospective enlargement of the EU to include the countries of Central and Eastern Europe will create further problems in complying with GATT commitments. This is most likely in respect of the commitments on subsidized exports. The exports of the CEE countries are projected to exceed significantly the limits permitted for those countries for export with subsidy, while the EU will be at or close to its own limits for subsidized exports. The accession of the CEE countries, at high support levels, will therefore result in surpluses in the enlarged EU considerably in excess of those allowed to be exported with subsidy.

It is projected that the enlargement with the first six CEE countries will bring a budgetary increase of 15 billion ECU making the case for urgent reforms because 'major policy changes introduced under pressure are rarely the right ones'. The re-emergence of these familiar productivist pressures recalls the debates of the early 1980s which led first to the introduction of milk quotas and then to Regulation 797/85. Whether the return of those days will be prevented by the recent CAP reforms and the commitment to allowing world prices to influence farm production in the EU remains an open question at this stage. Suffice it to conclude that a convincing and sustained determination to expose European agriculture to world prices has yet to be fully implemented and there are several hurdles to be surmounted before we can assume that the new era has indeed arrived.

Convergence of political and legal structures

The further evolution of European policies and institutions is clearly basic to the healthy development of environmental policies. Optimists here will point to the rapid development of these policies from a standing start in the last few years. The provisional budget for measures accompanying CAP reform for

1995 was 1.4 billion ECU which may seem large but it remains less than 4% of the total Guidance and Guarantee Funds in that year. Further, budgetary headings do not necessarily reflect all of the objectives on which funds are disbursed: a well recognized effect of the introduction of ESAs has been to support the incomes of the farmers who participate in such policies and it may well be that there are some environmental benefits from expenditure on production policies as well as their potential damage. The CAP has clearly not yet been thoroughly purged of its potential for environmental damage and much more diversion of funds to other headings will be needed before the lobbyists will feel ready to relax on this issue.

In the longer term we should expect that the issues of institutional convergence will reappear on the political agenda from time to time. The debate surrounding subsidiarity in connection with the Treaty of Maastricht provides an example of the type of issue which will arise. The dichotomous use of the term subsidiarity in the treaty is emphasized by several authors (for example Scott *et al.*, 1994) pointing out that it may be interpreted as a substantive principle that decisions should be taken in Europe as closely as possible to the citizens affected or as a mechanistic procedural criterion to determine how and where the EU should intervene. In the latter sense it has been most influential in the debates on Maastricht and was the basis for an eventual acceptance of the use of the term across several Member States. The result of such debates is to leave substantial uncertainty as to the meaning of the term until a closer measure of consistency can be reached. Meanwhile the term is being used by the UK to constrain the scope for EU intervention in its affairs while other states accept the idea of policies being implemented at the most efficient level in the system. In such a context it becomes similar to the Tinbergen concept of efficiency as deployed by Scheele (Chapter 1). In that sense subsidiarity may be seen to be working through the Regulation in that states are handling it in diverse ways. Thus Germany and Spain are devolving the implantation to the regional level (Höll and von Meyer, Chapter 5; Garrido and Moyano, Chapter 6), in the UK it is handled at the level of the constituent countries (Whitby, Chapter 11) whereas in Denmark it is a matter for local authorities to manage (Primdahl, Chapter 4).

Outlook for GATT

Within the next decade a new round of GATT discussions will be under way but before that we might expect some continuation of the modifications from the last round of discussions. International conferences scheduled for the next two years will shed light on the next round of GATT. At present it is expected to be concerned with the environment although precisely what that may mean in practice has yet to be determined. If there is to be a serious attempt to take account of environmental impacts in trade treaties, much analysis and discussion will be needed before this can happen. This in turn will require much more transparency in policies than has been achieved to date and resources devoted to careful examination of the policy options.

What is quite clear from this analysis is that substantial political uncertainties hang over the future of these arrangements and, while these market-simulating mechanisms are the main source of environmental protection of land under agriculture and forestry, their impact will be subject to the strength of the competing market and policy forces flowing through world food markets. The future of present arrangements depends very much on a series of conferences in the next few years. The expected Inter-Governmental Conference in 1996, the Conference on the MacSharry round of reforms in 1998 and the next GATT conference in 1999 will be key fora where the relevant policy issues will be considered and modified. The new world of 'minimal' agricultural price policies will be a more volatile one for commodity markets and events initiated outside the control of agricultural policies may put great pressure on existing conservation arrangements. Any attempt to strengthen the forms of conservation provided would do well to start from the basic question of the property rights of cultivators and owners of land.

The discussion in this chapter and, indeed, in the rest of this book, does not seek to encourage complacency amongst policy makers. Behind the obvious issues of enlargement and living within agreed GATT commitments whilst containing the budgetary pressures that still constrain the CAP, there are serious issues of policy design and implementation which require examination before Regulation 2078/92 can be hailed as an unqualified success. The limited experience of CAP Reform so far suggests that further evidence will appear rapidly as the policies evolve, including those aimed at conserving the agri-environment.

References

Adger, N., Brown, K., Shiel, R. and Whitby, M. (1991) *Dynamics of land use change and the carbon balance.* Department of Agricultural Economics and Food Marketing, University of Newcastle upon Tyne, ESRC Countryside Change Initiative Working Paper 15.

Agger, P. and Brandt, J. (1988) Dynamics of small biotopes in Danish agricultural landscapes. *Landscape Ecology* 1, 227–240.

Allanson, P. and Moxey, A.P. (1996) The Geography of Agricultural Land Use: England and Wales 1892–1992. *Land Use Policy Journal of Environmental Planning and Management* (in press).

Allen, P. and Van Dusen, D. (1988) *Global Perspectives on Agroecology and Sustainable Agricultural Systems.* University of California, Santa Cruz.

Allen, P., Van Dusen, D., Lundy, J. and Gliessman, S. (1991) Integrating social, environmental and economic issues in sustainable agriculture. *American Journal of Alternative Agriculture* 6, 34–39.

Alonso González, S. (1991) La política comunitaria de estructuras agrarias : objectivos y medios. *Revista de Estudios Agrosociales* 156, 169–184.

Alphandéry, P. and Deverre, C. (1994) La politique agri-environnementale communautaire et son application en France. *Recherches en Economie et Sociologie Rurales* 7, 2–3.

Andersen, E. (undated) Marginalisering i 1980erne. Unpublished working paper. Department of Economics and Natural Resources, The Royal Veterinary and Agricultural University, København.

Arbeitsgemeinschaft Umweltplanung (ARUM) (1989) *Umsetzung raumordnungspolitischer Ziele bei der Extensivierung und Stillegung landwirtschaftlicher Flächen.* Report to the Federal Ministry of Regional Planning, Building and Town Planning. ARUM, Hannover.

Bagnasco, A. (1977) *Tre Italie. La problematica territoriale dello sviluppo italiano.* Il Mulino, Bologna.

Baillon, J. (1993) Les nouvelles mesures. *Perspectives Agricoles* 184, 20–26.

Baldock, D. (1993) The implementation of the CAP reform accompanying measures. In: Dixon, J., Stanes, A J. and Hepburn, I. (eds) *A Future for Europe's Farmed Countryside.* Royal Society for the Protection of Birds, Sandy, pp. 171–182.

Baldock, D. and Beaufoy, G. (1992) *Plough On! An Environmental Appraisal of the Reformed CAP.* World Wide Fund for Nature, Godalming, 66pp.

Baldock, D. and Beaufoy, G. (1993) *Plough On! An Environmental Appraisal of the CAP.* A report to WWF UK. Institute for European Environmental Policy, London, 65pp.

Baldock, D. and Bennett, G. (1991) *Agriculture and the Polluter Pays Principle: a Study of Six EC Countries.* Institute for European Environmental Policy, London, 231pp.

Baldock, D. and Mitchell, K. (1995) *Local Influence – Increasing Local Involvement in the Development of Green Farming Schemes.* Council for the Protection of Rural England, London.

Baldock, D., Beaufoy, G., Haigh, N., Hewett, J., Wilkinson, D. and Wenning, M. (1992) *The Integration of Environmental Protection Requirements into the Definition and Implementation of Other EC Policies.* Institute for European Environmental Policy, London, 42pp.

Baldock, D., Beaufoy, G., Bennett, G. and Clark, J. (1993) *Nature Conservation and New Directions in the EC Common Agricultural Policy.* Institute for European Environmental Policy, London and Arnhem.

Barberis, C. (1979) *Famiglie senza giovani e agricoltura a mezzo tempo in Italia.* Franco Angeli, Milano.

Barrue-Pastor, M. (1994) Etat d'application de la disposition communautaire article 19 en France. In: Billand, J.-P. and Deverre, C. (eds) *Paper to the INRA/CNRS seminaire sur l'Agriculture et l'Environnement en Europe.* CNRS, Paris.

Bateman, I. (1994) Research methods for valuing environmental benefits. In: Bateman, I., Dubgaard, A. and Merlo, M. (eds) *Proceedings of a Workshop Organized by the Commission of the European Communities, Directorate General for Agriculture,* Brussels 7–8 June 1993. Wissenschaftverlag Vauk, Kiel.

Bateman, I., Willis, K.G., Garrod, G.D., Doktor, P., Langford, I. and Turner, R.K. (1992) *Recreation and Environmental Preservation Value of the Norfolk Broads: a contingent valuation study,* Unpublished report to the National Rivers Authority, Environment Appraisal Group, University of East Anglia, Norwich.

Bayerisches Staatsministerium für Ernährung, Landwirtschaft und Forsten (ByStMELF) (1994) *Bayerisches Kulturlandschaftsprogramm – Teil A (KULAP-A), Regulation B 4-7292-1661, from August 1, 1994.* ByStMELF, München.

Bayerisches Staatsministerium für Landesentwicklung und Umweltfragen (ByStMLU) (1986) *8. Raumordnungsbericht 1983/84.* Bayerisches Staatsministerium für Landesentwicklung und Umweltfragen, München.

Beaufoy, G., Baldock, D. and Clark, J. (eds) (1994) *The Nature of Farming: Low Intensity Farming Systems in Nine European Countries.* Institute for European Environmental Policy, London, 66pp.

Berlan-Darqué, M. and Kalaora, B. (1992) The ecologization of French agriculture. *Sociologia Ruralis* 32, 104–113.

Bernes, C. (ed.) (1993) *The Nordic Environment: Present State, Trends and Threats.* Nord1993:12. Nordic Council of Ministers, Copenhagen, Denmark.

Bernes, C. (ed.) (1994) *Biological Diversity in Sweden: a Country Study.* Monitor 14, Swedish Environmental Protection Agency, Solna, Sweden.

Bignal, E.M. and McCracken, D.I. (1992) *Prospects for Nature Conservation in European Pastoral Farming Systems.* Joint Nature Conservation Committee, Peterborough, 8pp.

Bignal, E.M., Curtis, D.J. and Matthews, J.L. (1988) *Islay: Land-use, Bird Habitats and Nature Conservation. Part 1: Land-use and Birds on Islay.* Nature Conservancy Council, Peterborough.

Bignal, E.M., McCracken, D.I., Pienkowski, M.W. and Branson, A. (1994a) *The Nature of Farming: Traditional Low-intensity Farming and its Importance for Wildlife.* World Wide Fund for Nature International, Brussels, 8pp.

Bignal, E.M., McCracken, D.I. and Curtis, D.J. (1994b) *Nature Conservation and Pastoralism in Europe.* Joint Nature Conservation Committee, Peterborough.

Bignal, E.M., McCracken, D.I. and Corrie, H. (1995) Defining European low-intensity farming systems: the nature of farming. In: McCracken, D I., Bignal, E.M. and Wenlock, S.E. (eds) *Farming on the Edge: the Nature of Traditional Farmland in Europe.* Joint Nature Conservation Committee, Peterborough.

Billaud, J.-P. (1992) L'article 19: une gestion agricole au nom de l'environnement? *Economie Rurale* 208/209, 137–141.

Bink, R.J., Bal, D., van den Berk, V.M. and Draajer, L.J. (1994) *Toestand van de Natuur 2.* IKC NBLF, Wageningen.

Bird Life International (1994) *Implementation of EU Agri-environmental Regulation 2078.* RSPB, London.

Blanc, G., Chabalier, F., Viallon, F. and Leyrissoux, C. (1993) Les mesures agri-environmentales, *Bulletin de la Société Languedocienne de Géographie* 116, 145–158.

Blanc, M. (1993) Les mesures agri-environnementales. *Bulletin de la Société Languedocienne de Géographie* 1–2, 145–158.

Bodiguel, M. and Buller, H. (1989) Agricultural pollution and the environment in France. *Tijdschrift voor Sociaal Wetenschappelijke Onderzoek van de Landbouw* 3, 217–239.

Bodiguel, M. and Buller, H. (1995) The regions and environmental policy in France. *Regional Policy and Politics* 4 (3), 92–109.

Boisson, J.-M. (1986) *Les Relations Agriculture–Environnement et la Politique Agricole Commune.* Institut d'Etudes Catalanes, Annuari de las seccio d'economica, Barcelona.

Bonnieux, P. and Rainelli, P. (1994) Les mesures agri-environnementales et le recours à l'evaluation contingente. *Reformer la Politique Agricole Commune, INRA Actes et Communications* 12, 247–261.

Bowers, J.K. and Cheshire, P.C. (1983) *Agriculture, the Countryside and Land Use.* Methuen, London.

Brandt, J., Holmes, E. and Larsen, D. (1994) Monitoring 'small biotopes'. In: Klijn (ed.) *Ecosystem Classification for Environmental Management.* Kluwer Academic Publishers, Dordrecht.

Braudel, F. (1986) *L'identité de la France.* Artaud, Paris.

Brouwer, F.M., Terluin, I.J. and Godeschalk, F.E. (1994) *Pesticides in the EC.* Agricultural Economics Research Institute (LEI-OLO), The Hague.

Brunet, P. (1992) *L'Atlas des Paysages Ruraux de France.* de Monza, Paris.

Buckwell, A.E., Harvey, D.R., Thomson, K.J. and Parton, K. (1982) *The Costs and Benefits of the Common Agricultural Policy.* Croom Helm, London.

Buller, H. (1991) *Agricultural Structures Policy and the Environment in France : Report to the Directorate General VI of the European Commission.* Groupes de Recherches Sociologiques, Paris.

Buller, H. (1992) Agricultural change and the environment in Western Europe. In: Hoggart, K. (ed.) *Agricultural Change, Environment and Economy.* London, Mansell, pp. 68–88.

Bundesministerium für Ernährung, Landwirtschaft und Forsten (BMELF) (1988) *Agrarbericht der Bundesregierung 1988. Bundestagsdrucksache 11/760.* BMELF, Bonn.

Bundesministerium für Ernährung, Landwirtschaft und Forsten (BMELF) (1991) *Agrarbericht der Bundesregierung 1991. Bundestagsdrucksache 12/70.* BMELF, Bonn.

Bundesministerium für Ernährung, Landwirtschaft und Forsten (BMELF) (1994a) *Statistisches Jahrbuch über Landwirtschaft und Forsten der Bundesrepublik Deutschland 1994.* Landwirtschaftsverlag, Münster-Hiltrup.

Bundesministerium für Ernährung, Landwirtschaft und Forsten (BMELF) (1994b) *Die Verbesserung der Agrarstruktur in der Bundesrepublik Deutschland 1991 bis 1993 (Agrarstrukturbericht).* Report of Federal State and Länder on the conduction of the Common Task 'Improvement of Agricultural Structures and Coastal Protection'. BMELF, Bonn.

Bundesministerium für Ernährung, Landwirtschaft und Forsten (BMELF) (1995a) *Agrarbericht der Bundesregierung 1995, Bundestagsdrucksache 13/400.* BMELF, Bonn.

Bundesministerium für Ernährung, Landwirtschaft und Forsten (BMELF) (1995b) Umweltgerechte Landwirtschaft auf 4,4 Million Hektar. Erste Ergebnisse der EG-Fördermaßnahme in Deutschland. In: *BMELF-Informationen Nr. 24.* BMELF, Bonn.

Burrell, A., Hill, B. and Medland, J. (1990) *Agrifacts: a Handbook of UK and EEC Agricultural and Food Statistics.* Harvester Wheatsheaf, London, 175 pp.

Campos Palacín, P. (1993) Valores comerciales y ambientales de las dehesas españolas. *Agricultura y Sociedad* 66, 9–41.

Casadei, E. (1991) Attività produttiva agraria e tutela del paesaggio: profili giuridici. In: *Agricoltura e Paesaggio.* Quaderni Vol. 4, Accademia dei Georgofili, Firenze, pp. 35–56.

Cauville, L. (1993) OGAF : La prime ou la friche. *L'Acteur Rural* 1, 42–45.

Cavalli, S., Moschini, R.and Saini, R. (1990) *I Parchi Regionali in Italia.* UPI, Roma.

CEC (1986) *Agriculture and Environment: Management agreements in the EC Countries. Annex 1 Les accords de Gestion in France.* Luxembourg.

CEC (1992) *Council Directive Concerning the Conservation of Natural Habitats and Wild Flora and Fauna.* EEC 43/92. Commission of the European Communities, Luxembourg.

CNASEA (1993) *Rapport d'Activité.* Paris.

Colman, D. (1994) Comparative evaluation of environmental policies, ESAs in a policy context. In: Whitby, M. (ed.) *Incentives for Countryside Management: the Case of Environmentally Sensitive Areas.* CAB International, Wallingford, UK.

Colman, D., Crabtree, R., Froud, J. and O'Carroll, L. (1992) *Comparative Effectiveness of Conservation Mechanisms.* Department of Agricultural Economics, University of Manchester.

Commission of the European Communities (1993) *The Agricultural Situation in the Community: 1992 Report.* Brussels, Luxembourg.

Commission of the European Communities (1994) *The Agricultural Situation in the Community: 1993 Report.* Brussels, Luxembourg.

Comolet, A. (1989) *Prospective à Long Terme de la Deprise Agricole et Environnement.* Institut pour une Politique Européene de l'Environnement, IPEE, Paris.

Crabtree, J.R. (1992) Effectiveness of standard payments for environmental protection and enhancement. In: *Actes du 30ème Séminaire de l'Association Européenne des Economistes Agricoles*, Chateau d'Oex, Suisse, 11–13 November, 1992.

Curtis, D.J. and Bignal, E.M. (1991) *The Conservation Role of Pastoral Agriculture in Europe.* Scottish Chough Study Group, Argyll, 8pp.

Curtis, D.J., Bignal, E.M. and Curtis, M.A. (1991) *Birds and Pastoral Agriculture in Europe.* Scottish Chough Study Group, Argyll, and Joint Nature Conservation Committee, Peterborough, 137pp.

Danmarks Statistik (1994) *Landbrugsstatistik 1993.* København.

de Putter, J. (1995) *The Greening of Europe's Agricultural Policy: the 'Agri-Environmental Regulation' of the MacSharry reform.* Ministry of Agriculture, Nature Management and Fisheries and Agricultural Economics Research Institute LEI-DLO, Netherlands.

Delpeuch, B. (1991) Politique agricole européenne et environnement: une intégration croissante. *Eau et Rivières* 78 (August), 11–17.

Delpeuch, B. (1992) PAC et environnement. *Perspectives* 1 (June), 17–19.

Delpeuch, B. (1994) Ireland's Agri-environmental programme in the European context. In: Maloney, M. (eds) *Agriculture and the Environment.* Royal Dublin Society, Dublin.

Department of the Environment (1983) *Wildlife and Countryside Act 1981. Financial Guidelines.* HMSO, London.

Der Rat von Sachverständigen für Umweltfragen (SRU) (1985) *Umweltprobleme der Landwirtschaft. Sondergutachten März 1985.* Kohlhammer, Stuttgart and Mainz.

Deverre, C. (1994) Rare birds and flocks: agriculture and social legitimization of environmental protection. In: Jansen, A. and Symes, D. (eds) *Agricultural Restructuring and Rural Change in Europe.* University Press, Wageningen, pp. 220–234.

Dillman, B.L. and Bergstrom, J.C. (1991) Measuring environmental amenity benefits of agricultural land. In: Hanley, N. (ed.) *Farming and the Countryside: an Economic Analysis of External Costs and Benefits.* CAB International, Wallingford, UK, pp. 250–271.

Dion, R. (1939) *Essai sur la Formation du Paysage Rural Francais.* 1991 edn. Durier, Paris.

Dixon, J. (ed.) (1992) *A Future for Europe's Farmed Countryside.* RSPB, Sandy, UK.

Dixon, J. (1995) Implementation of Agri-environment Regulation 2078. In: McCracken, D.I., Bignal, E.M. and Wenlock, S.E. (eds) *Farming on the Edge: the Nature of Traditional Farmland in Europe.* Joint Nature Conservation Committee, Peterborough, pp. 156–167.

Dixon, J., Stanes, A.J. and Hepburn, I. (1993) *A Future for Europe's Farmed Countryside.* Royal Society for the Protection of Birds, Sandy, 206pp.

Dombernowsky, L. (1988) Det landbrugs historie 1720–1810. In: Bjorn (ed.) *Det Danske Landbrugs Historie.* Landbohistorisk Selskab, Odense, pp. 211–394.

Drake, L. (1992) The non-market value of the Swedish agricultural landscape.

European Review of Agricultural Economics 19, 351–364.
Dubgaard, A., Bateman, I. and Merlo, M. (eds) (1994) *Economic Valuation of Benefits from Countryside Stewardship.* Wissenschaftverlag Vauk , Kiel.
Duby, G. and Wallon, A. (1976) *Histoire de la France Rurale,* Volume 4. Seuil, Paris.
Economist Intelligence Unit (1994) *Pocket Europe: Profile, Facts and Figures about Europe Today.* Penguin Books, London, 216pp.
Errejón, J.A. (1989) La política communitaria par la conservacíon de la naturaleza. *Revista de Estudios Agrosociales* 148, 31–60.
Etxezarreta, M. and Viladomiu, L. (1989) The restructuring of Spanish agriculture and Spain's accession to the EEC. In: Goodman, D. and Redclift, M. (eds) *The International Farm Crisis,* London, pp. 156–183.
European Commission (1985a) *Perspectives for the Common Agricultural Policy – The Green Paper of the Commission, Green Europe – News Flash.* Brussels.
European Commission (1985b) *Perspectives for the Common Agricultural Policy.* Com (85) 333.
European Commission (1988a) *The Future of Rural Society.* Com (88) 501.
European Commission (1988b) *Environment and Agriculture.* Com (88) 338.
European Commission (1990) *Proposal for a Council Regulation (EEC) on the Introduction and the Maintenance of Agricultural Production Methods Compatible with the Requirements of the Protection of the Environment and the Maintenance of the Countryside.* Com (90) 366.
European Commission (1991) *The Development and Future of the Common Agricultural Policy – Reflections Paper of the Commission.* Com (91) 100.
European Commission (1992) *Council Regulation on Agricultural Methods Compatible with the Requirements of the Protection of the Environment and the Maintenance of the Countryside.* EC 2078/92.
Fabiani, G. (1986) *L'Agricoltura Italiana tra Sviluppo e Crisi (1945–1985).* Il Mulino, Bologna.
Fédération des Parcs Naturels Régionaux (1994) *Mise en Oeuvre d'Opérations Agriculture – Environnement: Les Clés de la Réussite.* FPNR, Paris.
Felton, M. (1993) *Wildlife Enhancement Scheme Pilot Project Interim Review Report.* English Nature, Peterborough.
Ferro, O. (1988) *Istituzioni di Politica Agraria.* Edagricole, Bologna.
Focken, B. (1988) Wo liegen die derzeitigen Probleme und deren mögliche Ursachen? – Aus der Sicht der Wasserversorgung. In: *Sauberes Wasser und Moderne Landbewirtschaftung – ein Gegensatz?* Report to DLG colloquium on December 15, 1987. DLG, Frankfurt.
Frouws, J. (1988) State and society with respect to agriculture and the rural environment in the Netherlands. In: Frouws, J. and De Groot, W.T. (eds), *Environment and Agriculture in the Netherlands.* Centrum voor Milieukunde, Leiden, pp.39–56.
Frouws, J. (1993) *Mest en Macht-een Politiek-Sociologische Studie naar Belangenbehartiging en Beleidsvorming Inzake de Mestproblematiek in Nederland vanaf 1970.* Agricultural University Department of Rural Sociology, Wageningen.
Frouws, J. and Van Tatenhove, J. (1993) Agriculture, environment and the state : the development of agro-environmental policy-making in the Netherlands. *Sociologia Ruralis* XXXIII (2), 220–239.
Ganzert, C. (1988) Gedanken zur Wirkung staatlicher Naturschutzmaßnahmen auf das

Verhältnis von Landwirtschaft und Natur. In: Öko-Institut, (ed.) *Für eine umweltgerechte und sozialverträgliche Landwirtschaft.* Werkstattreihe Nr. 48, Freiburg.
Ganzert, C. (1995) The agri-environment programmes: are they effective? In: McCracken, D.I., Bignal, E.M. and Wenlock, S.E. (eds) *Farming on the Edge: the Nature of Traditional Farmland in Europe.* Joint Nature Conservation Committee, Peterborough, pp. 168–175.
Garrido, F. (1992) *Representación de intereses en la agricultura holandesa.* Trabajo Profesional Fin de Carrera. ETSIAM. Universidad de Córdoba (mimeo).
Garrido, F. and Moyano, E. (1994) *EC and National Regulations on Environment and Agriculture in Denmark, The Netherlands and Spain.* Report 1, EU SEER Project, Esbjerg.
Garrod, G.D. and Willis, K.G. (1992) Valuing good's characteristics: an application of the hedonic price method to environmental attributes. *Journal of Environmental Management* 34, 59–76.
Gavignaud, G. (1990) *Les campagnes en France au XXe siècle.* Ophrys, Paris.
Genghini, M. and Scalzulli, P. (1989) Vincoli paesistici e agricoltura. *Genio Rurale* 6, 21–27.
Goriup, P.D., Batten, L.A. and Norton, J.A. (eds) (1991) *The Conservation of Lowland Dry Grassland Birds in Europe.* Joint Nature Conservation Committee, Peterborough, 136pp.
Hackl, F. and Pruckner, G.J. (1994) *Subsidising external benefits: the case of agriculture and tourism in Europe.* Department of Economics, University of Linz, Austria.
Hansen, B. and Primdahl, J. (1991) *Miljofolsomme Omrader. Evaluering af MFO-Ordningens Evaerksaettelse og Betydning.* Betyning, DSR Forlag, Kobenhavn.
Hansen, H.O. (1994) *Landbrugets Placering i Samfundet.* Jordbrugsforlaget, København.
Harvey, D.R. (1996) Can markets sustain the rural economy? In: Allanson, P.F. and Whitby, M. (eds) *Blueprint for a Rural Economy.* Earthscan, London (in press).
Hees, E., Renting, H. and de Rooij, S. (1994) *Naar Lokale Zelfregulering: Samenwerkingsverbanden voor Integratie van Landbouw, Milieu, Natuur en Landschap.* Circle for Rural European Studies, Ministry of Agriculture, Nature Management and Fisheries, Agricultural University, Wageningen, Wageningen.
Hellekes, R. and Perdelwitz, D. (1986) *Nitratbelastung und Trinkwasser.* IIUG Report 86/2. Wissenschaftzentrum Berlin, Berlin.
Hervieu, B. (1993) *Les Champs du Future.* Bourin, Paris.
Hervieu, B. and Lagrave, R-M. (eds) (1992) *Les Syndicats Agricoles en Europe.* L'Harmattan, Paris.
Hodge, I.D., Adams, W.M. and Bourn, N.A.D. (1992) The cost of conservation: comparing like with like – a comment on Brotherton. *Environment and Planning A* 24, 1051–1054.
Hodge, I.D., Castle, R. and Dwyer, J. (1993) Covenants for conservation: widening the options for control of land. *ECOS* 13 (3), 41–45.
Höll, A. (1996) *Verminderung von Stickstoffüberschüssen aus der Landwirtschaft. Forderungen an Agrar-und Umweltpolitik.* Umweltstiftung WWF-Deutschland (ed.) WWF, Frankfurt (in press).
House of Commons (1985) *Operation and Effectiveness of Part II of the Wildlife and Countryside Act.* 1st Report. Environment Committee, HMSO, London. 1985–85.
House of Commons Environment Committee (1995) *The Environmental Impact of*

Leisure Activities. Volume 1. Fourth Report. HMSO, London.

INEA (1995) *L'applicazione del Regolomento CE 2078/95 in Italia. Quadro di sintesi: de programmi regionali.* Working paper, Istituto Nazionale di Economia Agraria, Roma.

INSEE (1993) *Les Agriculteurs.* INSEE, Paris.

Isermann, K. (1989) *Die Stickstoff- und Phosphoreinträge in die Oberflächengewässer der Bundesrepublik Deutschland durch verschiedene Wirtschaftsbereiche unter besonderer Berücksichtigung der Stickstoff- und Phosphor-Bilanz der Landwirtschaft und der Humanernährung.* Paper presented at a DLG workshop 'Umwelt- und Gesundheitspflege und spezielle Ernährungsfragen in der tierischen Produktion', 14 and 15 March, 1989.

ISTAT (1992) *IV Censimento Generale dell'Agricoltura,* Istituto Nazionale di Statistica, Roma.

ISTAT (1993) *Statistiche Ambientali 1993.* Istituto Nazionale di Statistica, Roma.

Jacques, A. and Blouët, K. (1995) *Valeur des paysages agricoles et tourisme rural.* Unpublished Report to the Direction de la Provision, Ministére de l'Economie et des Finances, Paris.

Janmaat, R., van Woerkum, C.M.J. and ter Keurs, W.J. (1995) Natuurbescherming, landbouw en Communicatie. *Landinrichting* 35 (1), 11–17.

Jordan, J. (1995) Inconsistencies of the CAP reform of 1992. Integration of different policies at regional level required. *Agra-Europe (Germany)* 95 (4), 23 January 1995. Sonderbeilage.

Just, F. (1990) *Co-operatives and Farmers' Unions in Western Europe.* South Jutland University Press, Esbjerg.

Kaule, G. (1988) *Arten-und Biotopschutz,* 2nd edn. UTB, Stuttgart.

Kayser, B. (1994) Angoisses et espérances de la France rurale. *Sciences Humaines* Hors Série 4, 16–21.

Kayser, B., Brun, A., Cavailhes, J. and Lacombe, P. (1994) *Pour une Ruralité Choisie.* Editions de l'Aube, Paris.

Kim, K.C. and Weaver, R.D. (1994) Biodiversity and humanity: paradox and challenge. In: Kim, K.C. and Weaver, R.D. (eds) *Biodiversity and Landscapes: A Paradox of Humanity.* Cambridge University Press, Cambridge.

Knickel, K., Hoffman, H-J. and Höll, A. (1994) *Ex-post Evaluation in Objective 5b Region (Hessen).* Report to the European Commission, DG VI. Institut für ländliche Strukturforschung, Frankfurt/Main.

Land Use Consultants (1995) *Countryside Stewardship Monitoring and Evaluation: third interim report.* Countryside Commission, Cheltenham.

Landbrugsministeriet (1994) *Jordbrug* 1993. København.

Le Goffe, P. and Gerber, P. (1994) *Coûts environnementaux et bénefices de l'implantation d'une sablière en zone périurbaine: le cas de Pacé.* Unpublished report, INRA, Rennes.

Ledru, M. (1994) *L'Espace Rural: Entre Protection et Contraintes.* Report to the French Government by the Conseil Economique et Social, Paris.

Leferink, J. (1994) *Financiele Ondersteuning Biologische Landbouwbedrijven.* IKC Informatie Akkerbouw en Vollegrondsgroente.

Leuba, M. and Simonet, E. (1992) L'article 19 et la dynamique de l'environment. *Amenagement et Nature* 105, 28–29.

Lévy, J. (1994) Oser le désert. *Sciences Humaines* Hors Série 4, 6–12.

Leynaud, E. (1985) *L'État et la Nature : l'Exemple des Parcs Nationaux Français.* Parc National des Cevennes, Florac.

Liefferink, J.D., Lowe, P.D. and Mol, A.P.J. (eds) (1993) *European Integration and Environmental Policy*. Belhaven, London.

Lockeretz, W. (1988) Open questions in sustainable agriculture. *American Journal of Alternative Agriculture* 4, 174–181.

Lowe, P. and Goyder, J. (1983) *Environmental Groups and Politics.* Allen and Unwin, London.

Lowe, P.D. (ed.) (1992) Industrial agriculture and environmental regulation. *Sociologia Ruralis* 32 (1), 1–188.

Lowe, P.D. and Buller, H. (1990) The historical and cultural contexts. In: Lowe, P.D. and Bodiguel, M. (eds) *Rural Studies in Britain and France.* Belhaven, London.

Lowe, P.D., Cox, G., MacEwen, M., O'Riordan, T. and Winter, P. (1986) *Countryside Conflicts: the Politics of Farming, Forestry and Conservation.* Gower, Aldershot, UK.

Lowe, P.D., Marsden, T. and Whatmore, S. (eds) (1994) *Regulating Agriculture.* Wiley, Chichester, UK.

McCracken, D.I., Bignal, E.M. and Wenlock, S.E. (eds) (1995) *Farming on the Edge: the Nature of Traditional Farmland in Europe.* Joint Nature Conservation Committee, Peterborough, 216pp.

Madelin, V. (1994) La remunération des externalités. *Economie Rurale* 220/221, 215–217.

MAFF (1990) *UK ESA Statistics.* HMSO, London

MAFF (1991) *Ministerial Information in MAFF (MINIM)* 1990. HMSO, London.

MAFF (1992) *Ministerial Information in MAFF (MINIM)* 1991. HMSO, London.

MAFF and Intervention Board (1995) *The Government's Expenditure Plans 1995–96 to 1997–1998.* HMSO, London.

MAFF (Ministry of Agriculture, Fisheries and Food) (1995) *Environmental Land Management Schemes : Consultation Document.* MAFF, London

MAFF Working Group on CAP Reform (1995) *European Agriculture – the Case for Radical Reform.* MAFF, London.

Mahé, L. and Ranelli, P. (1987) Les mesures agri-environnementales et les politiques agricoles sur l'environnement. *Cahiers d'Economie et Sociologie Rurales* 4, 9–31.

Marsh, J. (1991) *The Changing Role of the Common Agricultural Policy.* Belhaven, London.

Massot Martí, A. (1988) La reforma de la política estructural agrícola de la Comunidad. *Agricultura y Sociedad* 49, 49–119.

Mateu, E. (1992) Agricultura y medioambiente. *Revista Valenciana d'Estudios Autonomics* 14, 147–168.

Mazey, S. and Richardson, J.J. (1992) Environmental groups and the EC: challenges and opportunities. *Environmental Politics* 1 (4) 110–128.

Mazey, S. and Richardson, J.J. (1993) EC policy making: an emerging European policy style. In: Liefferink, J.D., Lowe, P.D. and Mol, A.P.J. (eds) *European Integration and Environmental Policy*. Belhaven, London, pp. 114–125.

Melman, T.C.P. and Buitink, G.H.M. (1994) De relatienota-uitvoering, kritiek en perspectief. *Landrichting* 34 (8), 15–20.

Merlo, M. (1991) The effects of late economic development on land use. *Journal of Rural Studies* 4, 445–457.

Merlo, M. and Boscolo, M. (1994) L'evoluzione nell'uso dei suoli fra intensificazione ed abbandono: una misura dell'indice di concentrazione della produzione agricola. In: Cannata, G. and Merlo, M. (eds) *Interazione fra Agricoltura e Ambiente in Italia.* Il Mulino, Bologna, pp. 11–32.

Merlo, M. and Della Puppa, F. (1994) Public benefits valuation in Italy: a review of forestry and farming applications. In: Dubgaard, A., Bateman, I. and Merlo, M. (eds) *Economic Valuation of Benefits from Countryside Stewardship: Proceedings of a Workshop Organised by the Commission of the European Communities.* Directorate General for Agriculture, Brussels, 7–8 June 1993, Wissenschaftverlag Vauk, Kiel.

Métais, M. (1992) The role of ESAs in nature conservation in France. In: Dixon J. (ed.) *A Future for Europe's Farmed Countryside.* RSPB, Sandy, UK, pp. 95–104.

Meyer, H.v. (1991) Natur und Agrarpolitik. *LÖLF-Mitteilungen* 91(1), 15–21.

Meyer, H.v. (1993) Agriculture and the environment in Europe. In: Group of Sesimbra (ed.) *The European Common Garden – Towards the Buiding of a Common Environmental Policy.* Group of Sesimbra, Brussels, pp. 69–86.

Ministère de l'Agriculture (1990) *Propositions d'Expérimentations en France de l'Article 19. Document d'Intention.* Paris.

Ministère de l'Agriculture (1991) *Article 19 : Etat d'Avancement des Projets Experimentaux au 1 Septembre 1991.* Paris.

Ministère de l'Agriculture (1992) *Note pour la Commission : Mise en Oeuvre du Réglement no. 2078/92.* Paris.

Ministère de l'Agriculture (1993) *Circulaire DEPSE 7010, 26 March 1993.*

Ministère de l'Agriculture (1994a) *Mise en Oeuvre des Mesures Agri-environnementales.* Circulaire DEPSE SDSEA C.94 No. 7004, 1 February 1994.

Ministère de l'Agriculture (1994b) *Gestion des Programmes Régionaux Agri-environment: Programmation.* Circulaire DEPSE/SDSEA 94-7046, 23 December 1994.

Ministère de l'Agriculture (1994c) *Mise en oeuvre des programmes 'agri-environnementales': lancement des opérations et souscriptions de contrats.* Circulaire DEPSE/SDSEA C.94 No. 7005, 1 February 1994.

Ministère de l'Agriculture (1994d) *Mise en oeuvre des programmes 'agri-environnementales': Cahiers des Charges.* Circulaire DEPSE/SDSEA C.94 No. 7006, 1 February 1994.

Ministère de l'Environnement (1989) *L'Etat de l'Environnement.* Paris.

Ministerio de Agricultura Pesca y Alimentacíon (MAPA) (1994) *Programas de Ayudas para Fomentar Métodos de Producción Agraria Compatibles con las Exigencias de la Protección y la Conservación del Espacio Natural.* Madrid.

Ministry of the Environment (1994) *The Danish Environmental Strategy.* Ministry of the Environment.

Montgolfier, J. (1990) *Coûts et Avantages d'une Agriculture Compatible avec les Exigences de l'Environnement.* CEMAGREF, Paris.

Moreux, I. (1994) Agriculture et Environnement. *Structures* 5, 45–50.

Mormont, M. (1994) La agricultura en el espacio rural europeo. *Agricultura y Sociedad* 71, 17–49.

Moussset, S. (1992) Protection de la nature et agriculture dans le Parc national des Cévennes: bilan d'une expérience insolite. *Annales du Parc National des Cévennes* 5, 223–243.

Moyano, E. (1988) *Sindicalismo y Política Agraria en Europa.* Serie Estudios del MAPA, Madrid.

Moyano, E. (1993) *Las organizaciones profesionales agrarias en la CEE.* Serie Estudios del MAPA, Madrid.

National Audit Office (1994) *Protecting and Managing Sites of Special Scientific Interest in England.* HMSO, London.

Neville-Rolfe, E. (1983) *The Politics of Agriculture in the European Community*. Croom Helm, London.

Norton, B.G. (1994) Thoreau and Leopold on science and values. In: Kim, K.C. and Weaver, R.D. (eds) *Biodiversity and Landscapes, A Paradox of Humanity.* Cambridge University Press.

Olwig, K. (1984) *Nature's Ideological Landscape.* George Allen and Unwin, London.

Oosterveld, E.B. (1994) *Beleid in Strijd – Knelpunten in het Milieu – en Natuurbeleid voor de Land-en Tuinbouw.* Centre for Agriculture and Environment, Utrecht.

Osti, G. (1992) Co-operative regulation: contrasting organizational models for the control of pesticides. The case of North-East Italian Fruit-Growing. *Sociologia Ruralis* 1, 163–177.

Plankl, R. (1995) *Synopse zu den umweltgerechten und den natürlichen Lebensraum schützenden landwirtschaftlichen Produktionsverfahren als flankierende Maßnahmen zur Agrarreform. Tabellarische Übersicht über die einzelnen Umweltprogramme gemaß VO (EWG) 2078/92.* Institut für Strukturforschung, Braunschweig/Völkenrode.

Poppe, K.J., Brouwer, F.M., Welten, J.P.P.J. and Wijnands, J.H.M. (1994) *Landbouw, Milieu en Economie.* Agricultural Economic Institute, The Hague.

Primdahl, J. (1991) Countryside planning. In: Hansen, B. and Jorgensen, S.E. (eds) *Introduction to Environmental Management.* Elsevier, Amsterdam, pp. 275–300.

Primdahl, J. (1992) Miljofolsomme omrader – en kortlaegning af milfofolsomme omrader, braklaegning og det vedvarende graesareal. In: Asbirk (ed.) *Naturen pa Landet.* The National Forest and Nature Agency, the Ministry of the Environment, pp. 73–79.

Primdahl, J. and Hansen, B. (1993) Agriculture in environmentally sensitive areas: implementing the ESA measure in Denmark. *Journal of Environmental Planning and Management* 36 (2), 231–238.

Rahmann, H. and Kohler, A. (eds) (1991) *Tier- und Artenschutz.* Margraf, Weikersheim.

Rasmussen, J.D., Jensen, S.P., Bjorn, C. and Christensen, J. (1988) Det danske landbrugs historie 1860–1914. In: Bjorn, C. (ed.) *Det Danske Landbrugs Historie bd III.* Landbohistorik Selskab, Odense, pp. 193–430.

Rebolledo, D. and Perez y Perez, L. (1994) *Valoracion contingente de bienes ambientales: aplicacion al Parque Nacional de la Dehesa del Moncayo.* Gobierno de Aragon, Departamento de Agricultura, Ganaderia y Montes.

Rennie, F.W. (1991) Environmental objectives in land-use practices. In: Curtis, D.J., Bignal, E.M. and Curtis, M.A. (eds) *Birds and Pastoral Agriculture in Europe.* Scottish Chough Study Group, Argyll and Joint Nature Conservation Committee, Peterborough, pp. 101–105.

Reus, J.A.W.A. (1994) Pesticide reduction strategies in Europe: towards a comprehensive and stimulating pesticide policy. In: Suckling, D.M. and Popay, A.J. (ed.) *Plant Protection: Cost, Benefits and Trade Implications.* New Zealand Plant Protection Society, Christchurch, New Zealand.

Riera, P. (1994) Amenity aspects of Spanish countryside policy: proceedings of a workshop organised by the Commission of the European Communities Directorate General for Agriculture, Brussels, 7–8 June 1993. In: Dubgaard, A., Bateman, I. and Merlo, M. (eds) *Economic Valuation of Benefits from Countryside Stewardship*

Wissenschaftsverlag Vauk Kiel KG, Germany.

Rijks Planologische Dienst (1993) *Ruimtelijke Verkenningen 1993.* Ministerie van Volkshuisvesting, Ruimtelijke Ordening en Milieu, The Hague.

Rodgers, C.P. (1992) Land management agreements and agricultural practice: towards an integrated legal framework for conservation law. In: Howarth, W. and Rodgers, C.P. (eds) *Agricultural Conservation and Land Use: Law and Policy Issues for Rural Areas.* University of Wales Press, Cardiff, pp. 139–164.

Saunders, C.M. (1995) *Subsidies without Farming.* Department of Economics and Marketing, Lincoln University, New Zealand 15pp.

Scheele, M. (1994) Reform der Gemeinsamen Agrarpolitik. Konsequenzen für die Agrarstruktur-und die Agrarumweltpolitik. In: Schriftenreihe der FAA (ed.) *Forschungsgesellschaft für Agrarpolitik und Agrarsoziologie.* Band 300, FAA, Bonn.

Schou, A. (1949) *Atlas of Denmark: The Landscapes.* The Royal Danish Geographical Society, Hagerup, Copenhagen.

Scott, A., Peterson, J. and Millar, D. (1994) Subsidiarity: A Europe for the Regions' v the British Constitution. *Journal of Common Market Studies* 32 (1), 47–67.

Semaille, M.L. (1994) *Les mesures agri-environnmentales: analyse socio-économique en province de Luxembourg.* Unité d'Economie Rurale, Université Catholique de Louvain.

Shoard, M. (1980) *The Theft of the Countryside.* Temple Smith, London.

Sillani, S. (1987) Land consolidation and environment. In: Merlo, M., Stellin, G., Harou, P. and Whitby, M. (ed.) *Multipurpose Agriculture and Forestry: Proceedings of the 11th Seminar of the EAAE, 28 April–3 May, 1986.* Wissenschaftsverlag Vauk, Kiel.

Struktur-og Planudvalget (1987) *1. Delbetænkning om gennemførelse af EF's socio-strukturelle foranstaltninger i Danmark.* Betænkning nr. 1122, Landbrugsministeriet, København.

Swedish Environmental Protection Agency (1993a) *Jordbruk och Miljö. Underlagsrapport till Aktionsprogram Miljö'93.* Rapport 4208. Solna, Sweden.

Swedish Environmental Protection Agency (1993b) *Markanvändningen och Miljön. Miljön i Sverige – tillstånd och Trender.* Rapport 4137. Solna, Sweden.

Swedish Environmental Protection Agency (1993c) *Strategy for Sustainable Development – Proposals for a Swedish Programme.* Report 4234. Solna, Sweden.

Swedish Environmental Protection Agency (1994a) *Fördjupad Utvärdering av Åtta Lånsstyrelsers Arbete 1990–1993 med LANDSKAPSVÅRD och NOLA.* LiM Rapport 4353. Solna, Sweden.

Swedish Environmental Protection Agency (1994b) *Livsmedelspolitikens Miljöeffekter. LiM-Projektets Rapport 1994 till Regeringen.* LiM Rapport 4364. Solna, Sweden.

Swedish Parliament (1990) *Regeringens Proposition 1989/90: 146 om Livsmedelspolitiken.* Stockholm, Sweden.

Tábara, D. (1995) Percepció pública i acció social en problemes de medi ambient. PhD thesis, Universidad Central de Barcelona.

Tacet, D. (1992) *Un Monde sans Paysans.* Hachette, Paris.

Tamames, R. (1991) *Introducción a la Economía Español.* Alianza Editorial, Madrid.

Termeer, C.J.A.M. (1993) *Dynamiek en Inertie Rondom Mestbeleid – een Studie naar Veranderingsprocessen in het Varkenshouderijnetwerk.* Erasmus University, Rotterdam.

Terrasson, F. (1988) *La Peur de la Nature.* Sang de la Terre, Paris.

Terwan, P. (1992) *Boeren met Natuur – een Verkenning van Kansen voor Natuur op Landbouwbedrijven.* Centre for Agriculture and Environment, Utrecht.

Tinbergen, J. (1964) *Economic Policy: Principles and Design.* North Holland, Amsterdam.

Tió, C. (1991) La reforma de la PAC, desde la perspectiva de las agriculturas del Sur de la CEE. *Revista de Estudios Agrosociales* 156, 41–66.

Trap, J.P. (1958) *Danmark. Bd. I, Landet og Folket.* G.E.C. Gads Forlag.

Trigilia, C. (1992) *Sviluppo senza Autonomia.* Il Mulino, Bologna.

Trommetter, R. (1994) La rationalisation des contrats entre pouvoirs publics et agriculteurs. *Reformer la Politique Agricole Commune, INRA Actes et Communications* 12, 307–323.

Umweltbundesamt (UBA) (ed.) (1994) *Stoffliche Belastung der Gewässer durch die Landwirtschaft und Maßnahmen zu ihrer Verringerung.* Erich Schmidt, Berlin.

van Bruchem, C. (1994) *Landbouw Economisch Bericht 1994.* Agricultural Economics Research Institute (LEI-OLO), The Hague.

van der Weijden, W.J. (1995) Policy instruments to integrate the environment into the CAP. In: Reus, J.A.W.A., Mitchell, K., Klauer, C.J.G.M. and Baldcock, D.M. (eds), *Greening the CAP: Report of the European Conference on the CAP, Environment, Nature Conservation and Rural Development, Ghent, 10–11 October 1994.* IEEP and CLM, Utrecht.

van Vuuren, W., Larue, B. and Ketchebaw, E H. (1993) *Logistic Estimation of Farm Practices on Rented Land. Reprint of: Serie Reserche No. 29, Groupe de recherche agro-alimentaire.* Department d'Economie Rurale, Université Laval.

Vera, F. and Romero, J. (1994) Impacto ambiental de la actividad agraria. *Agricultura y sociedad* 71, 153–181.

Viard, D. (1990) *Le Tiers Espace: Essai sur la Nature.* Méridiens Klincksieck, Paris.

Vieira, M. (1992) Environmentally sensitive areas in Portugal. In: Dixon J. (ed.) *A Future for Europe's Farmed Countryside.* RSPB, Sandy, UK, pp. 149–155.

Vourc'h, A. and Pelosse, V. (1985) Chasseurs et protecteurs: les paradoxes d'une contradiction. In: Cadoret, A. (ed.) *Protection de la Nature: Histoire et Idéologie.* L'Harmattan, Paris, pp. 108–123.

Walsh, R.G., Sanders, L.D. and Loomis, J.B. (1985) *Wild and Scenic River Economics: Recreation Use and Preservation Values.* American Wilderness Alliance, Englewood, Colorado, USA.

Ward, N. and Munton, R. (1992) Conceptualizing Agriculture–Environment Relations. Combining Political Economy and Socio-Cultural Approaches to Pesticide Pollution. *Sociologia Ruralis* 32 (1), 127–145.

Ward, N., Buller, H. and Lowe, P. (1995) *Implementing European Environmental Policy at the Local Level: The British Experience with Water Quality Directives.* 2 vols. Centre for Rural Economy Research Report, University of Newcastle upon Tyne.

Weaver, R.D. (1994) Market-based economic development and biodiversity: an assessment of conflict. In: Kim, K.C. and Weaver, R.D. (eds) *Biodiversity and Landscapes: a Paradox of Humanity.* Cambridge University Press, Cambridge.

Weaver, R.D. (1996) *Prosocial Behaviour: Private Contributions to Agriculture's Impact on the Environment* (in press).

Weaver, R.D. and Abrahams, N. (1994) *Information and Education: Their Roles in Adoption Decisions for Environmentally Beneficial Agricultural Practices (EBAPS).* Presented at the annual meeting of the American Agricultural Economics Association, Baltimore, 1993.

Weaver, R.D. and Kim, K.C. (1994) Biodiversity and humanity: toward a new paradigm. In: Kim, K.C. and Weaver, R.D. (eds) *Biodiversity and Landscapes: a Paradox of Humanity*. Cambridge University Press, Cambridge.

Whitby, M. (1989) Environmental application of Article 19 of Economic Directive 797/85 in UK. In: Dubgaard, A. and Nielsen, A. (eds) *Economic Aspects of Environmental Regulation in Agriculture.* Wissenschaftverlag Vauk, Kiel.

Whitby, M. (ed.) (1994) *Incentives for Countryside Management: the Case of Environmentally Sensitive Areas.* CAB International, Wallingford, UK.

Whitby, M. (1996) *Management Mechanisms for Countryside Conservation, Working Paper*, Centre for Rural Economy, The University of Newcastle upon Tyne.

Whitby, M. and Lowe, P. (1994) The political and economic roots of environmental policy in agriculture. In: Whitby, M. (ed.) *Incentives for Countryside Management: the Case of Environmentally Sensitive Areas.* CAB International, Wallingford, UK.

Whitby, M.C. and Saunders, C.M. (1994) *Estimating the Supply of Conservation Goods*, Centre for Rural Economy, Department of Agricultural Economics and Food Marketing, University of Newcastle upon Tyne.

Whitby, M.C., Coggins, G. and Saunders, C.M. (1990) *Alternative Payment Systems for Management Agreements.* Nature Conservancy Council, Peterborough.

Willis, K.G. and Benson, J.F. (1993) Valuing Environmental Assets in Developed Countries. In: Turner, R.K. (ed.) *Sustainable Environmental Economics and Management*. Belhaven, London.

Willis, K.G. and Garrod, G.D. (1993) Valuing landscape: a contingent valuation approach. *Journal of Environmental Management* 37, 1–22.

Willis, K.G. and Garrod, G.D. (1994) The ultimate test: measuring the benefits of ESAs. In: Whitby, M. (ed.) *Incentives for Countryside Management: the Case of Environmentally Sensitive Areas.* CAB International, Wallingford, UK.

Willis, K.G., Garrod, G D. and Saunders, C.M. (1993) *Valuation of the South Downs and Somerset Levels Environmentally Sensitive Area Landscapes by the General Public; Summary Report.* Centre for Rural Economy, University of Newcastle upon Tyne.

Withrington, D. and Jones, W. (1992) The Enforcement of Conservation Legislation. In: Howarth, W. and Rodgers, C.P. (eds) *Agriculture Conservation and Land Use: Law and Policy Issues for Rural Areas.* University of Wales Press, Cardiff, pp. 90–107.

World Wildlife Fund for Nature (1994) *Dossier Economia e Parchi.* WWF Italia, Roma.

Wulff, H. (1991) Danish environmental law. In: Hansen, P.E. and Jorgensen, S.E. (eds) *Introduction to Environmental Management. Developments in Environmental Modelling 18*. Elsevier, Amsterdam.

Appendix: Council Regulation (EEC) No. 2078/92

of 30 June 1992

on agricultural production methods compatible with the requirements of the protection of the environment and the maintenance of the countryside

THE COUNCIL OF THE EUROPEAN COMMUNITIES.

Having regard to the Treaty establishing the European Economic Community, and in particular Articles 42 and 43 thereof.

Having regard to the proposal from the Commission [1],

Having regard to the opinion of the European Parliament [2],

Having regard to the opinion of the Economic and Social Committee [3],

Whereas the requirements of environmental protection are an integral part of the common agricultural policy;

Whereas measures to reduce agricultural production in the Community must have a beneficial impact on the environment;

Whereas many factors affect the environment; whereas it is subject to very diverse pressures within the Community;

Whereas an appropriate aid scheme would encourage farmers to serve society as a whole by introducing or continuing to use farming practices compatible with the increasing demands of protection of the environment and natural resources and upkeep of the landscape and the countryside;

Whereas the introduction of an aid scheme to encourage substantial reductions in the use of fertilizers and plant-protection products or the use of organic farming methods can help not only to reduce agricultural pollution but also to adapt a number of sectors to market requirements by encouraging less intensive production methods;

Whereas a reduction in farm livestock or in animal proportion per hectare can help to avert environmental damage due to pressure from excessive numbers of sheep and cattle; whereas, therefore, the extensification scheme for various products provided for in Article 3 of Council Regulation (EEC) No 2328/91 of 15 July 1991 on improving the efficiency of agricultural structures [4] should be incorporated in the scheme introduced under this Regulation;

Whereas the production of products for non-food uses under a Community set-aside scheme must comply with the requirememts of environmental protection; whereas, therefore, this scheme must not apply to such products;

Whereas a scheme to encourage the introduction or maintenance of particular farming practices may help to solve specific problems related to protection of the environment or the

[1] OJ No C 300, 21. 11. 1991, p. 7.
[2] OJ No C 94, 13. 4. 1992.
[3] OJ No C 98, 21. 4. 1992, p. 25.

[4] OJ No L 218, 6. 8. 1991, p. 1.

countryside and thus contribute to environmental policy goals;

Whereas many agricultural and rural areas in the Community are increasingly threatened by depopulation, soil erosion, flooding and forest fires; whereas the institution of special measures to encourage the upkeep of land can reduce such risks;

Whereas because of the scale of the problems such schemes should be applicable to all farmers in the Community who undertake to use farming methods which will protect, maintain or improve the environment and the countryside and to refrain from further intensification of agricultural production;

Whereas the current set-aside scheme for arable land provided for in Article 2 of Regulation (EEC) No 2328/91 has been replaced by provisions in the regulations covering the common organization of the markets; whereas it appears nonetheless appropriate to introduce a scheme for long-term set-aside of agricultural land for environmental reasons and for the protection of natural resources;

Whereas the measures provided for in this Regulation must encourage farmers to make undertakings regarding farming methods compatible with the requirements of environmental protection and maintenance of the countryside, and thereby to contribute to balancing the market; whereas the measures must compensate farmers for any income losses caused by reductions in output and/or increases in costs and for the part they play in improving the environment;

Whereas the introduction by the Member States of codes of good agricultural practice can also help to make farming practices more compatible with the requirements of environmental protection;

Whereas the diversity of the environment, natural conditions and the structure of agriculture in the various parts of the Community call for the measures provided for to be adapted; whereas they should therefore be implemented within the framework of zonal programmes for the management of agricultural or abandoned land and possibly as part of national regulations;

Whereas both the Community and the Member States must increase their effort to educate farmers in, and inform them of, the introduction of agricultural and forestry production methods compatible with the environment, and in particular regarding the application of a code of good farming practice and organic farming.

Whereas, in order to guarantee the maximum effectiveness of such programmes, it is vital to ensure that the results are disseminated and monitored regularly;

Whereas such measures must contribute towards certain specific environmental goals set out in Community legislation;

Whereas, given that the Community is to contribute to the financing of the scheme, it must be able to ascertain that the implementing arrangements adopted by the Member States contribute towards the attainment of its objectives; whereas the structure of cooperation between the Member States and the Commission introduced by Article 29 of Regulation (EEC) No 4253/88 of 19 December 1988, laying down provisions for implementing Regulation (EEC) No 2052/88 as regards coordination of the activities of the different Structural Funds between themselves and with the operations of the European Investment Bank and the other existing financial instruments ([1]), should be used for this purpose;

Whereas the resources available for implementing the measures provided for in this Regulation must be additional to those available for the implementation of measures under the rules governing the Structural Funds, and in particular for measures applicable in regions covered by Objectives 1 and 5 (b) as defined in Article 1 of Regulation (EEC) No 2052/88 ([2]),

HAS ADOPTED THIS REGULATION:

Article 1

Purpose of the aid scheme

A Community aid scheme part-financed by the Guarantee Section of the European Agricultural Guidance and Guarantee Fund (EAGGF) is hereby instituted in order to:

— accompany the changes to be introduced under the market organization rules,

([1])OJ No L 374, 31. 12. 1988, p. 1.
([2])OJ No L 185, 15. 7. 1988, p. 9.

— contribute to the achievement of the Community's policy objectives regarding agriculture and the environment,

— contribute to providing an appropriate income for farmers.

This Community aid scheme is intended to promote:

(a) the use of farming practices which reduce the polluting effects of agriculture, a fact which also contributes, by reducing production, to an improved market balance;

(b) an environmentally favourable extensification of crop farming, and sheep and cattle farming, including the conversion of arable land into extensive grassland;

(c) ways of using agricultural land which are compatible with protection and improvement of the environment, the countryside, the landscape, natural resources, the soil and genetic diversity;

(d) the upkeep of abandoned farmland and woodlands where this is necessary for environmental reasons or because of natural hazards and fire risks, and thereby avert the dangers associated with the depopulation of agricultural areas;

(e) long-term set-aside of agricultural land for reasons connected with the environment;

(f) land management for public access and leisure activities;

(g) education and training for farmers in types of farming compatible with the requirements of environmental protection and upkeep of the countryside.

Article 2

Aid scheme

1. Subject to positive effects on the environment and the countryside, the scheme may include aid for farmers who undertake:

(a) to reduce substantially their use of fertilizers and/or plant protection products, or to keep to the reduction as already made, or to introduce or continue with organic farming methods;

(b) to change, by means other than those referred to in (a) to more extensive forms of crop, including forage, production, or to maintain extensive production methods introduced in the past, or to convert arable land into extensive grassland;

(c) to reduce the proportion of sheep and cattle per forage area;

(d) to use other farming practices compatible with the requirements of protection of the environment and natural resources, as well as maintenance of the countryside and the landscape, or to rear animals of local breeds in danger of extinction;

(e) to ensure the upkeep of abandoned farmland or woodlands;

(f) to set aside farmland for at least 20 years with a view to its use for purposes connected with the environment, in particular for the establishment of biotope reserves or natural parks or for the protection of hydrological systems;

(g) to manage land for public access and leisure activities.

2. In addition, the scheme may include measures to improve the training of farmers with regard to farming or forestry practices compatible with the environment.

Article 3

Aid programmes

1. Member States shall implement, throughout their territories, and in accordance with their specific needs, the land scheme provided for in Article 2 by means of multiannual zonal programmes covering the objectives referred to in Article 1. The programmes shall reflect the diversity of environmental situations, natural conditions and agricultural structures and the main types of farming practised, and Community environment priorities.

2. Each zonal programme shall cover an area which is homogeneous in terms of the environment and the countryside and shall include, in principle, all of the aids provided for in Article 2. However, where there is sufficient justification, programmes may be restricted to aids which are in line with specific characteristics of an area.

3. Zonal programmes shall be drawn up for a

minimum period of five years and must contain at least the following information:

(a) a definition of the geographical area and, where applicable, the sub-areas concerned;

(b) a description of the natural, environmental and structural characteristics of the area;

(c) a description of the proposed objectives and their justification in view of the characteristics of the area, including an indication of the Community environment legislation the objectives of which the programme seeks to fulfil;

(d) the conditions for the grant of aid, taking into account the problems encountered;

(e) an estimate of annual expenditure for implementing the zonal programme;

(f) the arrangements made to provide appropriate information for agricultural and rural operators.

4. By way of derogation from paragraphs 1, 2 and 3, Member States may establish a general regulatory framework providing for the horizontal application throughout their territory of one or more of the aids referred to in Article 2. That framework must be defined and, where appropriate, supplemented by the zonal programmes referred to in paragraph 1.

Article 4

Nature and amounts of aid

1. An annual premium per hectare or livestock unit removed from a herd shall be granted to farmers who give one or more of the undertakings referred to in Article 2 for at least five years, in accordance with the programme applicable in the zone concerned. In the case of set-aside, the undertaking shall be for 20 years.

2. The maximum eligible amount of premium shall be:

— ECU 150 per hectare for annual crops for which a premium per hectare is granted under the market regulations governing the crops in question,

— ECU 250 per hectare for other annual crops and pasture,

— ECU 210 for each sheep or cattle livestock unit by which a herd is reduced,

— ECU 100 for each livestock unit of an endangered breed reared,

— ECU 400 per hectare for specialized olive groves,

— ECU 1 000 per hectare for citrus fruits,

— ECU 700 per hectare for other perennial crops and wine,

— ECU 250 per hectare for the upkeep of abandoned land,

— ECU 600 per hectare for land set aside,

— ECU 250 per hectare for the cultivation and propagation of useful plants adapted to local conditions and threatened by genetic erosion.

The table for converting animals into livestock units is given in the Annex.

3. The maximum eligible amount for annual crops and pasture shall be increased to ECU 350 per hectare if the farmer has, at the same time and for the same area, given one or more of the undertakings referred to in Article 2 (1) (a) and (b), together with an undertaking as referred to in Article 2 (1) (d).

4. Where a premium is granted for the reduction of the number of livestock units:

— the aids provided for in Article 2 (1) (a) and (b) may not be granted for the forage area of the holding,

— the maximum eligible amount of premium for forage areas under Article 2 (1) (d) shall be reduced by 50%.

5. Subject to conditions to be determined by the Commission in accordance with the procedure laid down in Article 29 of Regulation (EEC) No 4253/88, the Community may also contribute to the premiums referred to in the preceding paragraphs which are granted by Member States in order to compensate for income losses resulting from the mandatory application of the restrictions referred to in Article 2 in the context of measures implemented in the Member States pursuant to Community provisions.

6. Member States may stipulate that a farmer's undertaking may be given in the context of an overall plan for the entire holding or for a part thereof.

In such cases, the amount of the aid may be

calculated as an overall figure taking account of the individual amounts and conditions in this Article and Article 5.

Article 5

Conditions of grant

1. In order to achieve the objectives of this Regulation in the context of the general rules referred to in Article 3 (4) and/or the zonal programmes, Member States shall determine:

(a) the conditions for granting aid;

(b) the amount of aid to be paid, on the basis of the undertaking given by the beneficiary and of the loss of income and of the need to provide an incentive;

(c) the terms on which the aid for the upkeep of abandoned land as referred to in Article 2 (1) (e) may be granted to persons other than farmers, where no farmers are available;

(d) the conditions to be met by the beneficiary to ensure that compliance with the undertakings may be verified and monitored;

(e) the terms on which the aid may be granted where the farmer personally is unable to give an undertaking for the minimum period required.

2. No aid may be granted under this Regulation in respect of areas subject to the Community set-aside scheme which are being used for the production of non-food products.

3. While ensuring that the incentive content of the measure is retained, Member States may restrict the aid to a maximum amount per holding and differentiate it according to holding size.

Article 6

Courses, traineeships and demonstration projects

1. Where no financing is granted under Article 28 of Regulation (EEC) No 2328/91, Member States may introduce a separate aid scheme for training courses and traineeships concerned with agricultural and forestry production practices compatible with the requirements of protection of the environment and natural resources and maintenance of the countryside and the landscape, and particularly with codes of good farming practice or good organic farming practice. The aid scheme shall include the grant of aid:

— for attendance of courses and traineeships,

— for the organization and implementation of courses and traineeships.

The expenditure incurred by the Member States in granting the aid referred to in the first subparagraph shall be eligible up to ECU 2 500 per person completing a full course or traineeship.

The measure concerned by this Article shall not cover courses or traineeships which are part of normal programmes or curricula of secondary or higher agricultural education.

2. The Community may contribute to demonstration projects concerning farming practices compatible with the requirements on environmental protection, and in particular the application of a code of good farming practice and organic farming practice.

The Community contribution referred to in the first subparagraph may cover assistance for training and education initiatives (including materials) organized by local or non-governmental organizations competent in this field.

Article 7

Programme appraisal procedure

1. Member States shall communicate to the Commission, by 30 July 1993 the draft general regulatory framework referred to in Article 3 (4) and the draft programmes referred to in Article 3 (1) and any existing or proposed laws, regulations or administrative provisions by which they intend to apply this Regulation.

2. The Commission shall examine the texts communicated in order to determine:

their compliance with this Regulation, taking account of its objectives and the links between the various measures,

the nature of the measures eligible for part-financing,

the total amount of expenditure eligible for part-financing.

3. The Commission shall decide on the approval of the general regulatory framework and zonal programmes, on the basis of the factors listed in paragraph 2 and in accordance with the procedure laid down in Article 29 of Regulation (EEC) No 4253/88.

Article 8

Rate of Community financing

The rate of Community part-financing shall be 75% in regions covered by the objective defined in point 1 of Article 1 of Regulation (EEC) No 2052/88 and 50% in the other regions.

Article 9

Detailed rules of application

Detailed rules for the application of this Regulation shall be adopted by the Commission in accordance with the procedure laid down in Article 29 of Regulation (EEC) No 4253/88.

Article 10

Final provisions

1. This Regulation shall not preclude Member States from implementing, except in the field of application of Article 5 (2), additional aid measures for which the conditions of granting aid differ from those laid down herein or the amounts of which exceed the limits stipulated herein, provided that the said measures comply with the objectives of this Regulation and with Articles 92, 93 and 94 of the Treaty.

2. Three years after the date of entry into force in the Member States, the Commission shall present to the European Parliament and the Council a report on the application of this Regulation.

Article 11

Transitional provisions

Application of the measures referred to in Article 39 of Regulation (EEC) No 2328/91 shall be extended with the following effect:

1. Article 3 of Regulation (EEC) No 2328/91, dealing with extensification of production, shall remain applicable until the entry into force of the zonal programmes referred to in Article 3 (1) of this Regulation or of the general regulatory framework referred to in the said Article 3 (4).
2. Articles 21 to 24 of Regulation (EEC) No 2328/91, dealing with aid in environmentally sensitive areas, shall remain applicable until the entry into force of the zonal programmes referred to in Article 3 (1) of this Regulation or of the general regulatory framework referred to in the said Article 3 (4).

   The maximum eligible amounts for the remaining annual payments shall be adjusted in line with the ceilings provided for in Article 4.

Article 12

Entry into force

This Regulation shall enter into force on the day of its publication in the *Official Journal of European Communities.*

This Regulation shall be binding in its entirety and directly applicable in all Member States.

Done at Luxembourg, 30 June 1992

For the Council
The President
Arlindo MARQUES CUNHA

Author Index
(first author only)

Subject Index

In preparing this index the following conventions have been adopted:-

Matters of importance for individual countries have been blocked together under the country's name – see France, Italy etc.

Some frequently recurring themes (e.g. ESAs, extensification and management agreements) are not fully indexed.

European Union (Community) legislation is all listed under the relevant Regulation. References to specific Articles within each Regulation (e.g. Article 19 of 797/85) are also linked with the Regulation entries.

In the specific case of Regulation 2078/92, the subject of this book, only references to its impact on specific countries are included, in the country blocks, where there is some particular local relevance.

3 8002 02274 211 0

Give Me Tomorrow

By the same author

Wives and Mothers
The Long Way Home
Oranges and Lemons
This Year, Next Year
The Lost Daughters
Thursday's Child
Eve's Daughter
King's Walk
Pride of Peacocks
All That I Am
The Happy Highways
Summer Snow
Wishes and Dreams
The Wise Child
You'll Never Know . . .
Should I Forget You
Falling Star
Too Late to Paint the Roses
True Colours

GIVE ME TOMORROW

Jeanne Whitmee

ROBERT HALE • LONDON

First published in Great Britain 2015

ISBN 978-0-7198-1584-3

Robert Hale Limited
Clerkenwell House
Clerkenwell Green
London EC1R 0HT

www.halebooks.com

2 4 6 8 10 9 7 5 3 1

Typeset in Palatino
Printed by Berforts Information Press Ltd

Chapter One

'EXCUSE ME, MADAM, I must ask you to accompany me to the manager's office.'

Karen stared in amazement at the dark-suited man standing at her elbow. 'Why? I don't understand,' she said.

The man put his hand on her arm. 'If you'd just come back into the store with me ...'

Karen shook the hand off indignantly. 'Just tell me what the problem is.' She felt her cheeks warming and her heartbeat quickened as the curious eyes of other shoppers turned towards her. 'What do you want? Who are you?'

The man leaned towards her and lowered his voice. 'Come along now, madam. We don't want to make a scene, do we?'

'I'm not making a scene.' Karen could hear the rising note of alarm in her voice. 'I just want to know why you're asking me to go to the manager's office with you.'

The store detective glanced round him. 'Very well. If you insist, I have reason to believe you are concealing goods for which you have not paid.'

'*What*?' Karen caught her breath. 'You're accusing me of shoplifting?' She thrust her handbag into his hands. 'Here, look for yourself. My sister will tell you ...'

She turned around to Louise, who had been following her out of the store, only to find that her sister was nowhere to be seen. Bemused, she looked around her. Where on earth had she gone? Turning back to the detective, she saw that he had unzipped her bag and to her horror, the hand he dipped inside emerged holding a fragment of black lace with a store label attached to it. He looked at her triumphantly.

'I think you had better accompany me at once,' he said. 'If you refuse I shall be obliged to call security.'

Karen stared at the man. 'But – that's not *mine*! I've never even seen it before. I didn't ...' Panic-stricken, she looked around again for Louise, who seemed to have vanished into thin air. Acutely aware of the eyes of other shoppers boring into her back, she followed the detective back into the store. As they walked through the ground-floor departments and through a door into a corridor, he was speaking on his mobile phone.

Neville Smith, the store manager, was a small man with an inflated awareness of his important position as manager of Hayward's department store. He wore a baggy grey suit and rimless glasses, and his thinning hair was carefully combed over his bald patch. He looked up sternly as Karen was presented to him.

'I must inform you here and now that it is the policy of Hayward's to prosecute shoplifters,' he said without preamble. 'We have suffered so much loss over recent months that we have to be stringent.'

'There is some mistake,' Karen said shakily. 'Those things in my bag – I didn't take them. Perhaps you should look at your CCTV footage.'

The manager gave a sardonic little smile and pushed his glasses further up the bridge of his nose. 'May I point out that the fact that they are in your possession is more than enough to prove your guilt.' He glanced at the store detective. 'Did you witness the theft, Marshall?'

The store detective cleared his throat. 'A member of the public alerted me,' he replied.

Karen turned to him. 'Then whoever that was has made a grave mistake.'

Neville Smith treated her to another of his scornful little smiles. 'I'm afraid that's an all too familiar line, madam.'

Karen turned to the store detective. 'So where is this member of the public?'

Smith looked at the store detective. 'I take it you asked her to wait?'

Marshall looked a little flustered. 'I – er – no.'

Smith sighed. 'How many times must I tell you that any witnesses should be asked to wait?'

'She may still be in the store,' Marshall said quickly. 'I think she was on her way to the coffee lounge. I could probably find her.'

'Then I suggest you do so immediately.' As the detective left the room, the manager turned his attention to Karen. 'In the meantime, madam, would you be good enough to empty your handbag onto the desk?'

'No. I would not be good enough to empty my bag,' Karen said stubbornly. 'Not until you have proved that I am guilty. As I suggested before, surely you only need to view the tape in your CCTV to clear this whole thing up.' She opened her bag and pulled out what she now saw was a set of black lace designer lingerie which she laid on the manager's desk. 'These items are not mine but I did not steal them. Anything else in my bag is private.'

'I must warn you that once we have verified that the witness actually saw you conceal the goods, I shall be obliged to send for the police.'

Karen sighed. Her heart was hammering in her chest but she was determined not to let it show. 'Perhaps I can be permitted to sit down,' she said. If this woman insisted that she saw her take the goods what was she to do? *Louise*. This was Louise's doing. It had to be. She'd caused plenty of problems in the past but this was the last straw. Where on earth *was* she? How dare she land her in a mess like this and then just disappear?

'You may be seated,' the manager conceded pompously.

Karen sat down gratefully and an uncomfortable five minutes passed as they waited. Then the office door opened to admit Marshall and a flashily dressed woman with magenta hair. Karen stood up and moved aside. Marshall introduced the woman.

'This is Mrs Jones, sir. She is a regular customer here at Hayward's.'

'Quite so.' Smith's beady eyes assessed the woman's appearance. 'I understand that you witnessed an act of theft this afternoon, Mrs Jones.'

'That's right,' the woman said self-righteously. 'I hate to see a lovely store like Hayward's being taken advantage of. When I saw it I felt I had to report it at once.'

'Very public-spirited, I'm sure.'

'But I don't really want to be involved in any legal action,' she added guardedly. 'I don't want to be called to give evidence in court nor nothing.'

'All I need you to do for now is to identify this person as the one you saw taking the articles.' The manager glowered at the store detective. 'Marshall – would you be good enough to step aside and allow Mrs Jones to *see* the accused?'

Marshall stepped smartly to one side and Karen and the woman came face to face. Karen looked directly into the other woman's eyes and immediately saw her confidence fade.

Mrs Jones bit her lip. 'Ah – well – she was certainly wearing a red coat like this person's got on,' she said. 'And I'm fairly sure that was the bag she stuffed them into – or one like it. But the woman I saw take the knickers was a blonde.'

The manager looked irritated. 'Are you now saying that you are unsure?'

'Well, I definitely saw *someone* take the knick – er – undies and stuff them into a bag,' she said. 'But – I'm sorry but I can't say for sure that it was this person, just someone in a red coat. Unless – er ...'

'Unless what, Mrs Jones?'

'I suppose she could be wearing a wig,' she said.

Karen looked round at the other three people in the office. 'Perhaps one of you would like to try to test my hair,' she suggested, leaning forward.

'That will not be necessary.' The little man at the desk seemed to have diminished in size. His face flushed an unbecoming beetroot shade as he glared at the store detective. 'Thank you, Mrs Jones,' he said through clenched teeth. 'Perhaps next time you could make sure you are reporting the right person.'

The woman looked apologetically at Karen. 'I'm sorry if I caused you any bother,' she muttered.

'Oh, please don't worry yourself about it,' Karen said, trying not to sound sarcastic. She resisted saying, *Maybe you should have gone to Specsavers.*

Mrs Jones left the room hastily and Karen turned to the manager. 'Perhaps you would like to confirm what the witness has said by viewing your CCTV tape as I suggested?'

'There will be no need,' Smith said. His colour had now faded to a sickly grey. 'It is obvious that there has been a mistake. You are free to go.'

'Thank you.' Karen stood her ground. 'But first I'd like an apology,' she said. 'It's not every day one is named as "the accused" or threatened with prosecution.'

Smith cleared his throat. 'I think you will agree that it was a natural mistake.' Karen waited, saying nothing and Smith cleared his throat again and continued, 'Please accept my apologies for any inconvenience and – er – embarrassment caused.'

Karen picked up her handbag and as she left the office, she heard Smith upbraiding his store detective.

'Marshall! How many times must I tell you to make sure those tapes are regularly changed and running!'

In the car park Karen climbed gratefully into her car. It was only then that she realized that she was still shaking. What a horrible experience. Suppose she had been seen being apprehended by one of her pupils' parents? It would only take one! Simon was going to be furious about this. If the store had insisted on pursuing the case and the papers had got hold of it, they could both have lost their jobs. It was so irresponsible of Louise – and so typical.

Louise Davies, or Louisa Delmar as her half-sister liked to call herself, was fourteen years Karen's senior. Although they shared the same father they were as different in temperament and character as it was possible to be. Louise was an actress. Touring around the country and never in the same town for more than a few months, she always turned up on her married half-sister's doorstep whenever she was between jobs. Recently, the summer show she had been appearing in down in Devon had come to a close and she was waiting for her agent to arrange a pantomime for her.

Delving into her handbag, Karen took out her phone and clicked on Louise's number. The call went straight to voicemail. Furious, she left a message: '*Louise*! Where the hell are you? What did you think you were playing at? You've just almost had me arrested for shoplifting. Please ring me as soon as you get this.'

She started the car and backed out of her space. She was almost halfway home when she heard her phone ringing in her bag. Pulling over, she stopped and took the phone out. It was Louise calling.

'Louise!'

'Darling, I've just got your message. What's the matter?'

'Don't play the innocent with me. You know bloody well what's the matter. You put those things in my bag in Hayward's, didn't you?'

'Things? What things?'

'Designer lingerie ring any bells? Simon and I could have lost our jobs through this. I could be ringing you from the police station right now for all you know.'

Louise gave a maddening little giggle at the other end of the line. 'Oh, come off it, sweetie, don't exaggerate. You're not at the police station, are you?'

'No, no thanks to you. More to the point, where are *you*?' Karen demanded. 'And where did you disappear to? You left me to face the music on my own.'

'Look, it's all too silly for words. I can explain everything.'

'Then I think you'd better start. Where are you, Louise?'

'I'm in the park. It's such a lovely day I thought I'd have—'

'Whereabouts? I'm coming right now.'

'The little café by the lake. Look, there's no need to be so stuffy.'

Karen didn't reply, she just ended the call and thrust her phone back into her bag, then, turning the car, she headed back in the opposite direction.

The park was looking lovely in the autumn sunshine. The trees were turning to gold and the sky was blue and cloudless. The sound of children's voices filled the air as they played on the swings, making the most of their half-term holiday. But Karen hardly took any of it in. She had only one aim in mind: to find her half-sister and let her have both barrels. She spotted Louise as soon as she came through the yew archway at the lakeside. She was sitting on a bench, her elegant legs crossed and her short skirt pulled up, displaying several inches of thigh as she calmly nibbled at a choc-ice. Karen had never in her life wanted to hit anyone as much as she did now.

Louise looked up and gave a cheery wave. 'Hello, darling! There you are. Would you like one of these?' She held up the half-eaten choc-ice. 'They've always been my weakness.'

'I remember,' Karen said drily, sitting down on the bench next to her. 'I don't want a choc-ice. It would probably choke me. What I want is an explanation – *now*, please.'

'Oh, all right.' Louise sighed and popped the last of the ice cream into her mouth and screwing the wrapper into a ball, tossed it into the bushes.

'You put those things in my bag when I went to the loo, didn't you?'

Louise shrugged. 'OK – yes, but why not? I'm sure Hayward's can afford to lose a few pounds a damned sight more than I can.' She glanced sideways at Karen. 'I couldn't resist those sexy, black lace undies. Gorgeous, weren't they? What a pity you had to go and get yourself caught.'

'It's *theft*, Louise. Against the law, and it was despicable to involve me in it and then just disappear.'

Louise frowned. 'I knew you were going to be bloody boring about it. You're so damned buttoned up, Karrie.'

'I don't think you understand how serious this is. If I'd been charged, Simon and I could both have lost our jobs.'

'Oh, don't be so melodramatic!' Louise said dismissively.

'I'm not. It's true. Do you imagine that the school governors would allow either of us to continue after shoplifting charges and all the publicity that would have ensued? Why did you do it, Louise? You were seen taking the things and if it hadn't been for the difference in our hair colour, I'd be in custody right this minute.'

'But you're not. So what are you going on and *on* about?'

'You still haven't explained why you took these things.'

Louise shrugged. 'I fancied them – simple as that.'

'You *fancied* them!' Karen shook her head. 'So why put them in *my* bag?'

Louise grinned impishly. 'I couldn't resist winding you up. You always take the bait so beautifully. Oh, come on, darling. I couldn't know you'd get nabbed, could I?'

Karen's fingers itched to smack her sister's face. It was useless arguing. They could go round and round for ever. It was obvious she was going to get no more sense out of Louise. She could be completely amoral at times. She tried a different tack. 'Have you heard from that agent of yours lately?'

Louise chuckled. 'You're not terribly subtle, are you, darling? What you're trying to say is that you've had enough of me.' She pouted. 'Don't tell me I've outstayed my welcome.'

Karen shrugged. 'What do you think? After this afternoon – yes, you have!'

'Pity.' Louise sighed. 'I'll call Harry tomorrow morning if you're really going to insist on punishing me.' She glanced at Karen. 'I haven't said anything but there could be something quite exciting in the pipeline.'

Karen raised a cynical eyebrow. 'Oh yes? I've heard that one before. But then I suppose there's bound to be a panto coming up in some remote corner of the country.' She looked at her sister. 'You could always get a proper job, you know. The department stores will be taking on extra staff for Christmas soon.'

'Thanks for the vote of confidence, darling. Anyway, as I told you, there's a possibility of something really big in the offing.'

'Well, let's hope it comes off this time.'

'OK, don't go on about it. I'll get on to it in the morning.'

'Good,' Karen said without conviction.

Louise looked at her half-sister sheepishly. 'Look, there's no need to mention any of this to Simon. You know what he's like. He's so po-faced. He wouldn't see the funny side.'

Karen raised an eyebrow. 'Strangely enough, I know the feeling.' She turned to Louise. 'I'll make a bargain with you. I won't tell if you promise you actually will move out.'

Louise stared at her. 'You *what*? Where else can I go?'

'You could find yourself a flat or a room somewhere. I've had enough, Lou. Whenever you're around there's always trouble. This afternoon is the last straw.'

For a moment Louise looked crestfallen. 'I suppose I could try Susan.'

Karen looked at her. 'Mum? You know she only has a one-bedroom flat.'

'I could doss down on the sofa.'

'Why are you always so hard up? Where did all the money go to that you got for the house?'

Louise sighed. 'Trust you to bring that up again.'

'So – you should have quite a healthy bank balance. You could easily afford to rent a flat.'

'Perhaps.'

'I hope you think it was worth chucking Mum out for.'

'It was my house,' Louise said stubbornly. 'You were married and I had no one. Dad left it to me and I needed the cash. Susan wanted to downsize anyway. The place was far too big for her.'

'You got a good price for it so you can't be short of cash.'

'A girl has to put something away for a rainy day.'

'Really? Well, this is it, Lou. This is the rainiest day I've seen for a long time, so get your umbrella out.' Karen got up from the bench and began to walk away. Louise jumped up and followed her.

'I suppose I'd better come home with you and start packing, then?'

'I suppose you better had.'

Chapter Two

KAREN WATCHED FROM THE window as Louise walked down the path and out through the gate and into the waiting taxi without a backwards glance. Why did she always feel so guilty? Louise took advantage quite outrageously and this afternoon she had really overstepped the mark, but still, perversely, she couldn't help feeling bad about telling her to go, especially when she knew she'd be landing herself on poor Mum.

'You still haven't told me what that was all about.' Simon had come into the room and was standing behind her.

'You don't want to know,' Karen said without turning round.

'I do actually.' Simon dropped an arm across her shoulders. 'Whenever she's around, you look really stressed but when you came in this afternoon, I could see you were at the end of your tether.'

Karen turned to look at him. Most of the time, Simon put up with Louise's visits without saying a word, although he had never liked or approved of her. In her turn, Louise made no secret of the fact that she considered Simon dull and boring. She was always making cutting remarks to Karen about his clothes and hairstyle, his taste in music and cars, most of all, his job as head teacher of St Luke's Primary School, which she considered mind-numbingly dreary and bourgeois.

'She overstepped the mark,' she said non-committally. 'Suddenly I'd had enough.'

'What did she do? Come on, you might as well tell me. I'll find out in the end anyway.'

Karen took a deep breath. 'She shoplifted some underwear in Hayward's.'

Simon shrugged. 'Why doesn't that surprise me?'

'But she put the stuff in *my* bag and then disappeared when the store detective stopped me.'

'Oh my God!' Simon's eyes widened in horror. 'What happened?'

'Luckily I was able to prove that I didn't do it. But not before I had the humiliation of being stopped and frogmarched through the store to the manager's office.'

'My God!' Simon's face flushed angrily. 'Do you realize what that might have meant for me?'

'For *you*? I think you mean *us*! Of course I do. I eventually found Lou calmly eating ice cream in the park. She made light of the whole thing as though it was some hilarious prank. That's when I saw red and asked her to leave.'

Simon frowned. 'For God's sake! When I think what the outcome could have been. I would never have worked again if you'd been charged. So where's she off to now?'

'To Mum's, unfortunately. I rang ahead to warn her and I was hoping she'd say no but you know what a soft touch she is. I just hope Lou doesn't take advantage.'

'That's a fond hope. Seriously, Karen, something will have to be done about your sister. We can't allow her to keep on scrounging on us, especially as she must be rolling in cash since she sold the house. It's quite preposterous. No one else would stand for it.'

'Well, let's hope Susan is able to knock some sense into her.'

'I doubt that.' He looked thoughtful. 'Speaking of sense – have you given any more thought to what we talked about last night?'

'About me giving in my notice at school? No, I haven't. I've only been back at work for two terms. What's that going to look like? The governors will think I can't cope.'

'It doesn't matter what they think. You know my feelings

on the matter. I've always thought you should be at home with Peter at least until he goes to school.'

Karen sighed. 'We've been through all that. He's quite happy with Mum and in a few months, he'll be old enough for nursery.'

'Well, I happen to think that your mother isn't a fit person to be bringing him up. She spoils him rotten with ice cream and chocolate and all the wrong food.' He spread his hands. 'And frankly, Karen, since you've been back at school the house is a tip.'

She stared at him. 'A *tip*, is it? Well, if you gave me a hand now and again, maybe it wouldn't be.'

He looked outraged. 'Gave you a hand! My job doesn't finish at half past three as you well know. Even when I get home, there's a pile of paperwork to be done and then at the weekend there's the garden. I can't possibly do more.'

'Of course I realize all that, but I have preparation to do too – *plus* the housework and Peter.'

He smiled smugly. 'My point exactly. You're taking on too much.' As she opened her mouth to protest he ran an exasperated hand through his hair. 'Oh, for Christ's sake, Karen, why don't you just admit it and give in? Peter's almost two. In three years he'll be off to school. The time will fly by.'

'Other new mothers go back to school even sooner than I did,' she protested. 'As I said before, he'll be going to nursery soon and I'd be twiddling my thumbs all day.'

'You could always fill the time by doing some housework and cooking some decent meals. I'm getting a bit tired of supermarket ready-meals.'

She gave him a withering look. 'Well, I'm not giving my job up just because you say so. If you want me to go you're going to have to fire me.' She smiled defiantly up at him. 'And I don't think you'd like what the governors would make of *that*!' She turned towards the door. 'I'm off to bathe Peter now. As you don't seem to be up to your neck in paperwork at the moment, perhaps you'd like to peel some potatoes for tea.' She was halfway up the stairs when she heard his infuriating parting shot:

'As it's half-term next week, perhaps you could spend it tidying up a bit. I'm told that's what other working mothers do. And by the way, it's *dinner* – not tea!'

Chapter Three

Susan replaced the receiver and sighed. Karen had sounded apologetic.

'I'm really sorry, Mum, but Louise is on her way over to you. We've had a bit of a falling-out and I'm afraid she's going to ask if she can stay with you. You don't have to say yes,' she added hurriedly. 'If you've got anything planned don't put it off for her. She's got absolutely no consideration, so stick to your guns.'

Susan sighed. 'Yes, dear. I will.'

'I thought I'd just warn you.'

'Thanks. Just leave it with me. It'll be fine.'

She closed the balcony doors and drew the curtains reluctantly. Soon it would be winter. Next weekend the clocks would go back and the nights would lengthen. She sighed. It was no fun, being on your own. Sometimes she was glad she had no one to please but herself, but at others – when she felt down or depressed, she longed to have someone to talk to, to share a meal or a cup of tea with, someone who'd be nice to her when she was feeling low. Not that Louise exactly filled that description, but she was company at least, and Susan didn't see her stepdaughter very often.

She looked around her at the things she'd managed to salvage from her marital home. It was easier, of course, having a smaller place to look after. It was quite a nice flat and she could just about afford it on her pension. But it had been a shock when

Frank died and she found that he'd left the house to Louise. An even greater shock when her stepdaughter announced that she was in desperate need of the money and intended to sell it. At first she had felt as though the rug had been pulled out from under her feet. Karen had been furious and offered her a home with her and Simon, but Susan valued her independence too much to accept. After all, she wasn't in her dotage yet and Frank hadn't exactly left her destitute. He'd left her a small legacy which she'd invested in the hope that she'd be able to afford a better home in the not too distant future.

Frank had always spoilt Louise. It was understandable in a way. His first wife – Louise's mother – had walked out when her daughter was barely more than a toddler and Frank had never stopped trying to make it up to her.

Susan had been twenty-four when she'd met Frank. She worked as manageress at the Blue Bird Café. Frank worked for the local council as a highway surveyor. His office was just a few doors away from the café and he was in the habit of dropping in for his morning coffee. They got to chatting and eventually he asked her out. She learned that his wife had walked out and left him with a four-year-old daughter and her heart had gone out to him. By that time, Louise was ten and for the last six years Frank had had no social life at all; hurrying home from work each evening to pick Louise up from the childminder or school, make her tea, and put her to bed.

'But now she's beginning to get a bit more independent,' he told her with a smile. 'So maybe I'll be able to get a bit of time to myself.'

Not that he ever did. Louise was very demanding. And she always came first. Often when he and Susan had arranged to go out he'd have to ring her and cancel; the reason invariably being something to do with Louise. He'd got into the habit of giving in to her and he couldn't let go.

It was a whirlwind courtship. She and Frank married just six months after their first date. Susan had no family and she looked forward to having a home and family of her own; someone to care for who would care for her in return. But sometimes it seemed that Frank cared only for his daughter.

When Karen arrived things improved. She was such a sweet-natured little baby and Susan was overjoyed to have a baby to love and care for. Louise took little interest in the baby, though Susan told herself that it was understandable. The fourteen-year age gap meant that the girls had little in common.

Karen met Simon at college, and Susan and Frank were delighted when they got engaged. He was a pleasant, steady young man with good prospects in the teaching profession and when baby Peter came along a couple of years later, Susan had been overjoyed. Naturally, Karen wanted to return to her work as a teacher as soon as she could and Susan was thrilled to be asked to step in and care for her little grandson on a regular basis once Karen's maternity leave came to an end. It made her feel useful and needed again now that she was on her own. There was plenty of time for her to do other things in the school holidays and catch up with her hobbies and social life, such as it was.

Susan knew that Louise often visited Karen when she was between acting jobs and although Karen never complained, Susan felt that she took advantage. Not that there was anything she could do about it. Karen never said anything, although she did let drop the occasional telling remark. Privately, Susan thought that although she would never admit it, Louise was envious of Karen's settled lifestyle. She'd made her choice when it seemed she had acting talent and begged to go to drama school. Susan often wondered just how much talent Louise actually had. She certainly hadn't got very far in her chosen career, picking up the odd part in a touring play, a place in the company of a seaside summer show and then of course the inevitable pantomime for a few weeks at Christmas. It all added up to less than a third of the working year and when she was 'resting' as she liked to call it, she fell back on her family's generosity and came to visit.

Susan switched on the electric fire. The evenings were getting chilly now. Karen had been looking forward to spending time at home with Peter. Not that she'd been able to enjoy it as she'd hoped. Louise had landed on her doorstep a week last Monday and showed no sign of moving on any time soon. Her stepdaughter hadn't paid her a visit, but then that was par for the course. Susan sometimes felt that she was at the bottom of Louise's list when it came to priorities.

Picking up the *Radio Times,* she saw that there was a gardening programme on in ten minutes' time. She missed her garden sadly, and she still hoped that one day she'd be able to afford a little bungalow with a bit of garden for her to tend. She switched on the TV and settled down to watch. But the opening credits had barely finished when there was a ring at her doorbell. Picking up the entry-phone, she heard Louise's voice.

'Hi, Susan. It's me. I'm here on a flying visit and I couldn't leave without coming to see you.'

'Oh, how nice.' Susan pressed the door release. 'Come on up.'

Louise appeared a few minutes later and flung out her arms in her usual affected manner. 'How lovely to see you and how well you're looking.' She enveloped Susan in a hug. 'It's so *good* to see you.'

'It's good to see you too, Louise,' Susan said, extricating herself from Louise's embrace. 'It must be almost a year,' she added pointedly.

Louise looked shocked. 'No. Really? Time flies so, doesn't it? And of course I've been so busy.'

Susan knew perfectly well that Louise had been to stay with Karen at least twice since Louise had last visited her but she let it pass. 'Well, now that you're here, come in and tell me all your news.' She led the way through to the sitting room and Louise took off her coat and sat down on the settee.

'Karrie has been telling me how marvellous you are with little Peter,' she said. 'Rather you than me. Babies have never been my thing as I'm sure you know.' She laughed. 'As a friend of mine puts it – noisy, messy and smelly. It's so good of you to be a free childminder for her. She's so lucky.'

'I'm not a *free* childminder,' Susan corrected. 'Karen insists on paying me the going rate, even though as Peter's grandmother

I would willingly do it for nothing. He's such a dear little boy and I adore looking after him.'

'Well, anyway, she's lucky to have you.'

'You must sometimes feel that you'd like a more settled, conventional life, Louise.'

'Me? Good heavens, no!' Louise gave a brittle little laugh. 'Life in suburbia looks like a living death to me. I love living on the edge; the excitement of never knowing what's round the corner.'

'I see, and is there anything round the corner for you at the moment?'

'I'm expecting to hear from my agent any day now,' Louise said. 'There might be something exciting coming up for me.'

Susan had heard all this before. 'Really, so are you going to tell me about it?'

Louise shook her head. 'Not till it's confirmed. I'd hate to jinx it.' She cleared her throat and glanced sideways at her stepmother. 'Actually, Karrie has a bit of a crisis on at the moment. I can see it's a problem for her – me staying there, and I don't want to be a nuisance. So I wondered – could you put me up for a few days?'

Susan took a deep breath. Karen had already prepared her but she had always been so bad at saying no. She cleared her throat. 'You realize that I only have one bedroom, don't you, Louise?'

Louise shrugged and patted the seat beside her. 'This sofa is very comfy. I could manage beautifully on it for a few nights.'

'What do you call a few nights?'

Louise's eyebrows rose. 'Susan, just say if it's inconvenient. I just thought that with you being family…'

Susan took a deep breath. Karen would be so cross with her if she caved in so easily. 'Forgive me for reminding you, Louise, but you're not hard up, are you? Surely you could afford to rent a small flat for a few weeks, until your agent comes through with this wonderful job?'

'It wouldn't be worth taking on a flat,' Louise told her. 'I'm expecting a call from him any day.' She frowned. 'And actually

I *am* quite hard up,' she said. 'Money goes nowhere these days.'

'But you got a good price for the house.'

'You always have to bring that up, don't you? Then you wonder why I don't come to visit you. You don't understand, Susan, an actress has to keep up appearances. I need to buy clothes. Not just any old clothes but designer fashion. I have to be seen around town in all the right places – have my hair done and watch my looks. It would be fatal to let myself go.'

Susan took in what Louise was wearing and told herself that if that was designer wear, standards must have dropped. 'It all sounds quite exhausting,' she said dryly. 'But of course, none of us are getting any younger.'

Louise bridled. 'I'm thirty-five but I think I look at least ten years younger if I do say it myself.'

Susan knew perfectly well that Louise would be forty next birthday but she kept her mouth shut on the subject. 'So you're saying that all of the money you got for the house has gone on clothes and keeping up appearances?'

Louise bridled. 'Quite a bit of it, yes. I do have a nest egg of course but I don't want to touch that. No one knows what the future might bring.'

'Indeed,' Susan said. She didn't bother to add that that was exactly how she had felt when Frank died and the house had been sold from under her. 'I take it you sign on for Jobseeker's when you're not working,' she said.

Louise nodded reluctantly. 'Yes, but you know what a pittance that is.'

'I suppose there's always temporary work.'

'You sound just like Karrie,' Louise said. 'What do you expect me to do, *scrub toilets*?' When Susan smiled and shook her head she added, 'Anyway, I told you, there's a strong possibility of something really exciting coming up any day now. I don't want to miss it because I've taken on some mundane job, do I?'

'I suppose not.'

'Right, so getting back to the point – can you put me up?'

Susan frowned. 'You mentioned that Karen is going through a crisis. What's happened? Should I be worried?'

'No, nothing like that.' Louise paused. 'OK – we've fallen out. It was all about something trivial and I'm sure it will soon blow over, but in the meantime ...'

'If, as you say, it's only for a few days.'

'At the most.'

'Then I suppose you'd better go and collect your stuff.'

'As a matter of fact, I left my case outside on the landing.'

Susan loved order and she always felt that when Louise was staying, it was as though the flat was full of people. Although she was sleeping on the sofa, she didn't get up until halfway through the morning. When she did, she used up all the hot water and made full use of the washing machine. The flat was strewn with her belongings and the bathroom was festooned with her drying tights and underwear. It never occurred to her to give a hand with the cleaning or cooking and in the evenings, she hogged the TV remote control, dismissing Susan's taste in programmes as 'boring'. She made absolutely no attempt to help with the shopping, nor did she offer to contribute towards the food bills.

Once the half-term week was over, Susan suggested to Karen that she babysat Peter at her house. It was a relief to get out of the flat although she dreaded the state the flat would be in on her return each evening. She also tried not to think about the heavy telephone bill Louise was running up. She used the landline all the time. It never seemed to occur to her to use her mobile. The worst of it was that most of her calls were long-distance ones to her agent and various other people.

When she ventured to ask if there was any news about the so-called 'exciting prospect', Louise merely shrugged and muttered something vague about 'these things taking time'.

Susan asked Karen why she and Louise had fallen out but her daughter was cagey, shrugging it off as a 'storm in a teacup'.

'If she's taking advantage, Mum, just kick her out,' she said, and although Susan agreed that she would, they both knew that it was easier said than done.

*

Louise had been at the flat for ten days when Susan began to have a suspicion that the marvellous job she had been so effusive about was either a non-starter or all in her mind. She became quiet and preoccupied and jumped every time the phone rang, racing to be first to pick it up. Then suddenly one evening, her mobile phone trilled out. Rummaging for it in her bag, she listened in breathless anticipation to the voice at the other end and slowly a look of excited relief lit her face. The call seemed to last forever but eventually Louise said goodbye and ended the call and looked at her stepmother, her face wreathed in smiles. Her whole demeanour had changed.

'I'll be off tomorrow morning, Susan,' she said cheerily. 'It's been lovely staying with you and I really appreciate it. Thanks a lot.'

Susan's heart leapt. She was going at last! 'Was that your agent?' she asked. 'Was it about the job?'

Louise nodded. 'Well, fingers crossed. There are some details to iron out yet. But anyway, I'll be out of your hair tomorrow.'

She left soon after breakfast, before Susan had left for Karen's.

'Thanks for everything, Susan. I'll be in touch,' she said, pecking her stepmother on the cheek.

'Where will you be staying?' Susan asked.

'With Dianne, an old friend from drama school. She's got a flat in Earl's Court and she's always happy to put me up for a few days. I rang her last night. Would you just ring a taxi for me while I gather my stuff together?'

As she disappeared into the bedroom, Susan lifted the receiver. One more call and it would all be over.

Ten minutes later, she watched from the window as Louise stepped into a taxi down in the street.

She was bursting with her news as she took off her coat later at Karen's.

'She's gone,' she said briefly, knowing there was no need to elaborate.

'Louise – gone? Just like that?'

'Just like that. She had a phone call yesterday evening and

announced right after it that she was leaving this morning. She's gone already.'

'So – is it a job, or what?'

'I asked that. She just said something about an exciting possibility.'

'I take it she'll be sponging on her friend, Dianne, again.'

'She did say she'd be staying with her, yes.'

'Oh well, at least you've got your home to yourself again. I'm sorry, Mum. I'm afraid it was partly my fault you got saddled with her.'

'I wish you'd tell me what you quarrelled about.'

'Maybe I will one of these days,' Karen said, putting her coat on. 'For now, just be thankful that she's gone. If you've got any sense you'll say no next time she asks to stay. I'm going to. It's time we left her in no doubt about how we both feel, Mum. She's always ridden roughshod over us and it's time we called a halt to her freeloading. After all, she can't be hard up.' She kissed her mother's cheek briefly. 'Thanks, Mum. I'll have to dash now.'

What Karen said made good sense. It was high time they told Louise a few home truths. But she had the distinct feeling that Louise would find a way to be the injured party when they did. Somehow she always did.

Chapter Four

It was my fourth birthday that day. I'd had a party and I was really sleepy when I went to bed but something woke me. Rising up the stairs, I heard my parents' raised voices. I couldn't make out the words, just the frightening tone of their voices, sharp and ugly with anger; Mum's shrill and Dad's hoarse and rough. They were rowing again. I felt my heartbeat quicken and my tummy churned sickeningly. Pulling the covers over my head, I stuck my fingers in my ears. I must have gone to sleep again because suddenly it was morning and I was awake. Slipping out of bed, I padded through every room on my bare feet, searching the whole house for Mum. When I couldn't find her, a wild panic filled my chest.

'*Mum*!' I heard myself screaming. 'I want Mum!'

'Lou – *Lou*, wake up!'

Someone was shaking my shoulder.

'Lou, you're dreaming again. Wake up now, love, everything's all right.'

I opened my eyes to see Dianne standing beside the bed. My heart was still pounding but I took a deep breath and sat up, shaking my head. 'Sorry, Di. Did I wake you? Was I shouting again?'

'Yes. I don't know what those nightmares are about but they certainly seem to terrify you.' Dianne sat down on the side of the bed.

'I've had them on and off for most of my life,' I admitted. 'They usually recur when I'm nervous about something.' I pulled myself up into a sitting position. 'My parents used to have these horrendous rows when I was little and I'd wake and hear them.'

'They must have scared you pretty badly if they're still affecting you.'

'They did. Then one morning my mum wasn't there any more. I never saw her again.' I shuddered, my stomach still churning at the memory of the dream. 'When I was about eight I started getting bullied at school. The other kids said my dad had killed her and buried her body in the garden. That gave me even worse nightmares. Then, on my ninth birthday I got a card from her so I knew it wasn't true. She'd written her address on it. I managed to keep the card but I never got to see the letter she enclosed with it; Dad tore it up and wouldn't let me read it.'

Dianne laid a hand on my shoulder. 'Shall I make you a cup of tea?'

I shook my head. 'No. I'll be OK once the dream has faded.'

'You're seeing this mystery guy tomorrow, aren't you?' Dianne observed. 'Is that what you're nervous about? You didn't tell me much about it last night.'

'There's a lot hanging on it.'

'You've never really told me much about him – or this project he's planning.'

'It's not really a mystery. I haven't said anything because I didn't want to jinx it. You know how superstitious I am. I didn't mention it to anyone at home. They don't have a very high opinion of me or my talent and they probably wouldn't have believed me anyway. When I was down in Devon in the summer with the Sunshine Follies, Harry came down and brought this guy along. His name is Paul Fortune. He's a musician and composer and he's written this musical called *Oh, Elizabeth*. It's based on Jane Austen's *Pride and Prejudice*. Harry brought him backstage afterwards and apparently Paul had been quite impressed by me in the show. He said that he saw me as the perfect actress to play Elizabeth Bennet.'

'Wow!' Dianne's eyes widened. 'So is it him – this Paul Thingummy that you're meeting tomorrow?'

'Yes, I'm having lunch with him and Harry. There was some ground-laying work he had to do before going ahead, like booking venues and rehearsal rooms – setting up advertising and so on, which is why it's taken so long, but now it seems it's all systems go. The show kicks off in the provinces and then – fingers crossed – eventually moves into the West End.'

'It sounds like a great opportunity,' Dianne said. 'And I do wish you the best of luck. You're sure there are no strings attached, aren't you?'

I laughed. 'No, of course there aren't. Harry's hinted that he might ask me to put up a small amount of the money, but that's not a problem. I've still got most of the cash I got for the house stashed away.' I looked at Di. 'You know how old I am and how long I've been waiting for a break. This could be it; my last chance at making the big time.'

Dianne looked doubtful. 'Mmm, I'd be a bit careful if I were you. How much do you know about this guy?'

'Basically just that he's someone my agent knows. But Harry and I go back a long way and I trust him.'

'I see. What was your initial gut feeling when you met him – this Paul Fortune?'

'Frankly, wow, what a hunk!' I giggled. 'The three of us went out to dinner after the show and Paul and I hit it off like a house on fire right from the off.'

'Are you saying you slept with him?'

'Oh come on, give me *some* credit. They were only there the one night. Even I don't work that fast.'

Dianne wrinkled her nose. 'Just be careful, Lou. I know how much you want to grab what looks like a great opportunity and I also know how impulsive you can be. Just how much of your money is he after, if you don't mind me asking?'

'I've no idea, he might not want much at all,' I hedged. 'Anyway, it seems that every cast member will be putting a share in and we'll all get it back, plus dividends when the show takes off.'

'*If* the show takes off.'

'Oh, come on, Di, don't pour cold water on it. What makes you such a cynic? I don't understand you. You were one of the most talented students in our year,' I told her. 'With your looks and talent, you could have been a star by now if you hadn't given the idea of a stage career the elbow.'

'Or I could have been penniless and homeless. The theatre is far too precarious for my liking, Lou, and in spite of what you say I never really felt I had any real talent. As it is I've got a good steady job that I enjoy, a nice flat and a reliable income and I still enjoy a bit of amateur work.'

I grinned at her. 'Bor-ing. Still, whatever floats your boat, I suppose.'

'Do you think it's a subconscious anxiety about this project that's causing you to have these nightmares about your childhood?'

'No. I get them from time to time. As for this new musical, I can't wait.' I paused, wondering whether to voice the thought that was in my mind. 'I think I know what is causing them, though. Since Dad died I've been thinking a lot about my mother, wondering if she's still alive and what she's doing. I've had this really strong feeling that I'd like to find her again.'

'Why, when she let you down so badly? It's an awful long time ago and after all, you've got Karen and your stepmum.'

I shook my head. 'They've never felt like a real family to me. I've always felt like the odd one out. As a kid I felt resentful that they'd stolen Dad away from me. I think he understood that. It was probably why he left me the house. And I've never been able to shake off the feeling that the split might have been my fault.'

Dianne shook her head. 'You were only a little kid. How could it have been your fault?'

'I don't know. It's just a feeling. I suppose all kids from broken marriages feel like it. Anyway, I'd like to know the real reason for the split and only she can tell me.'

'OK, so if you want to try and find her why don't you go ahead? It shouldn't be that difficult.'

'I wouldn't know where to start.'

'They say that the Salvation Army is good at finding people. You could try them for starters.'

'Well, maybe,' I said. 'But she could have married again – have a different name. Thinking about it and doing it are two different things. Anyway, my main priority right now is my meeting with Paul tomorrow.'

'Have you ever talked to anyone about it?'

'Trying to find my mother, you mean?' I shook my head. 'Susan and Karen wouldn't understand. Neither of them ever met my mum. Dad married Susan when I was ten and there are too many years between Karen and me for us to feel like sisters. Anyway, she's as different from me as chalk from cheese. She and her precious Simon are so bourgeois and horrendously smug with it. They always make me long to shock them out of their complacency.' I bit my lip at the memory of the shoplifting episode. I'd never gone over the top quite that far before. It was a lot worse than a wind-up and I baulked at telling Dianne about it. She'd never understand. 'I'm afraid I can't resist rattling her cage,' I said lightly. 'Old habits die hard.'

'Mmm.' Dianne smiled ruefully. 'I remember the practical jokes you used to play at drama school; quite merciless, some of them. But going back to your family – Susan's always been a good stepmum, hasn't she?'

'I suppose so. She's a bit of a pushover though. She never comes out with what she's really thinking – just makes bland comments.'

'Which – if I know you – makes you even more secretive?'

I couldn't help grinning. 'You're right. I try to keep in touch. I visit whenever I can but quite frankly, I don't think any of them would give a damn if they never saw me again.'

'I'm sure that's not true.' Dianne glanced at the alarm clock. 'Look, it's past one. Maybe we should both get some sleep now.'

Dianne looked in before she left for work at eight o'clock the following morning.

'Here, I've brought you a black coffee.' She put the cup on the bedside table. 'Don't go back to sleep.'

I emerged from under the duvet and tried to force my eyes open. 'Thanks, Di, you're an angel. I'll get up and have a cold shower when I've drunk it. That should wake me up.'

'Well, good luck.' Dianne hesitated in the doorway. 'Lou – if you get a few spare minutes before you leave, could you tidy up a bit?'

I bit my lip. 'Sorry, Di. I know I'm a slob. If I get time I'll put some of my stuff away.'

'Great. And while I remember, there isn't a scrap of food left in the fridge so maybe on your way home you could pop into the supermarket and get something for tea. I'll have to dash now. Bye for now.'

When I'd showered I did my best to tidy up the stuff I'd left lying about in Di's living room. It wasn't that bad and I felt slightly resentful that she'd actually asked me to do it. I'd have sorted it eventually, and this morning I had other things on my mind.

Once I'd stuffed the last pair of tights down the back of the sofa and hung the towels back on the bathroom rail, I set about preparing for the coming meeting. In spite of what I'd told Susan, I actually only possessed one decent designer outfit, a classic black suit by Chanel. I washed my hair and borrowed Di's hairdryer to blow-dry it. Twisting it up into the French pleat I peered at the roots. I should really have had them touched up but it was too late to worry about it now. It would have to do. Five years ago, a brightening rinse would have been enough to revive its vitality but now the dreaded greys were all too evident and I was till trying to ignore the faint lines that had begun to appear round my mouth. Sitting at the dressing table, I half-closed my eyes and regarded my reflection, thinking about Jane Austen's Elizabeth Bennet, the part I was hoping to get. It would mean dropping almost twenty years. The one and only time Paul Fortune had seen me on stage I was wearing a glamorous evening gown and full stage slap, then later at the restaurant seated opposite him at a candlelit dinner table. When he saw me in the cold light of day, would he change his mind about offering me the part?

I took a deep breath and reached into my bag for the little sapphire necklace that I'd relieved Susan of. It had belonged to my mother. I remembered seeing her wearing it and it should have been mine by right anyway, so when I saw it lying on her dressing table I had no feeling of guilt about taking it. I decided to wear it now – for luck. Fastening it round my neck, I slipped it inside the neck of my top. Then I laid out the collection of concealers, foundations and blushers that would hopefully bring about the necessary transformation.

The table for lunch had been booked at an Italian restaurant in the Strand. In my eagerness I arrived too early. Anxious not to appear too keen, I disappeared into the ladies' and lost some time reapplying my lipstick and fussing with my hair. When I decided I was just late enough to appear relaxed yet not impolite, I gathered up all my courage and made my way to the bar.

I spotted them at once and sighed with relief. At least they had turned up. Pausing, a little out of their line of vision, I assessed them both. They looked an unlikely pair; bald, bespectacled Harry in his usual formal dark-grey suit and tie; Paul in jeans and a black roll neck sweater, his dark hair longish and flopping slightly over his brow. I took a deep breath and breezed in.

'Hello, you two. I'm so sorry I'm late. I had such a job finding a taxi.'

Paul stood up and offered his hand. 'How nice to meet you again, Louise. What can I get you to drink?'

I hardly tasted my lunch. In fact I can't even remember what I ordered. I was too busy assessing Paul's thoughts and wondering if he still felt the same about offering me the part. As soon as the dessert had been served, Harry made his excuses.

'I hate to eat and run but I'm afraid I'm going to have to leave you two now,' he said, getting up from the table. 'I've got an important meeting at two. But I'm sure you can manage without me.' He smiled. 'Let me know how your meeting goes,' he whispered in my ear as he bent to drop a formal kiss on my cheek.

'I'll be in touch.' I smiled up at him.

When he'd gone Paul asked if I'd like a liqueur. I shook my head. 'I won't, thanks. It's a bit early in the day. But don't let me stop you.'

'I don't want any more to drink either,' he said. 'I prefer to keep a clear head.' As he glanced at his watch I wondered if, like Harry, he was about to make an excuse to leave, but he looked at me enquiringly. 'Look – if you're not doing anything else this afternoon, would you like to come back to the flat with me? I could play you some of the numbers I've written for the show and you could have a copy of the script to take home and read. What do you say?'

What did I *say*? I was absolutely thrilled. He hadn't written me off after all. 'That would be lovely,' I said, trying not to appear too keen. 'I'd love to hear some of your music.'

His flat was on the top floor of a fashionable block in Kensington. He opened the door into the living room, which contained a grand piano and a couple of black leather sofas. I tried not to let him see how impressed I was. Going over to the piano he picked up a copy of the music, complete with words, and handed it to me.

'Have a seat. I'll run through a couple of songs and then perhaps you'd like to try them out.'

My heart was thumping as he ran his fingers over the keys. If this was an audition it was the most unusual one I'd ever had. He ran his hands over the keys and then broke into the first theme. I don't read music so I tried hard to memorize the tune. Luckily, it was quite a simple melody and the words seemed to fit well. When I got up to stand beside him at the piano, my knees were trembling but I got through the first verse without stumbling and he looked quite pleased.

'That was great, Louise. Shall we try another? There's a nice duet between Elizabeth and Darcy. I'll take Darcy's part if you like, though I warn you, I'm no Pavarotti.'

The duet was pleasant and although I was nervous it seemed to go quite well. When it was over Paul smiled up at me.

'I know it's early days but I've got a really strong feeling that we're going to have a hit on our hands.'

My mouth was dry as I asked, 'Is this an audition, Paul?'

He looked surprised. 'I think we've passed that stage. I already knew you could sing. The part is yours – if you want it, that is. I thought that was a foregone conclusion.'

I felt my colour rise. 'I see. Well, of course I want the part.'

'Right, that's settled, then.' He got up from the piano. 'Shall we break for a cup of tea?'

'I'd love one. Can I do anything to help?'

The kitchen was very modern and minimalist with every labour-saving gadget imaginable. While he put the kettle on I found cups and saucers and laid a tray.

'This is a beautiful apartment,' I remarked.

He nodded. 'It's nice and central. It suits me very well.'

'I love the songs,' I said. 'What I've heard so far, that is. Have you gone very far with the casting?'

'It's almost complete,' he told me. 'I'm producing myself as well as being musical director, naturally. I do still have to get a director but I have a very good one in mind. I can't tell you who yet. It's always difficult to get a big-name director when you don't plan to put a big star in the leading role.'

I looked at him. 'Yes, I was going to ask you about that.'

'*You* are going to be my star, Louise,' he said. 'If the show really takes off, as I believe it will, your face is going to be on the front of all the magazines.' He left me for a moment to absorb this heady prospect and for a second I wondered if he could hear my excited heart drumming from across the room. Then he asked casually, 'Did I tell you that I've arranged a read-through for next week?'

'No,' I said, my mouth dry. It was all gathering momentum even faster than I'd expected.

'I hope you'll be free to come.'

'Of course I will. Just tell me where and when.'

Over the tea and biscuits he asked me about myself and I outlined my career, trying to boost it up and make it sound better than it actually was. He asked me about my family too, which surprised me. I told him how my mother had walked out when I was very young and how I intended to try to find her again. He was sympathetic.

'Well, once you're a West End star I've no doubt she'll want to know you,' he said with a smile. As he poured a second cup of tea for both of us he asked, 'Louise – there was some mention of you putting some money into the show …'

I nodded. 'Yes. I'm not wealthy but I could manage a little.'

'The rest of the cast are chipping in,' he went on. 'It's because at the moment I don't have a backer. Again it's the start thing, but I hope to get someone interested very soon. It's just to get the ball rolling and the loans will all be paid back as soon as possible – with interest of course.'

'Naturally.'

He paused, biting his lip. 'I hate to ask, but have you any idea how much your input might be?'

I did some quick sums in my head. 'I take it there'll be no salary during the rehearsal period,' I said tentatively.

'Well, no, I'm afraid not.'

I gave him the figure that was in my head. 'Don't take that as positive,' I added. 'I'll have to work out what I'll need to live on. I'll do it later this evening when I get home.'

He looked pleasantly surprised. 'That – or something like it – would be very generous,' he said. 'But think of it as an investment. Once we get into the West End the cash should start rolling in.'

I couldn't wait for Di to get home. I was longing to tell her that all her misgivings were unfounded and that the new musical was going to be a hit. The moment she got in I began.

'I've had the most exciting day,' I told her. 'It was perfect. After lunch, Paul took me back to his flat. You should see it, Di. It's in Kensington; the last word in luxury and it …'

'Great,' she interrupted as she took off her coat. 'So what did you get for tea?' She shook the raindrops off her coat and hung it up without looking at me. 'I'm starving. I didn't have time for lunch and …' She turned and caught my expression.

Tea! I'd completely forgotten about the supermarket. I bit my lip. How could she expect me to think about mundane things like food after the exciting day I'd had?

'Oh – I'm sorry, Di,' I mumbled. 'I forgot.'

She stared at me. 'You forgot! Oh, really, Lou. Surely it's not asking too much for you to pop into Tesco on your way home for a few bits.'

I was shocked at her response. 'I *said* I'm sorry. Surely you can rustle up something just for tonight.'

'Well, no, I can't actually. The fridge is completely empty, thanks to you and your constant snacking while I'm out at work. Perhaps you could pop out now and get something. I don't care what, just as long as it's edible.'

I stared at her. 'But it's raining!'

'Yes, oddly enough I noticed that on the way home when I got soaked,' she said with unnecessary sarcasm.

'You're serious, aren't you?'

'Never more so.'

Furious, I reached for my coat. 'There's a mini-market up the road,' I snapped. 'Will that do for you?'

She shrugged. 'Like I said – as long as it's edible and *you're* paying, it's fine.'

On the way to the corner shop my temper cooled a little. It was good of Di to let me stay and if she decided to throw me out it would make life difficult; that, I had to admit. Maybe I should have made more of an effort; after all, if I had to find other accommodation now before the provincial tour began, it would be expensive and with my contribution to the show I was going to have to watch the pennies. Reluctantly, I acknowledged that I was going to have to pocket my pride.

At the corner shop I bought a packet of pasta and a jar of sauce; a tin of peaches and a carton of cream. Then, on second thoughts, I bought a box of chocolates. Keeping on the right side of Di seemed the diplomatic way to go under the circumstances.

Back at the flat she seemed to have calmed down too. My purchases seemed to satisfy her and when I gave her the chocolates and apologized for forgetting to shop earlier, she shook her head.

'Thanks. It's all right, Lou. I'm sorry too if I was a bit sharp but I've had a pig of a day at the office.' She put a pan of water on for the pasta and opened the jar of sauce. 'I know you're dying to tell me all your news but there are a couple of things I'd like to clear up first.'

'OK, like what?' I said, my back to her as I began to lay the table. I had the distinct feeling I wasn't going to like what she was about to say.

'Well, if you're planning to stay on for a while that's fine. You know you're always welcome, but I'm going to have to ask you to contribute to the bills – food and so on.'

'Oh – OK then.' Fair enough, I thought. Maybe I had been taking advantage a bit.

'And – maybe a little towards the rent.'

'Right.' *The rent*! What else was she going to ask for?

'And – I hate having to mention this, Lou, but do you think you could try and be a bit tidier around the place?'

I bit back a sharp retort. I'd always thought of Di as my best friend. Now all at once she was behaving more like some sort of nanny. But I didn't really have a choice but to agree to her terms if I wanted to stay. I forced a smile. 'OK, Di,' I said through clenched teeth. 'Sorry if I've been making your home look like a tip.'

She didn't argue, seeming to miss the irony. Smiling, she said, 'Well, that's got that out of the way. Now – are you going to tell me about your day while we wait for the pasta to cook?'

Somehow she failed to see that she'd just taken most of the shine off it.

Chapter Five

'COME INTO THE KITCHEN and sit down, Mum. I've got something to tell you.'

Susan had just brought Peter home from a trip into town to see Father Christmas. It hadn't been a success. To her horror he'd been terrified and ran screaming from the scarlet-clad figure into his grandmother's arms, without even waiting for his gift. All the way home on the bus he'd been inconsolable and now he stood clutching Susan's hand, his lip trembling. She was shocked that Karen hadn't even noticed his distress.

'Can we just get Peter his tea first?' she asked. 'He's a bit upset.'

For the first time Karen looked down at her son. 'What's the matter with him?'

Susan frowned and shook her head in an attempt to play the situation down but Karen persisted. Crouching down to the little boy's level she asked, 'What's the matter, baby? Don't you feel very well?'

Peter shook his head. 'Don't like man,' he said, a tear rolling down his cheek.

Karen looked up accusingly at Susan. 'What man? What happened?'

Susan sighed. 'I took him to Harvey's to see Santa but Peter was afraid of him.'

Karen picked Peter up and dried his tears. 'Really, Mum,'

she said over his head. 'What were you thinking? He's far too young for that kind of thing.'

'I remember taking you at that age and you loved it. I'm sorry. I did tell him beforehand where we were going and he seemed to be looking forward to it.'

Peter soon cheered up at the promise of a boiled egg and soldiers and it was while Karen was tying on his bib that Susan asked, 'So – what was it you wanted to talk to me about?'

'Oh, yes.' Karen looked up, a finger of wholemeal bread halfway to Peter's open mouth. 'I just wanted to say that we won't need you to babysit Peter after next week.'

'Oh!' Susan was stunned. 'So have you decided to go along with what Simon wants?'

'What do you mean?'

'I take it you're giving up work till Peter is older?'

'Good heavens, no, I'm not!' Karen said firmly. We're getting an au pair.'

Susan's eyebrows rose. 'An *au pair* – from abroad, you mean? To live in?'

'That's what au pair means, yes. She's Dutch. I always think they're very homely people, the Dutch, don't you?'

'I don't think I know any,' Susan said faintly. 'Where did you find her?'

'Online, but of course we've spoken on the telephone.'

'Since when have you been able to speak Dutch?'

'I don't, of course, as you very well know, Mum. Adrey – that's her name – speaks perfect English. I'd hardly be employing someone Peter couldn't understand, would I?'

'So I'm to be surplus to requirements after next week, then, am I?'

Karen sighed. 'Oh dear, I did hope you wouldn't take it like that, Mum.'

'How else did you think I'd take it? I thought you were happy with our arrangement. Have I done something wrong?'

'No, of course you haven't and I – *we* are really grateful for all you've done. It's just that Simon has been really stuffy about me going back to work so soon. As you know, he's always been adamant about Peter going to a nursery while he was still so young and he's got all these archaic ideas about a woman's place.'

'So you're making a stand?'

'I suppose you could call it that, yes.'

'And Simon's all right with this au pair thing, is he?'

Karen wiped Peter's mouth and replaced his empty plate with a bowl of mashed banana which she began to spoon into his eager mouth. 'We've come to an agreement. He's been complaining that the house isn't as spick and span as it was before, and he says he's fed up with convenience food although we only have ready-meals very occasionally. Adrey will do the housework and some of the cooking as well as taking care of Peter, so he could hardly argue on those counts.'

'Can you afford it?'

'As long as I'm working, yes.'

'And what about your privacy?'

'We're going to make the spare room into a bedsit.'

'And suppose you don't like each other or she proves to be unsuitable? Once she's here it'll be difficult to get rid of her, won't it?'

Karen had the grace to look slightly embarrassed. 'As a matter of fact, she came over last weekend for that very reason and we all got along very well. She's a lovely girl, very practical and down to earth. And most important, Peter seemed to take to her right away.'

'I see, so it's all set in stone, then?'

'Oh, don't be offended, Mum. I didn't want to tell you until we were sure it was the right way to go.'

'Until *you* were sure it was the right way to go,' Susan corrected. 'I'm still not convinced that Simon approves of your plan. It doesn't sound like his kind of thing at all to me.'

'Well, it'll have to be and that's that!' Karen said. 'Anyway, I reminded him that if the spare room was occupied we wouldn't be able to accommodate Louise any more.' She smiled. 'It was my trump card and it seemed to seal the deal.'

'I suppose that means *I'll* have to put her up.'

Karen wiped Peter's mouth, took off his bib and lifted him out of his highchair. 'There you are, darling. Off you go and play for a while till bath time.' As he scampered away she looked at Susan. 'I had a text from Louise last week. It seems she got that part she was hoping for in this new musical show so I don't think we'll be seeing her again for a while.'

'A text, eh? That's more than I got – not so much as a phone call or a postcard. Did she mention anything about Christmas?'

'Not a thing.'

'Well, at least that's a relief.' Susan stood up. 'I'd better be getting home. It's dark so early these evenings and the buses get full up at the rush hour.'

'I'd run you home but Simon's not in yet and I can't leave Peter.'

'Of course you can't. I wouldn't dream of putting you out.'

Karen shot a quick look at her mother as she walked to the door with her. 'Mum, please don't think we're not grateful for all you've done since I went back to work. I couldn't have managed without you. But you know, you're not getting any younger and an energetic toddler must be tiring for you.'

Susan bridled. 'I'm not *ninety*, Karen. If Peter had been too much I'd have said.'

Karen bit her lip. 'I'm not sure that you would. Anyway, you know what I mean, Mum.' She laid a hand on Susan's arm. 'You should be enjoying life – making new friends, joining things.'

'Like bingo or a sewing circle, you mean?'

Karen chuckled and gave her arm a push. 'If that's what turns you on!'

Susan laughed in spite of herself. 'Well, I hope this new scheme of yours turns out well. But if it doesn't you know where I am. I'm not one to bear grudges. At least, not where my grandson is concerned.'

Karen kissed her cheek. 'Thanks, Mum. You're a real treasure.'

Susan queued in an icy drizzle for fifteen minutes before a bus finally turned up. As she climbed aboard, she saw that it was standing room only and she gave a resigned sigh as she grabbed a strap.

'Please, do have my seat. I'm getting off soon.'

A distinguished-looking man with thick silvery hair was easing himself out into the crowded aisle beside her. He wore a belted trench coat and carried a furled umbrella. Susan looked up into the smiling brown eyes and felt herself blush.

'Oh – well, thank you very much but there's really no need. I'm not going very far.'

He inclined his head. 'I insist.'

Susan sank gratefully into the seat and smiled up at him 'Thank you.'

Although he didn't speak again, Susan was acutely aware of him standing next to her in the crowded bus and to her surprise as she stood up at her stop, she saw that he was already alighting onto the pavement. Turning and noticing her, he held out an arm to help her down.

'Please, allow me. Do you have far to walk?' he asked.

Susan shook her head. 'I live in the flats; Snowden House. It's on the next corner.'

'Then please share my umbrella. I'm going the same way. I'll walk along with you. You can't be too careful after dark these days.' He looked down at her with a wry smile. 'Listen to me! No doubt you're wondering why you should trust a man you've never set eyes on before.'

Susan felt herself blushing again and was grateful for the dusky half-light.

'Not at all. You're quite right and I'm very grateful. You hear of so many muggings and handbag snatches that I don't venture out much in the evening, this time of year.'

'Very wise. I don't normally use the bus but my car is in for servicing today so I don't suppose you've ever set eyes on me before.' He smiled down at her.' I've noticed you on several occasions though,' he confessed as they walked along the pavement. 'I've seen you with a little boy in a pushchair, at the corner shop and occasionally in the park.'

'That's Peter, my little grandson,' she told him. 'I take care of him while my daughter is at work. She's a teacher.' Suddenly she remembered that her services had just been discontinued and added, 'Well, that is to say I *used* to take care of him.'

'Used to?' He looked down at her. 'But surely he's too young for school?'

'My daughter and son-in-law are getting an au pair; a Dutch girl. Karen, my daughter, thinks I'm past it.'

He laughed out loud. 'Good heavens, these young people! *Past it*, indeed. The very idea!'

They reached the corner of the street and the entrance to Snowden House.

'This is me,' Susan said. 'Thank you so much for seeing me home. There aren't many gentlemen around nowadays, more's the pity.'

He looked up at the small block of flats. 'It looks very nice but I'm afraid I couldn't bear to live in a flat,' he said. 'I love my garden too much.'

'I've only been here a year,' she told him. 'And I miss my garden too. One of these days I'm going to buy myself a nice little bungalow with a garden.' Remembering Karen's bombshell she added, 'Now that I won't be taking Peter out, I won't get enough exercise. I don't like going for walks by myself and I hate those jolly hockey sticks keep-fit classes.'

'I run a gardening club at the local college,' he told her. 'I know you don't have a garden yet but you'd be very welcome to join – ready for when you get one.' Reaching into an inside pocket, he took out a card and handed it to her. 'The details are on there along with my telephone number. Think about it and give me a ring.'

Susan stared down at the card, unable to make out any details in the rapidly fading light. 'Thank you. It sounds really interesting. But surely you'll be closing for the Christmas break soon?'

'We still have a couple of sessions to go. If you're interested you could treat it as a taster. No charge of course. Well, I mustn't keep you standing here in the cold any longer,' he said. 'Give it some thought.' He held out his hand. 'I'm Edward

Mumford by the way – Ted.'

Susan took the proffered hand and found it large and strong. A typical gardener's hand, she mused. 'Lovely to meet you. I'm Susan,' she told him. 'Susan Davies.'

Chapter Six

THE VENUE FOR THE read-through was in a run-down church hall in Stoke Newington and I had a hell of a job to find it. When I finally tracked it down, in a scruffy back street, I was fifteen minutes late, hot and out of breath. The door creaked like something out of a budget horror film as I let myself in, and the atmosphere of damp mustiness nearly took my breath away. I needn't have worried about being late. Paul clearly hadn't arrived yet. In one corner of the large empty space, an assorted bunch of out-of-work actors sat on a semicircle of chairs next to an ancient upright piano; all of them half-hidden behind newspapers. They barely looked up as I entered and I didn't recognize any of them.

As I crossed the hall in my high-heeled shoes, my footsteps echoed embarrassingly on the bare floor and one of the assembled group looked up from her paper.

'Hi there! Are you here for the read-through?' I forced a smile at the middle-aged woman, taking in the tatty fake-fur coat and jeans. Her hair, an unlikely flame colour, was tied back with a purple scrunchie. She didn't look much like Jane Austen material.

'Yes. I'm playing the leading part actually,' I told her. What part could *she* possibly be playing? I asked myself. Although it was only a read-through, I'd made a special effort with my appearance this morning but she looked as though she'd thrown on the first thing she'd picked up off the floor.

'Really? Good for you. Come and sit down. I'm Carla Dean and I'm playing Mrs Bennet.' She looked me up and down critically and chuckled. 'Odd, that, isn't it? Seeing that we're obviously about the same age.'

I chose to ignore the barbed remark. 'Is there any coffee?' I asked through clenched teeth.

She laughed. 'Coffee? You must be joking, darling. Don't know if you've noticed but this is hardly the Ritz. I'm afraid you'll have to wait till after. I think there's a café round the corner.' Her voice was deep and throaty and she exuded a powerful odour of tobacco.

'Is Paul here?' I asked as I took a seat.

'Not yet.' Carla opened her bag and took out a packet of cigarettes. 'I'm going to slip outside for a drag.' She offered me the packet. 'Join me?'

'No, thanks. I don't smoke,' I told her stiffly. '*I* try to look after my voice.'

'OK, suit yourself,' Carla said good-naturedly as she stood up. 'Looks as though His Lordship's going to be late.' She crossed the hall, her scuffed trainers making no sound on the bare boards. As she reached the door, it opened to admit a man. Everyone perked up with a rustle of newspapers, but when they saw that it wasn't Paul they relaxed again. As he came closer my heart gave a leap of recognition.

'*Mark*!' I said. 'Mark Naylor!'

His face broke into a smile. 'If it isn't little Lou Davies. What a lovely surprise. Bloody hell! I haven't set eyes on you since drama school. How long is it – twenty years?''

'Nowhere near! Don't exaggerate.' I glanced around, hoping no one else had heard his crashingly tactless remark. 'And actually, it's Louise Delmar nowadays,' I added, lowering my voice.

He pulled a comically apologetic face. 'Whoops – sorry – on both counts.' He fetched a chair from the stack in the corner and sat down beside me. 'Well, this *is* a surprise,' he said. 'How did you get involved in this little epic?'

'In the usual way – through my agent,' I told him. 'I'm playing Elizabeth.'

His eyebrows rose. 'The lead, no less. Wow! Good for you.'

'What about you?'

'Wickham,' he said. 'Not much of a singing part but then I'm not much of a singer.' He nudged my shoulder. 'So – what have you been up to all these years? I must say you look as if you've done all right. Come on, tell me all about yourself.'

'I haven't done too badly,' I told him non-committally.

He glanced down at my hands. 'Married?'

'Good heavens, no! You?'

'Need you ask? You blighted my love life forever. After you turned me down, I never looked at anyone else.'

I laughed. 'I don't believe a word of it.' Mark had been besotted with me when we were at drama school – used to follow me around like a lost puppy. I'd been fond of him too. He was always such fun, but he didn't have a lot going for him. He wasn't blessed with looks and he had neither cash nor influence, all of which were important to me back then. Well, still are. When he asked me to marry him and I turned him down, he insisted in his over-the-top, flamboyant way that I'd broken his heart. To be honest, I was never all that certain that he was serious. Most people thought he was gay, although I knew from experience that he most definitely wasn't. Once we left our paths hadn't crossed again – till now.

I opened my mouth to answer him but before I could reply, the door creaked opened to admit Carla, accompanied by Paul Fortune. Feet shuffled and newspapers were hastily folded away as the rest of the cast came to life. Paul apologized for his lateness.

'Sorry, folks,' he said. 'Had a string of phone calls just as I was about to leave. Are we all here?' He looked round and his eyes alighted on me. 'Ah, Louise. Glad you could make it.'

Mark nudged me. 'Ooh! Looks like you're well in there, sweetie.'

'He saw me in a show I was in a couple of months ago and offered me the part on the spot,' I told him, massaging the truth slightly. 'We've had lunch together a couple of times – along with Harry, my agent, of course.'

'Oh, of *course*.' He treated me to his quizzical, lopsided grin. 'You'll be telling me next that you haven't been to his flat on your own.'

I gave him my enigmatic smile. 'Just the once.'

'Snap!' he said with a flash of his sharp blue eyes.

I felt my eyebrows shoot up. '*You've* been too?' I wanted to ask more but Paul was handing round the scripts.

'With your permission, I'll run through the songs for you before we start reading. Just so that you can get an idea of the melodies,' he said. He pulled a face as he lifted the lid of the piano to expose the discoloured keyboard. He placed his music on the dusty stand. 'I'd better apologize in advance,' he said. 'I've got a feeling this old girl isn't exactly a Steinway.'

'Or even a Yamaha,' Carla quipped. A half-hearted ripple of laughter went round the rest of the cast and I nudged Mark. 'She's playing Mrs Bennet,' I whispered. 'She thinks she's Judi Dench!'

Mark smothered his splutter of laughter behind his hand. '*Behave*!' he whispered back.

The read-through went off reasonably well and when it was over, Paul seemed quite pleased.

'Sorry, people, but I'm going to have to rush,' he said as he crammed his music into his briefcase. 'I've got all your addresses so I'll post you a rehearsal schedule as soon as. Do please start learning your lines. We'll start working on the songs once we all get together with a decent piano. It won't be till after Christmas now. Have a good holiday, all of you.'

As we packed up our scripts I turned to Mark. 'I don't know about you but I'm spitting feathers. Do you fancy a coffee?'

He nodded. 'You bet.'

I glanced across to where Carla Dean was deep in conversation with another member of the cast. 'Quick,' I said, grabbing Mark's arm. 'Let's get the hell out of here before we get stuck with her.'

We found a decent-looking pub and ended up ordering lunch from the tempting-looking menu. As we waited with our much-needed drinks, Mark took his coat off and for the first time I noticed his expensive, well-cut trousers and cashmere sweater. In our student days, he wore tattered jeans and T-shirts. As he reached out a hand for his drink, I also noticed the Rolex watch on his wrist.

'You look very prosperous,' I remarked. 'Have you been successful, or just lucky?'

He grinned and lifted his whisky and soda with a flourish. 'Here's to us, darling.' He took a sip and looked thoughtful. 'Successful or lucky? Mmm, I guess I've had a bit of both. I've had a few small parts in touring plays. You don't get rich on that but it was enough to keep me ticking over. No, the best break I had was when an uncle of mine died. He'd made a lot of money in his day – out of pet food, would you believe. He'd never married or had a family and he didn't leave a will, so as his next of kin I copped for the whole shebang.'

'Well, well! Pet food, eh?' I raised my gin and tonic. 'Congratulations. Here's to my very own pedigree chum!'

He laughed. 'Nicely put. I see you haven't lost your razor sharp wit!'

'So – I expect you live in a stately home in Surrey or somewhere.'

'No.' He shook his head. 'Part of the legacy was my uncle's apartment in Stanmore. It was nicely furnished and equipped so I just moved in there. It's quite handy for the Tube. I can get up to the West End in half an hour.' He took a sip of his drink.

I studied him over the rim of my glass. 'So – now that you're in the money, why on earth are you still bothering to work?'

He gave me a whimsical smile. 'This business gets under your skin. You know; *the sound of the greasepaint, the smell of the crowd*! I know I'll never have my name up in lights. I'm not very good. I've always known that, but the whole thing – atmosphere, excitement – it's very seductive.' He smiled wryly. 'Being offered the part in this show has been a huge break-through for me.' He smiled. 'The West End, eh? I only hope I can hack it. I'm sure you remember my flair for fluffing lines.'

I laughed. 'I remember your genius for ad-libbing. More than once you were the cause of sheer chaos. By the way, what did you think of the rest of the cast?'

He grinned. 'Bit of a motley crew. Your Darcy's not bad-looking, though I'd swear he's wearing a wig. Nobody's hair is that perfect; either that or he spends a fortune on hairdressing.' He glanced at me. 'Speaking of which, have you been asked to put cash into this show?' When I nodded he asked, 'So how did you come by the necessary readies?'

'My dad died and left me the family home.'

'Wow!' His eyebrows rose. 'That must have been a blow to the rest of the family.'

'Oh, they're all right,' I told him airily. 'My sister is married to a guy with a fabulous job and Dad left my stepmother enough to keep her comfortable. They were fine about it.'

'Good! So you and I are in the money at last? A far cry from those hard-up drama-school days.'

'You could say that.'

As our food arrived, I took in Mark's appearance again. He'd changed quite a bit. The old Mark with his sense of fun and his flamboyant manner was just the same, but the brown hair that had once straggled down to his shoulders was now cut in a crisp, short style and frosted at the temples in the very best romantic novel fashion. His wiry, stick-thin body had broadened into quite a presentable physique and the few lines on his face actually improved his looks. He turned and smiled at me as the waiter walked away.

'Well, this looks pretty good to me. I'm starving.' He picked up his napkin and tucked it under his chin.

Yes, he's certainly improved with age, I told myself. Not to mention the fact that he'd come into money too. Maybe it was fate, our meeting up again.

'Where do you live?' he asked.

'Earl's Court. A little flat I keep for when I'm up in town,' I told him glibly.

'Right, so will you be staying there for Christmas?'

I shook my head. 'Oh, no. I expect I'll be going home. I've had several invitations but you know how it is; Christmas is the time for families, isn't it? They'd be so disappointed if I cried off, especially Peter, my little nephew.'

He looked wistful. 'I see. You're so lucky. Unfortunately both my folks died a few years ago – within weeks of one another. And my sister lives in Australia, so for me it'll just be a solitary frozen dinner-for-one and *The Great Escape* on the telly.'

'Why don't you book into a hotel?' I asked, wishing I hadn't been quite so quick off the mark with the self-boosting lies.

He shook his head. 'I tried that once but the place was full of sad, lonely bastards like me; very depressing.' His eyes brightened into the warm smile I remembered so well. 'The best Christmas present I could have had is meeting you again,' he said, squeezing my hand.

As we parted on the pavement outside, we exchanged mobile numbers.

'See you when we start rehearsals,' Mark said. 'Take care and have a lovely time with your folks.'

I watched wistfully as he walked off down the street towards the Underground station. Why did I have to come out with all those bloody lies? I asked myself. If we were going to be seeing a lot of each other, the truth was bound to come out sooner or later, and I was going to have to think up a lot more fibs to cover myself. They rolled off my tongue without my even thinking. It was a defence method I'd learned as a child getting mercilessly bullied at school, and somehow it had become a habit I couldn't shake off. I never even stopped to think about the consequences – even fleetingly believing in my own fantasies at times. It had landed me in trouble more than once in the past. As I turned towards the bus stop, I mentally kicked myself.

'Why is it you never learn, you silly cow?' I muttered.

It really had been lovely meeting Mark again. No one except my dad had ever really loved me as he had and I'd chucked it back in his face. I'd been rotten to him back in our student days but he obviously held no grudges. As I boarded the bus, I resolved to make it up to him in the months that followed.

*

'I take it you're going home for Christmas?' Dianne asked as we prepared the evening meal together later. She hadn't even asked me about the read-through when she got in from work and I was feeling a bit miffed; too proud to bring it up myself. The only thing that seemed to concern her was that I'd remembered to do some shopping on my way home. I thought she'd have thanked me but she just seemed to take it for granted.

She glanced at me. 'Are you, then – going home, I mean?'

'I haven't actually been invited,' I replied. 'I had a text from Karrie to say that they've got this Dutch au pair living with them so that's obviously a hint that there won't be room for me.'

'Can't you stay with your stepmother?'

I shrugged. 'Her sofa isn't exactly what you'd call comfy. It's only a two-seater and lumpy with it. Anyway, I expect she'll be going to Karrie and Simon's for the day.'

'You could go too. And the sofa can't be all that bad – just for a few days.'

I looked at her. 'I thought you and I would be spending Christmas together,' I confessed. 'I was quite looking forward to it and I don't want to push in where I'm not really wanted.'

She looked uncomfortable, her head bent over the potatoes she was peeling. 'The thing is, Lou, my parents really want me to go home. My brother is getting engaged and they're planning a party on Boxing Day.'

'Oh, well you must go of course. I'll be OK here on my own.' Privately, I thought she might have suggested taking me along too.

Dianne frowned. 'I'm not allowed to sublet the flat, Lou,' she said.

I laughed. '*Sublet*? I'll only be staying here for a few days on my own, surely that doesn't constitute subletting?'

She dropped the potatoes into a saucepan of water and lit the gas under it, turning to me with a determined expression. 'To be brutally honest, Lou, I'd rather not leave you in the flat on your own.'

'Why not? I'll be OK.'

'Oh, I'm sure you would be, but to be frank, you're not the tidiest of house guests. Take today, for instance. I came home from a hard day at work to find your breakfast washing-up still in the sink, including a burnt porridge saucepan, and I found several pairs of your tights stuffed down the back of the settee when I was vacuuming the other day. You never even *think* of taking your turn with the cleaning and you leave wet towels all over the bathroom floor. If you're here for a week on your own, I dread to think what state the flat will be in when I get home.'

'Speak your mind, why don't you?' I sniped. I stared at her. 'Anyway, since when have you been so house-proud?'

'I'm not ...'

'You sound positively paranoid to me!'

'I do like some kind of order.'

'OK, I'll get out of your hair,' I told her. 'As a matter of fact, I turned down an invitation just today because I didn't want to disappoint you.'

She looked slightly relieved. 'Well, maybe it isn't too late to change your mind.'

I turned to walk out of the kitchen. 'Well, we'll just have to keep our fingers crossed, won't we? Otherwise it looks as if I'll be spending Christmas in a cardboard box!'

I waited until Dianne had left for work the following morning, then I got my phone out and clicked on Mark's number. He sounded sleepy when he answered.

'Who the hell is this ringing me in the middle of the night?'

'It's me – Lou,' I told him, laughing. 'And as a matter of fact, it's half past eight.'

'Like I said – the middle of the night.' He cleared his throat. 'Only joking, Lou. It's good to hear from you any time. What can I do for you?'

'It's more what I can do for you,' I told him. 'How would you like me to come and cook you a traditional slap-up Christmas dinner in your own home?'

I could almost hear him blinking. 'Sorry, it's a bit early for riddles,' he said. 'I could have sworn you offered to cook me

Christmas dinner in my own home. You did say you were Lou Davies, didn't you – not meals on wheels?'

I laughed. 'No, it's me all right – and it's *Louise Delmar*, cloth-ears! I'll explain – I had a call from my stepmum last night; they're all off to Sweden for a Scandinavian Christmas. Of course, they wanted me to join them but I don't fancy it. My flatmate is off home so I thought – why don't Mark and I team up? It could be fun.'

'That would be great, Lou!' He sounded fully awake now. 'A home-cooked Christmas dinner plus *your* company! How lucky could I get?' There was a pause then he said, 'Flatmate? You never said you had a flatmate.'

I bit my lip hard and forced a laugh. 'Didn't I? Well, it's only a recent arrangement. She had nowhere to go so I offered her a room.' Suddenly I remembered that he knew Di from our time at drama school, but I decided not to mention that.

'Right.' He lowered his voice. 'Hey – I hope you're a good cook.'

'The best,' I lied, crossing my fingers and thanking God for Aunt Bessie.

Chapter Seven

'YOU'RE NOT GOING TO let *her* cook the Christmas dinner, I hope.'

Simon tutted irritably as he straightened the bottom sheet and punched his pillow into shape. 'Look at this. She doesn't even know how to make a bed properly!'

'Oh, stop finding fault with the poor girl. You haven't stopped since she arrived.' Karen slapped cleansing cream onto her face and whipped a tissue out of the box on the dressing table. 'She's an absolute treasure with Peter. He adores her. She's so patient and creative with him.'

'So she might be but you can't say the same about her cooking. It's abysmal,' Simon complained. 'Tasteless stodge in watery gravy.'

'Shhh! Keep your voice down. She'll hear you.'

'That's another thing,' he hissed. 'The only place we can actually have a private conversation is in bed and even then it has to be conducted in whispers. And as for doing anything else in bed ...'

'*Simon*! Keep it down, for God's sake.' She turned to him. 'Look, if I keep working we'll be able to afford a bigger house in a couple of years. Surely it's worth a few sacrifices.'

'That's a matter of opinion. And I thought you said she spoke fluent English.'

'She does.'

'When *we're* around, yes, but when she's on her own with Peter she obviously speaks to him in Dutch. When I spoke to him the other day he came out with a mouthful of it and when I asked her about it, she said it was a nursery rhyme. I was horrified. It's coming to something when I can't even understand what my own son is saying.'

'Isn't it an asset for him to be growing up bilingual?'

'If it was German or French, yes, but where or when is he going to need Dutch?'

'You never know.' Karen drew back the duvet and climbed into bed. 'Why can't you think about all the advantages of having Adrey with us? You have to admit that the house is spotless and your shirts are beautifully ironed.' She turned to look at his unconvinced face and decided to play her trump card. 'Best of all, Louise won't be joining us this Christmas because we simply haven't a spare room any more.'

'Well, that is a plus, I suppose,' he said grudgingly.

'And aren't you pleased that I'm not so tired these days?' She reached across to kiss him and her hand crept under the waistband of his pyjamas. 'In fact I'm feeling really sexy tonight.'

He grasped her hand and firmly removed it. 'Well I'm *not*. How do you expect me to work up any enthusiasm when there's just a thin partition between us and her?'

'She's probably asleep. Anyway, we don't go in for all that noisy stuff.'

He turned his back to her and switched off his bedside lamp pointedly. 'I told you, I can't work up any enthusiasm and if I can't – well, surely I don't have to draw you a picture?'

'There's no need to be crude.'

'Just go to sleep, Karen, or you'll look like nothing on earth in the morning.'

Deeply hurt, she switched off her own lamp and turned over, a lump in her throat. Presently, a large tear slid down her cheek and she brushed it away with a corner of the duvet. After a few minutes she felt Simon turn towards her.

'I'm sorry, love. I didn't mean that,' he said quietly. When she made no reply he reached out a hand to rest on her waist. 'I've got such a lot of stress at school as you know, what with the coming festivities and everything. I know how hard you work too and I do appreciate it.' The hand crept up to cup her breast. 'Know what?' he whispered in her ear. 'I think I can hear Adrey snoring.'

Karen couldn't conceal the chuckle that rose in her throat. Taking his hand she moved it slowly down her body and lifted her face for his kiss.

Chapter Eight

SUSAN SAID GOODNIGHT TO the rest of the group as they left the classroom in twos and threes and began to put on her coat. It was the second meeting of the Green Fingers Club she had attended and the last before Christmas. She was sorry. She had enjoyed the meetings so much. The other members were pleasant company and she had learned a lot from Ted's expertise. Her only regret was not having a garden of her own on which to practise some of the new skills and tips that she had scribbled down in her notebook

'Can I give you a lift home, Susan?' Ted stood at her side, winding a scarf round his neck. 'It's freezing outside tonight.'

She smiled up at him. 'Thank you. That would be lovely.'

They walked along the corridor together and out into the frosty night air. As they walked across the car park he turned to her.

'I don't suppose I'll see you again until after Christmas, so would you like to come for a festive drink with me?' His brow furrowed. 'Or perhaps you don't drink and you'd prefer a coffee?'

Susan laughed. 'I do drink – moderately of course, but either would be nice. Thank you.'

They opted for the Coach and Horses, a comfortable pub almost next door to the college. Inside the lounge bar it was warm and comfortable, tasteful evergreens decorated the walls and there were red candles on every table. They chose a table near the log fire and Ted went to the bar for drinks, a pint of bitter for him and a gin and tonic for Susan. 'I got you ice and a slice,' he said as he joined her at the table. 'I hope that's all right.'

'Lovely. Thank you.' She raised her glass. 'Here's to a happy Christmas.'

Putting his glass down, he looked at her. 'So – do you think you'll be joining us next term, or did you find it unutterably boring?'

'Oh, no!' Susan said quickly. 'I mean about it being boring. I was fascinated. It was so interesting. And yes to your first question; I'd love to sign up next term. My only complaint is that it's going to be so frustrating, not being able to try out all your useful tips and advice.'

He leaned towards her. 'I've been thinking about that,' he told her. 'My garden is quite small. Only really big enough for a lawn and a few flower beds, so I grow my own vegetables on an allotment. The chap next to me is giving his up in the New Year. How would you feel about applying to the Council to rent it?'

Susan looked doubtful. 'An allotment? Do they let women have them?'

He laughed. 'Of course they do. Why not? The only people who are not allowed are those who have a bad track record for neglect.'

Susan sipped her drink thoughtfully. 'How big would it be? I'm not sure I could manage on my own.'

'You wouldn't be on your own. As I said, the plot is next to mine so I'd always be there to help you out.' He paused. 'Look, sorry if I'm going too fast for you. Maybe I'm being presumptuous – taking too much for granted. If you hate the idea just say. It was only a thought.'

He looked so embarrassed that Susan reached out an involuntary hand to pat his arm. 'No, no, it's a lovely idea and so kind of you to think of me.' She sat back with a smile. 'Just fancy, home-grown organic vegetables. I could keep Karen supplied and still have lots left over.'

'You could.' Very softly he placed his hand over hers on his sleeve. 'And there's a farmers' market every other week in town. If we get a glut, some of us club together and rent a stall. It's great fun and quite profitable too.'

'It sounds it.' Susan felt her cheeks flush and she didn't know whether it was the gin and tonic, the heat of the fire, or the warm feel of Ted's hand on hers. Then a thought suddenly occurred to her. 'Oh, but where are these allotments? Could I get there easily? And what about gardening tools? I mean, I did have some when I had a garden before but I gave them to Karen and Simon.'

'You could always ask for them back,' he said, his eyes twinkling. 'I have a lock-up shed on my allotment and you could keep your tools in there with mine.' He took a drink of his beer. 'As for getting there – it's only fifteen minutes in the car; you could come with me.'

She smiled tentatively. 'How can I refuse?'

'Of course you can refuse!' His smile vanished. 'Please don't feel obliged to agree if you don't like the idea. I'd hate to bulldoze you into it.'

'You've given it so much thought.'

He gave her a wry grin. 'To be honest I've thought of little else.' He looked away. 'It's been a bit of a pipe dream. Meg and I never had children and although I've been on my own for five years I still can't get used to my own company.' He turned to look at her. 'I don't make friends easily. When I started the Green Fingers Club I thought I might meet some like-minded folk and I have, but they come in couples or pairs of friends and anyway, there isn't anyone who I'd say was on the same wavelength as me.' He raised an eyebrow at her. 'Does that make sense or does it make me sound stuffy and difficult?'

Susan shook her head. 'Of course it makes sense. I know what it is to be widowed and like you, I've never been someone who's happy as part of a group. I prefer one-to-one friendships.'

'Somehow I knew that instinctively. But there is a big difference. You have a family.'

'That's true.' On impulse she asked, 'Ted, what are you doing for Christmas?'

He shrugged. 'Sitting in front of the box and wishing it was spring.'

'Then come with me to Karen and Simon's for Christmas Day?' she invited.

'Oh, I couldn't impose on people I don't even know.'

'You wouldn't be imposing, and you'd soon get to know them. I always contribute my share of the food and I help with the cooking; we all muck in. You'd be very welcome and I'd like you to meet my family.' She took in his hesitant expression. 'But it's only an idea. Don't say yes just because I've asked you.'

He smiled. 'I'd love to come. Thank you.'

'Really, Mum! Who is this old man and what on earth made you invite him without asking me first?'

Susan was shocked by Karen's reaction. 'To start with, he's not an *old man*; he's about the same age as me. He's on his own and he's lonely. I had no idea you'd be put out about it. One more can't make all that difference.'

'It could be very awkward, having some stranger in the house,' Karen said. 'Christmas is a family occasion after all.'

'What happened to *goodwill to all men?* And anyway, what about Adrey? She's not family.'

'That's different. She's away from her own family for the first time and I want to make it special for her. I've asked some of the neighbours for pre-lunch drinks so that she can get to know them.'

'*I* wanted to make it special for Ted too. He's been on his own for five years and he's terribly lonely.'

'That's hardly my problem. Anyway, what do you know about this Ted person? You want to be careful, Mum, picking up strange men on the bus. He could be some kind of conman.'

'Well, he's *not*!' Susan bristled. 'And I didn't *pick him up* as you so delicately put it. I haven't lived as long as I have without being able to tell a genuine person from a crook. And if he's not welcome then I'm not coming either.'

Karen looked shocked. 'Oh really, Mum! Don't be so ridiculous.'

'I never ask you to do anything for me, Karen,' Susan said as she marched down the hall. 'But if you can be mean-minded enough to turn down a lonely person at Christmas, then you're not the daughter I thought you were.'

'Mum – *wait*!' Karen caught up with Susan at the front door and took her arm. 'Please, don't go like this. Bring your friend, Ted, if you like. I'm sorry I was snappy. It's just that there's so much to do, what with shopping – cards and presents and everything at home and all that's going on at school. You know how stressful Christmas is …'

'I don't want you to wear yourself out on my account.' Susan firmly detached her arm from Karen's grasp. 'We'll be fine. I'll invite Ted to spend Christmas with me at the flat and you can concentrate on making it a special Christmas for your Dutch au pair and the neighbours.'

'So, I hope you don't mind, Ted, but I thought we'd have Christmas Day on our own at my flat,' Susan said when she rang Ted the next morning.

There was a slight hesitation at the other end of the line. 'Susan – my dear, I hope you're not giving up the day with your family on my account.'

'Not at all,' she assured him. 'When I thought about it I realized that it might all be a bit overwhelming for you. They're having friends in for the evening and I know you don't like crowds of strange people. I don't know any of them so I'm not very keen either.'

'You're absolutely certain about this?'

'Absolutely.'

'Well, if you really are, then the answer is a resounding yes. As a matter of fact the prospect of it just being the two of us is very pleasing,' he said. 'It'll be an opportunity to get to know one another better.'

'Yes, it will.' Susan smiled, feeling a little flutter of anticipation. It was so long since she had planned a Christmas of her own. 'Right, then,' she said. 'I'd better get down to the supermarket before the last-minute rush starts.'

'Do you mind if I come with you?' Ted said. 'I absolutely insist on sharing all the expense with you.'

'There's no need, but it would be lovely to have your company,' Susan said. 'Then I won't have to guess what you like and don't like to eat.'

'And on the day I insist on helping,' he said. 'I'm a dab hand with Brussels sprouts.'

Susan smiled to herself, serene in the knowledge that she had made the right decision.

Chapter Nine

MARK WAS WAITING FOR me as I stepped off the train, laden with bags of shopping.

'Wow!' he exclaimed as he took a couple of them from me. 'Looks as if you've bought enough to feed an army. It must have cost you a fortune. Come on, I've brought the car.'

'You've got a car?' I said, following him into the street.

'Of course. I don't use it for going up to town – too difficult to park but I do like it for holidays and days out and so on.'

The car turned out to be a Ferrari – sleek, black and shiny, and Mark's apartment was in a smart block of luxury flats built conveniently handy for the main shopping centre and the Underground station. Mark drove into the basement car park where he had his own numbered space, and then whisked me up to the penthouse apartment in the fastest lift I'd ever experienced.

The apartment was gorgeous, even more lavish than Paul's. The kitchen had every modern convenience. The furnishings were a little old-fashioned maybe, but I had to remind myself that an elderly man had lived here and Mark clearly hadn't updated anything.

The large living room had massive sliding doors, leading out onto a spacious balcony, and I stood in the middle of the room and spread out my arms.

'Oh, Mark, it's lovely. I have to say, you certainly fell on your feet, inheriting all this.'

He smiled. 'It is rather nice, isn't it? I suppose I really ought to change things round a bit – bring it up to date but it always feels like too much hassle. What do they say – if it ain't broke, don't fix it?'

'Well, there's nothing broke about this place.' I kicked off my shoes and threw myself onto the enormous corner settee. 'I'm surprised you can ever tear yourself away from all this to go on tour.' I sat up and looked at him. 'Er – what are the arrangements – where do I sleep?'

He laughed. 'Don't tempt me. Seriously, there are two bedrooms so you can take your pick. They both have their own en suite so we won't bump into each other in the nude first thing in the morning.' He grinned impishly. 'More's the pity!'

'In your dreams!' I said, laughing as I got up and followed him to the spare bedroom. Like the rest of the apartment, it was luxurious but I hazarded a guess that I wouldn't be staying in it for long.

I made spag bol for supper – the only dish I can actually cook from scratch – and we sat and ate it at the kitchen table. I'd hastily unpacked all the frozen Christmas fare I'd stocked up with at the supermarket and hidden them in Mark's massive freezer when he wasn't looking. The turkey was already defrosted (at least I know that much) and I slipped it onto a large serving dish and put it in the fridge. Luckily there was plenty of room. All it contained was a pint of milk and some cans of beer. Obviously Mark was no Gordon Ramsay and I guessed that he existed mainly on takeaways.

After clearing his plate, Mark leaned back in his chair and took a sip of the wine he had produced from a well-stocked wine rack. At least he didn't stint himself on that. 'That was delicious,' he said with satisfaction. 'This is really great. How long can you stay?'

'I think you could have timed that question a little more tactfully,' I told him.

He laughed. 'You know what I mean. Who would have thought a week ago that I'd be sitting here, looking forward to spending Christmas with the love of my life I'd given up hope of ever seeing again.' He drained his glass and refilled it, holding the bottle enquiringly towards my empty glass.

'Yes, please.' I took an appreciative sip. 'You're right. We never know what's round the corner, do we?'

'So – how long *can* you stay?'

Quickly, I calculated. Di was away till the day after Boxing Day. I'd give her a day to feel flat and miss me. 'Four days,' I said. 'That is if you can put up with me for that long.'

'I can put up with you for as long as you like,' he said.

'You might not be saying that a few days from now,' I warned him. 'You're still wearing those rose-coloured glasses you wore twenty years ago.'

'And very comfy they are too,' he said, holding up his glass. 'Here's to our meeting again and to our renewed acquaintance.'

'And to the new show.' I touched my glass to his. 'To it being a hit!'

Suddenly Mark was serious. 'About the show,' he said, putting his glass down. 'Isn't it usual for a show like this – heading for a West End theatre – to be backed by a consortium of people; you know, impresarios?'

'It will be,' I told him. 'Paul said that borrowing money from us to get off the ground is only temporary. He'd got someone lined up. And it's only been difficult because he hasn't booked a star attraction for the lead role.'

'OK, but where's the director?'

'He's been searching for the right person,' I explained. 'He has a really big-name guy interested.'

'Oh, yes – who?'

'He didn't tell me.'

'Well, I hope you're right.'

'I'm sure everything is in place,' I assured him. 'Harry Clay, my agent, is in on the whole thing. I've been with him for years and I trust him.'

Mark took a reflective drink of his wine. 'I was invited up to his flat to audition me for the part,' he said. 'My agent was as surprised as me. He – Paul – showed me the sketches for the sets; very impressive. He said they were already being built up in Yorkshire somewhere. After that we had tea and cakes and he asked me for money.'

'It was similar for me,' I told him. 'Although he'd already seen me in a show, I was in back in the summer. At his flat he played some of the songs for me and I sang one or two.'

'Then you had tea and he asked you for money?'

'Well – yes.'

'Are you with me in wondering if it's all completely kosher?'

'No. I told you; Harry, my agent, is in on it too. He's very shrewd. He'd never risk his money if he had any doubts and he certainly wouldn't let me be taken for a ride.'

Mark was shaking his head. 'It's all a bit odd,' he said. 'I mean, who's ever heard of Paul Fortune anyway – or any of those weirdos we met at the read-through the other day?'

'I told you, there are no big names.'

'Mmm.' He stroked his chin. 'You have to admit, Lou, it's one hell of a risk.'

'Well, that's up to Paul, isn't it? It's his risk and he seems confident enough.'

'I hope you're right.' He took a deep breath and smiled. 'Let's not be pessimistic. It's Christmas, you're here with me and tomorrow I'm going to have the first home-cooked Christmas dinner I've had in years with my first love cooking it for me.' He raised his glass. 'Here's to us!'

I clinked my glass to his. 'To us! And to being optimistic about the show.'

'Absolutely!' Mark said. 'What do I know anyway?'

Later, as I lay in bed – surprisingly alone – I couldn't help thinking about Mark's words. He was wrong, of course he was; he *had* to be. Paul had promised to make me a star. *Your face will be on the cover of all the magazines,* he'd said. He had to be on the level. I couldn't bear it if he wasn't. This was my very last chance.

Christmas dinner was a success – as much to my surprise as anyone else's. Though you have to be a complete loser to mess up a frozen, pre-cooked meal. Mark was delighted. If he suspected that it wasn't *exactly* home-cooked he didn't mention it. His pessimistic mood from the previous night had gone and instead he was on form in the style of the old Mark I remembered so well. After lunch, we watched TV and dozed in front of the realistic living-flame electric fire. After a couple of bottles of champagne, Mark grew amorous and we ended the day in bed together. He'd always been a good lover and he certainly hadn't lost his skills, making me ever so slightly curious about whom he'd been practising on in my absence.

Chapter Ten

KAREN WAS UP EARLY on Christmas morning. Peter had wakened them at six, bouncing on the bed and dragging a pillowcase full of presents.

'Open, Mummy!' he demanded.

Simon groaned and turned over, squeezing his eyes tightly shut as Karen switched on the bedside lamp. 'Take him back to bed, for God's sake. It's the middle of the night.'

Karen slipped out of bed and put on her dressing gown. 'Come downstairs with Mummy, darling,' she said, taking Peter's hand and picking up the pillowcase. 'We'll leave grumpy old Daddy to sleep.'

When all the presents were opened, she left Peter playing with his new toys in the living room and went into the kitchen to make a pot of tea. Adrey was already up, having risen early to make what she called *kerststol,* which turned out to be some kind of fruit loaf traditionally eaten at Christmas for breakfast.

'Peter loves his new teddy,' Karen told her. 'It looks expensive. You really shouldn't have.'

Adrey turned with a smile. 'He's such a good little boy. I wanted to give him something nice for *Sinterklaas.*' She reached into the pocket of her dressing gown and produced a small, brightly wrapped parcel. 'I get this for you too. You and Simon have been so kind and welcoming.'

Karen was surprised. Opening the package, she found a tiny

brooch in the shape of a Dutch clog, encrusted with crystals. 'Oh, Adrey, how sweet,' she said. 'It'll look great on my black dress. Thank you so much. There's a little something from Simon and me under the tree. We'll be opening those later.'

Putting the kettle on, she thought how pretty Adrey was. In her red dressing gown and with her fresh complexion and her long blonde hair hanging down her back in a thick plait, she looked like a Christmas angel. Karen turned and gave her a quick hug. 'I know you must be missing your family today,' she said. 'And you must feel free to telephone them.'

'Thank you, Karen. I would like to do that very much.'

The kitchen door opened to admit Simon in his old navy dressing gown. His hair tousled and his jaw dark with stubble, he still looked grumpy. Glancing at the two women he enquired, 'Any tea going?'

'I'm just making it,' Karen said, reaching for the teapot.

Adrey touched Karen's arm. 'I go now to make the telephone call to my family, if it is permitted?'

'Of course, help yourself – and take as long as you like,' Karen added.

'Thank you. Then I take Peter upstairs to wash and dress.'

Simon slumped at the table. 'Her family does live in Holland, you know,' he growled.

'I am aware of that.'

'Now you've given her carte blanche she'll probably be on the bloody phone all morning.'

'No she won't. She's not the type to take advantage.' Karen poured two cups of tea. 'Cut the poor girl some slack, Simon. This is her first Christmas away from home.'

Simon looked at the loaf, cooling on a wire rack on the worktop. 'What's that thing?'

'It's called *kerststol*. Adrey got up early specially to make it. It's a kind of fruit loaf. They eat it for breakfast in Holland at Christmas.'

Simon sniffed. 'Do they? Well, it smells OK anyway.'

'I'm going to put the turkey on in a minute. If you want, you can help with the vegetables.'

He snorted. 'No way. You know I'm all thumbs when it comes to domestic stuff.'

'All right, then, if Adrey is going to help me, you can take Peter to the park after breakfast.'

When everyone else was upstairs getting dressed, Karen took the telephone into the kitchen and dialled Susan's number.

'Mum – it's not too late to change your mind,' she said. 'There'll be plenty for all of us – including your friend.'

'That's perfectly all right,' Susan said. 'I have everything ready here. Ted and I are going to have a lovely day, thank you.'

'Oh, Mum, you're not still cross, are you? You know I'm only thinking of you.'

'I don't think that's quite true, Karen, but please don't worry. Maybe I'll see you sometime in the New Year.'

'Please, Mum, don't be like that,' Karen begged. 'Look, why don't you come round for a drink tomorrow – bring Ted too. I could make a brunch with some of the leftovers.'

'I'll eat my own leftovers if it's all the same to you,' Susan said. 'I'm not cross, Karen, just a little disappointed by your attitude. But don't worry. I'm perfectly all right and looking forward to having someone to cook for again.'

'But, Mum ...'

'No, I'm not being awkward, I mean it, Karen, I'm fine. Ted and I are going to have a really nice Christmas and I hope you do too. Goodbye, dear. Give my love to Simon and little Peter.'

'Oh – thank you for the presents, Mum.'

'I'm glad you liked them. Thank you for yours. The scarf will go beautifully with my new coat. Happy Christmas, dear.'

'Happy Christmas, Mum.' Karen said faintly. There was a lump in her throat as she switched the phone off. She hated falling out with her mother. Sometimes lately it seemed that she couldn't do right for doing wrong.

Chapter Eleven

TED ARRIVED ON THE dot of twelve. Susan answered his ring on the entry-phone and pressed the button that released the main door to the flats. Ripping off her apron, she took a quick look in the hall mirror to check that her newly set hair was still in place.

When she opened the door she found him beaming outside, a huge bunch of chrysanthemums in one hand and a carrier bag containing a bottle of sherry and another of champagne in the other. He handed both to her. Susan blushed with pleasure.

'Oh, Ted, how thoughtful. But you really shouldn't have.'

'Not at all, my dear. It's so good of you to invite me.' He took off his overcoat and hung it on one of the pegs inside the door. Susan saw that he wore his best dark-grey suit with a pristine white shirt and tasteful blue tie. She thought he looked very handsome.

'Happy Christmas, Ted,' she said. 'Come through and make yourself comfortable. Everything's almost ready so we can have a drink and relax for half an hour.'

'Oh, but I thought I was going to help.'

Susan smiled. 'It's all right. I haven't been up since dawn. I did most of the preparation yesterday afternoon.'

Ted followed her through to the living room and looked around appreciatively while Susan fetched glasses and poured the sherry. A cosy fire was burning in the hearth and in the centre of the room, the table was laid with her best glass and cutlery, set off by red napkins and crackers by each place setting. A Christmas tree stood in one corner, its coloured lights twinkling, and holly was draped around the mirror above the fireplace. There were red candles on the mantelpiece, their flickering flames reflected in the mirror.

'It all looks very festive,' he said, rubbing his hands and holding them out to the fire. 'Nice and warm too.' Susan handed him a glass of sherry.

'Here's to a happy Christmas,' she said.

Lunch was a great success. It was some years since Susan had cooked Christmas dinner for a man and she had loved every minute of it. Her reward was seeing him clear his plate with obvious relish.

'I'd forgotten how good home-cooked food tasted,' he said, pushing his chair back from the table and sighing. 'I don't think I'll need to eat again for at least a week!'

Susan laughed. 'I'm sure you will,' she said. 'I hope so anyway. I've got Christmas cake and mince pies lined up for later on.'

He smiled almost impishly. 'Oh well, I suppose I'll have to force them down.'

He insisted on helping her with the washing-up and then they settled down to see the Queen's speech on TV. When it was over, Ted went out into the hall and came back with a small package which he handed to her.

'Just a little token of my appreciation,' he said almost shyly.

Susan blushed. 'Oh, Ted, you shouldn't have done this. I haven't got you anything.'

'Indeed you have,' he said stoutly. 'This is the best Christmas Day I've spent for years. All the work and the planning you've put into it makes my little offering look meagre. Please open it and see if you like it. If not, please feel free to change it.'

Somewhat flustered, Susan quickly unwrapped the gift. Inside a black velvet box she found a single crystal threaded onto a fine gold chain. In the light it glittered and flashed like a diamond. She took it out of the box and held it up.

'Oh, Ted. It's beautiful!' Standing in front of the mirror she held it out to him. 'Will you fasten it for me?'

He clipped the chain around her neck and the crystal lay winking in the light at the base of her throat. She looked up at him with shining eyes.

'Thank you,' she said. 'It's a long time since anyone gave me anything as nice.' She stood on tiptoe to kiss his cheek but his hands cupped her face, turning it gently to kiss her lips.

For a moment his eyes held hers then he said softly, 'I'm so happy that I found the courage to speak to you that day on the bus, Susan.'

She smiled. 'So am I, Ted. And this is the nicest Christmas I've had for many a day too.'

'I can't wait for the next few weeks to pass so that we can start working on the allotments together.'

'Neither can I.'

They sat down side by side on the settee in front of the fire and Ted's hand found hers, squeezing it lightly. 'I can't remember when I've felt so happy,' he said.

'Neither can I.' For a moment they looked into the firelight together then Susan said, 'Tell me about your wife, Ted. Meg, wasn't it?'

His brow clouded. 'I'm afraid it wasn't the happiest of marriages,' he said. 'If you don't mind, I'd rather not talk about it at the moment. Let's not spoil a perfect day.' He smiled at her and squeezed her hand. 'What about you? All I know is that you have two daughters. I take it your marriage was happy.'

She nodded. 'Frank was a good man. Louise is my step-daughter. Frank's wife walked out on them both when Louise was a toddler. When he and I married she was ten and she resented me dreadfully – thought I'd stolen her beloved dad away, which wasn't the case at all. I tried my best but nothing seemed to work and I'm afraid she still sees me as an outsider.'

'But you have your own daughter.'

'Yes. Karen made up for a lot, but she has her own family now and lately even she seems to be drifting away from me.'

She looked at him. 'I suppose we've always been what they call a dysfunctional family.'

'Sometimes I wonder if there's any other kind,' Ted said with a wry smile. 'Do you know, I think it's time to open that bottle of champagne. I can't think of a better occasion.'

Chapter Twelve

WE WERE THREE DAYS into the New Year when the rehearsal schedule arrived. I rang Mark as soon as I'd opened it and scanned through.

'It doesn't look very intensive,' I said. 'Only two rehearsals a week – and it seems we're still in that draughty old hall in Stoke Newington.'

'So I see,' Mark replied. 'Still, at least things are moving. In the enclosed letter, Paul mentions extra sessions for the music and some with a choreographer. How's your dancing?'

I laughed. 'I think I can hoof it with the best.'

'I'm sure you remember my galumphing efforts. Let's hope Wickham doesn't have anything too energetic to do.'

'I see the first one is the day after tomorrow,' I said, looking at the schedule. 'How are you getting on with the lines?'

'Not too bad, how about you?'

'Oh, OK.' I was remembering Di's cool response when I'd asked her to test me. She'd treated the request as though I were some tiresome kid rehearsing for the school play. No one would ever have thought she'd once longed to become an actress herself. It was more the kind of reaction I'd have expected from Karen.

'See you on Wednesday, then,' Mark was saying.

'What? Oh yes, fine.'

*

Since Di had returned from her Christmas break, she'd been a bit distant. Apparently the engagement party had gone well and she'd had a great time with her family, but I could tell there was something brewing.

To tell the truth I was a bit jealous of Di's happy family life. I'd had a card from Karen and Simon and one from Susan, but no presents and certainly no invitations to visit or enquiries about the new show. I know it was partly my own fault. The trick I played on Karen last time I was staying with her was pretty much over the top, but there is something about Karen that stirs up the devil in me. One look at that sanctimonious, smug little face and I can't help myself. I realize now that what I did could have had disastrous consequences. I just hadn't thought it through and she was right to be angry with me. Nevertheless it hurt to be so completely excluded, especially at Christmas. Now that Dad has gone, I feel I have no one who really gives a damn about me.

Inevitably my thoughts turned again to my mother. I wondered if she regretted what she'd done just as I did. Was she lonely? Maybe she had married again. I could have half-siblings I didn't know about; a whole new family! The thought excited me and I promised myself that I would look into ways of finding her again just as soon as the rehearsals had got underway.

I was right about Di having a bee in her bonnet. When she got in from work that evening, I showed her the schedule that Paul had sent but as she set about unpacking her briefcase, I could see that there was something else on her mind. Eventually she stopped me in mid-sentence.

'Lou – look, can we just sit down a minute? There's something I need to tell you.'

My heart sank. I could tell from her face that whatever it was, it wasn't going to be to my advantage. 'OK, Di, what's on your mind?' I asked. 'I've been trying really hard to keep the place tidy and I'm not behind with my share of the rent, am I?'

'No, nothing like that.' She took a deep breath. 'It's just that – well, while I was at home this time I met up with an old flame of mine again. Mike and I used to go out together when we were teenagers but we drifted apart when I came up to London to drama school and he went off to study law. He married someone else during that time but they divorced last year.' She glanced at me. 'We found we still liked each other – quite a lot actually, the old spark was still there. Over the holiday we met a few times. Mike has just landed himself a fabulous job up here with a law firm. He said he was going to look me up when he started – ask me to help him find a place to live, so it was lucky our meeting again.' She glanced at me again. 'So – I thought – I said ...'

'You told him not to bother looking because he could move in with you,' I completed the sentence for her. I laughed at her expression. 'Christ, Di, I thought you were never going to spit it out. You could have said it all in a few words: 'I've met an old boyfriend and we still turn each other on so we're going to move in together – oh, and by the way, I want you to move out.'

She had the grace to wince. 'You make me sound like a ...'

'I make you sound like what you are,' I told her. 'A good friend who's put up with me for far too long. I know I've outstayed my welcome by a mile. It's high time I was out of your hair. Good luck with the renewed relationship, Di. When do you want me to leave?'

She looked so relieved that I thought she might faint. 'It's very good of you to take it so well,' she said. 'There's really no hurry about leaving. Where will you go?'

'Oh, don't worry about me,' I said lightly. 'I've got some contacts. I'll be fine.' Secretly, I have to admit that I'd never have been so generous if it hadn't been for Mark. I felt confident that he'd be more than delighted to have me move into the apartment with him. As for me, I was thrilled. The sex was fantastic and I wouldn't have to pay any rent.

The first rehearsal was a bit of a shambles. Paul brought this guy along who he said was an experienced director. He introduced him as Marvin Nash. Neither Mark nor I had ever heard of him. Not that either of us was that clued up about West End directors but he was certainly no Cameron Mackintosh, anyone could see that. When we were finally dismissed, Mark and I went along to the Prince of Wales, the same pub we'd lunched at before, and sat down with stiff drinks to pool our opinions.

'I daresay it'll all come together once we get the music and choreography sorted,' Mark said optimistically.

I shrugged. 'I hope you're right. They're like a bunch of amateurs, and that Carla is the worst. God only knows what her singing voice is like.'

Mark grinned. 'Well, old Ma Bennet isn't exactly supposed to be a diva, is she?'

'She is as played by Carla bloody Dean.' I took a quick swig at my gin and tonic. 'And I can't wait to see her dance!' I put my glass down on the table and took a look at Mark's face. This seemed like a good moment to spring my proposition.

'Talking of movement, I've got something I want to run past you,' I said, looking at him from under my lashes in the way I knew turned him on.

'Oh?' He raised his eyebrows. 'Am I going to like it?'

'I hope so.' I toyed with the stem on my glass. 'Since Christmas I've been thinking a lot – about us. We had a great time, didn't we?'

He nodded enthusiastically. 'The best.'

'I was a fool to be so rotten to you all those years ago,' I said. 'We have so much in common and I just couldn't see it. We make each other laugh. We're really compatible in every way, aren't we?'

'I've always thought so, yes.' He leaned towards me, his eyes twinkling. 'You're not trying to propose, are you? Because if you are I'm going to have to insist that you get down on one knee.'

'In your dreams!' I gave his shoulder a push. 'No, I'm not proposing – not marriage anyhow. What I am suggesting is that it might benefit both of us if I moved in with you. The lease on my flat is about to run out and my flatmate would like to take it over from me. Of course it would only be until we go on tour with the show and …'

'*Lou!*' He had grasped my hand so hard that the pressure

made my eyes water and stopped the words in my throat. His grin spread from ear to ear. 'Bloody hell, that would be marvellous. Just fancy – home-cooked food every day!'

I kicked him under the table. 'On your bike! That was a Christmas one-off. We share the cooking and the chores or the deal is off.'

He gave me a mock salute. 'Yes, *ma'am*! Anything you say. When can you move in?'

'Well, I'll have to break the news to my friend that I'm leaving sooner than expected. I daresay she'll be a bit upset but she'll get over it.'

'Are you sure? I'd hate you to lose a friend because of me.'

I smiled at him. 'I'm sure. I'll find a way to make it up to her somehow.'

'OK – so when?'

'Give me a few days. I'll let you know at the next rehearsal.'

I started packing that night. I'd told Di that I'd found somewhere else to live and she seemed pleased. A bit too pleased to be flattering, actually. I overheard her talking to Mike on her mobile later and arranging for him to move in the week after next. She certainly didn't intend to waste any time. But it didn't matter, I told myself. Two weeks from now I'll be living in the lap of luxury in Mark's lush apartment.

The day of the next rehearsal was grey and wet; a typical January day. The walk from the bus stop seemed endless. My heels skidded on the greasy pavement and my umbrella blew inside out so many times that I eventually gave up and put it down. After that, the icy rain dripped relentlessly down the back of my collar all the way to St Mary's Hall. Pulling the heavy door closed behind me with relief, I took off my wet mac and shook as much of the water off it as I could. Mark was already there. He was standing with his back to one of the lukewarm radiators and I joined him.

'Brrr! Budge up and let the dog see the rabbit,' I said, giving him a playful shove. 'I've got good news. I can move in at the weekend if it's OK with you.'

He turned to me with an expression like a whipped puppy and my heart sank. I knew right then that it wasn't going to be my day.

He began haltingly, 'Lou, darling – I don't know how to tell you this but I'm very much afraid that our plans are going to have to be put on hold – for the time being at least.'

I stared at him. 'You *what*? Why? What's up?'

He took a deep breath and looked at the floor. 'Last night I got an email from Cathy, my sister in Australia. She and her husband have split up and she's flying home today.'

I frowned. 'Home?'

'To England – and to me as there's no one else.'

'She's landing herself on you? What a cheek.'

'Not just her but her two kids as well. I couldn't say no, could I, Lou? She's the only family I've got and she's going through a crisis.' He looked at my bemused face. 'I've only got the two bedrooms so it's going to be a bit of a squeeze.'

'Yes, but I'll be sharing yours.'

'I know, but ...' He looked uncomfortable. 'She'll want to talk. She's bound to need some advice and moral support.'

'So? I won't get in the way.'

He shook his head. 'I'm sorry, Lou. It just won't work; not with five of us in the apartment. It won't be forever,' he added hurriedly. 'I'm sure she'll want to find a place of her own quite soon. Obviously she'll need to find schools for the kids and settle them down into some kind of normality as soon as possible, poor little devils.'

'Oh, yes – *tough*!'

He reached for my hand but I pulled it away. 'I've already told Di that I'm moving out.'

'But she won't mind if you change your plans temporarily, will she? After all, it's your flat and it'll give her all the more time to find someone new to share with. And you said she'd be devastated that you were leaving.'

There, I'd done it again, I told myself. What's the phrase – hoist with my own petard? (What *is* a petard anyway?) Mark was looking at me, his face creased with concern.

'I'm so sorry, Lou. I was really looking forward to the two of us being together.'

'So disappointed that you let your sister call all the shots,' I said, shaking off his hand. 'She and her kids have to come first, I suppose. After all, who am I?'

'Please – don't be like that. I told you, it's only temporary.'

'Does she know you'll be going on tour with the show in a few weeks' time?'

'I did mention it, yes.'

'Then she's not going to be in any hurry looking for a place of her own, is she?' I gave a dry little laugh. 'You'll have the perfect flat-sitter. Just let's hope that you don't come back to find the kids have wrecked the place!'

'They're good kids actually,' he said.

I stared at him. 'How would you know? They live in Australia. How many times have you seen them?'

'Well, not many admittedly.'

'I just hope you know what you're letting yourself in for.'

Paul chose that moment to arrive with Marvin in tow and there was no more time to discuss our ruined plans. I poured my heart and soul into the rehearsal, putting special effort into my scenes with Darcy, who seemed delighted by my enthusiasm. Whilst close to him, I endorsed Mark's opinion that he wore a toupee. I also discovered that he had terminal halitosis which didn't auger well for our love scenes. After the rehearsal, Marvin came up to me and congratulated me on my talent and the hard work I'd put in learning the lines. I felt it was no more than I deserved.

'Paul's done a good job of the casting,' he said. 'Especially in your case. Just wait till we get to the West End.' He grinned at me. 'Fancy your name in lights, do you?'

'I certainly do.'

'We'll have to get you a really good press agent,' he said. 'And a photographer. Some great pics for the magazines.' He looked at me, his head on one side and his eyes half closed. 'That wicked mixture of sexiness and innocence. I can see it now. You'll knock 'em dead, Louise.'

His comments made up a little for the day's disastrous start. I was slipping my arms into the sleeves of my still-damp mac when Mark sidled up. 'Shall we go to the Prince of Wales for lunch?'

I glanced round casually at him. 'Not today,' I told him. 'I'm not hungry – or in the mood.' I turned up my collar and picked up my bag. 'See you soon, Mark. Goodbye.' I noticed that Phil, the actor playing Darcy, was just about to leave. I called out to him.

'Oh, Phil! Wait for me. Do you fancy a drink?'

He looked surprised and delighted. 'Great! Yes, I'd love to.'

I turned to Mark, seeing with satisfaction his downcast expression. 'Have to rush. See you soon. Bye!'

'What do you mean, your plans have fallen through?' Di clearly wasn't pleased with my news. 'I've told Mike he can move in next week.'

'I can't help it,' I told her. 'The friend who was going to put me up has had a sudden family crisis.'

'Oh?' She failed to look sympathetic.

'I promise I'll keep out of your way,' I said. 'Presumably he'll be sharing your room and I'll keep to mine. It's only temporary anyway.'

'Actually we hadn't got as far as sharing a bed,' Di said stiffly. 'Mike thinks he's having my spare room.'

I couldn't disguise the little smile that lifted the corners of my mouth. '*Oh*! I thought you said you found the *old spark* was still there.'

'I didn't mean we'd jumped straight into bed together,' Di said icily. 'We're not all as promiscuous as you, you know.'

I laughed. I couldn't help it. '*Promiscuous*! Come off it, Di. How stuffy can you get?'

She bridled, and the red patch on her neck that always appeared when she was furious flared angrily above the neck of her sweater. 'Right, you can move all your stuff out of the spare room right now,' she said. 'And from next week on you'll be sleeping on the sofa.'

I picked up my bag, inwardly regretting the fact that I'd laughed at her.

'OK,' I said over my shoulder. 'As long as you realize that means I'll have nowhere to go to keep out of your way!'

I only slept on the sofa for three nights. I think it was partly the fact that I was always around playing gooseberry and partly the rekindling of that *old spark* they shared that had Mike moving in with Di so that I had my old room back again.

Mike and I disliked each other on sight. I thought he was a charmless weed and he made no bones about making me feel like a shameless scrounger. Personally, I couldn't see what she saw in him with his specs and his pedantic way of talking, but that was up to her. She always did have strange taste in blokes. I kept to my side of the bargain, spending my evenings shut away in my eight-by-ten bedroom while they canoodled on the sofa, recently vacated by me. But their audible lovemaking penetrated the thin walls into the small hours, making sleeping all but impossible for me, specially when I thought of what Mark and I could have been enjoying.

Rehearsals continued and we had one or two sessions with the choreographer and two more with Paul and the musical score. It wasn't going too badly. Mark and I were barely speaking. He told me that Cathy, his sister, and her two adorable angels had settled in happily. I don't know if that was supposed to make me feel better but I couldn't see why he expected me to work up any enthusiasm, seeing that this woman and her two brats had completely scuppered my plans. It seemed that I was doomed to be the outsider – the unwanted – surplus to everyone's requirements. Di had her beloved Mike; Mark had his sister and her kids and as for my so-called family, they had each other and for all they cared, I could go to hell.

Lying awake as dawn was breaking one morning, I made my decision. I'd find somewhere else to live. After all, it would only be for a few weeks. I made up my mind about something else too: I'd start looking for my mother. Maybe she'd been waiting for me to find her all these years. We might strike up a good relationship together. Getting together again might be just what we had both been waiting for. Who knew? But if I didn't try I'd never find out, would I?

A couple of days later, after scouring the local paper I found myself a bedsit in Stoke Newington. It was pretty grim, shared bathroom and kitchen, and the occupants of the other rooms looked a weird lot. But it was cheap and close to the rehearsal venue. It was only temporary, I told myself, so I could stick it out for a few weeks. I told Mark that I'd found a really nice flat; Di too, and I think she believed me. I say believed but she didn't try very hard to conceal the fact that she didn't really care one way or another. I was about to leave her and Dream Lover Mike to share their little love nest in peace. I moved into the bedsit and later made that all-important call to the Sally Army. If I was going to find Mum it was better to do it before the tour started.

Chapter Thirteen

AT THE FIRST SIGNS of spring, Ted began to work on his allotment. Susan had applied to the local council before Christmas and to her delight, she heard soon after that she had been allotted the vacant plot next to his. She lost no time in donning the wellingtons she had thought she would never need again, and began to accompany Ted to the huge plot of allotments on the outskirts of town, the car boot loaded with their assorted gardening equipment.

As the weeks passed, Susan could feel herself growing fitter. Her skin took on a healthy glow and seeing the weeds vanish from the ground to be replaced by rows of tiny seedlings gave her immense satisfaction.

Her relationship with Ted had grown from friendship to a comfortable affection. Sometimes she thought they were like an old married couple, sharing their love of working with nature. On the days when they returned tired from several hours of gardening, Susan would go home with Ted and cook a meal for them both. As the weather grew warmer, Ted would drive them both to the coast on Sundays where they would treat themselves to a leisurely lunch then take a bracing walk along the sea front. For Susan it was a whole new life, and Ted confessed that it was the same for him.

The first time Ted had tentatively suggested that Susan might stay the night after one of their outings she had reservations. It was a long time since she had shared her bed with a man. There had only been one man in her life and that of course was Frank. One of her uncertainties was that it would feel like a betrayal. But Frank had always urged her not to be alone when he had gone. *You're a woman who needs a man,* he had fondly told her once. *You need to be loved and cared for.* The fact was that Frank had never really known how strong she was. And maybe not letting him see that had been the secret of their happy marriage. But now there was Ted. She recognized that they had reached a milestone in their relationship and she was wise enough to know that the decision had to be mutual. Ted had insisted that it must be; he had urged her not to do anything she wasn't absolutely sure about but she felt that things would never be quite the same between them if she declined.

For a while she held back, thinking carefully about the difference intimacy might make to the pleasure they already had in each other's company. Ted did not pressure her and when at last she made up her mind, he asked her over and over if she was sure. By then she was.

So one Sunday in late March, she took an overnight bag with her when they drove to the coast and later, after they had spent a pleasant evening watching television, they retired to bed together as though they had been doing it for decades.

Ted was so sweet and gentle. He was kind and loving, arousing in her feelings and sensations that she had expected never to experience again. By morning she knew that she had made the right decision. Their new-found intimacy had brought them even closer than before and she felt happy and contented.

As the weeks went by, it occurred to Susan that Ted might ask her to marry him. She searched her mind, asking herself what she really wanted from their relationship. She had always vowed that she would never marry again and really, why change that? she asked herself. Nowadays, no one seemed to mind people being together without the benefit of a marriage licence. They were happy as they were. But as the weeks continued to fly, a tiny doubt nagged at the back of her mind. Ted was a conventional kind of man. Surely he would want to put their relationship onto a more formal basis. Then, one Sunday evening as she was taking her overnight case from the car, something happened to answer her questions in a way that was to shatter her world.

Mrs Freeman was the elderly widow who lived in the bungalow next door to Ted's. She was often to be seen twitching her net curtains on the days when Susan visited Ted. They had even laughed about it. On this occasion, she was watering her front garden when they returned from one of their Sunday outings. Ted had already gone indoors to put the kettle on while Susan took her case out of the car boot. She slammed the lid down, only to see Mrs Freeman's sour face inches from hers on the other side of the fence.

Susan smiled. 'Good evening.'

The woman glared at her. 'Good evening *indeed*!' she grunted. 'You do know he's married, I suppose? Or are you the kind of woman who doesn't care about little details like that? I've seen you sneaking off first thing in the morning. Ought to be ashamed of yourself!'

Susan frowned. 'Mr Mumford is a widower,' she said. 'Not that it's any of your business.'

'Huh!' The old woman gave a mirthless laugh. '*Widowed*! Is that what he told you? Well, you can take it from me that he isn't. His wife's still alive. Shoved into a home and left to rot while he enjoys heaven knows what with loose women. I think you should know that the disgraceful way you two are behaving is getting this neighbourhood a bad name!' She looked Susan up and down. 'I'd have thought a woman of your age would have had a bit more decency about her! Downright depraved, that's what it is ...'

Susan didn't wait to hear any more. Picking up her case, she hurried in through the open front door of the bungalow and slammed it behind her.

In the kitchen Ted was humming happily, his back towards her as he waited for the kettle to boil.

'I'm ringing for a taxi,' Susan said, fishing her mobile out of her handbag. 'I'm going home, Ted.'

'Going home? Why ...' He turned to see Susan standing in the doorway. Her face was white and she was trembling. '*Susan*! Darling, what's the matter?' He took a step towards her.

'You might well ask,' Susan said, holding out one arm to prevent him coming any closer. 'You should hear some of the things that woman next door just said to me. You lied to me, Ted. You told me your wife was dead and I believed you.'

'Susan, I never said any such thing. I ...'

'I want to go home.' Her voice was shrill and trembling. 'I won't wait for a taxi. I'll walk!'

'No, please! You don't understand. What did she say? Was it Mrs Freeman?' Susan turned and walked down the hall without another word but he caught her up, grasping her arm. 'Don't go like this. Please – let me explain.'

She shook off his hand. 'There's nothing to explain. Goodbye, Ted. Please don't try to get in touch again.'

'Well, at least let me drive you. You're in no fit state ...'

'*No*!' Trying not to look at his wounded expression she picked up her case and left.

Afterwards she didn't remember walking home. Once inside the flat, she gave way to the tears that had threatened all the way home, collapsing onto the bed to sob into the pillow. How could she have been taken in by him? They say there's no fool like an old fool and she'd been taken in good and proper. For days she didn't leave the flat. When the telephone rang she didn't answer it and she saw that there were several missed calls on her mobile, most of them from Ted. There were texts too but she deleted them all without reading them.

By the end of the week, she was obliged to venture out to the supermarket. Donning a pair of sunglasses and tying a scarf over her head, she went during the lunch hour when she felt she was unlikely to run into anyone she knew but she was just rounding the end of the fruit and veg section when a voice hailed her.

'*Mum*!'

Her heart sank. She'd forgotten that Karen often popped into the local supermarket in her lunch break. There was no escape.

She turned and forced a smile. 'Oh, hello, dear. How are you?'

'More to the point, how are *you*?' Karen looked cross. 'I've been trying to ring you for days on your mobile and your landline but no reply. I've been really worried. In fact, I was going to come round after school today to see if you were all right.'

Susan felt a pang of guilt. 'I'm so sorry, darling, I didn't think.'

'That's not like you, Mum.' Karen looked closer, narrowing her eyes. 'Are you really OK? Why are you wearing sunglasses in here?'

'My eyes felt a bit strained. The lights are very harsh in here.'

Karen laid a hand on her arm, her face concerned. 'What's wrong, Mum? You don't look well at all.'

Susan opened her mouth to speak but suddenly and without warning, her eyes filled with tears. She swallowed hard but she was totally powerless to stop the sobs escaping her. Just a few caring words and suddenly here she was, sobbing her heart out like an idiot, right here in the middle of Tesco with everyone staring at her. Karen grasped her arm.

'*Mum*! What is it?' She glanced round at the curious customers trying to pretend they weren't looking. She took the handle of the trolley from Susan's hands. 'You can finish your shopping later. Come on, leave it for now. Come and have a coffee.'

There was no escape. Susan meekly allowed herself to be led by Karen to the cafeteria and deposited at a corner table, while she fetched the coffees. In a moment of wild panic, she contemplated making a run for it while Karen was otherwise occupied, but suddenly she found that she hadn't the strength, either of purpose or purely physically. Days of eating scratch meals, plus the shock of what had happened at the weekend, had sapped all her energy.

Karen returned to the table with a determined look on her face. She set down the tray and began to unload it. 'I've got you a sandwich too,' she announced. 'Prawn, you like prawns, don't you? You look to me as though you haven't been eating properly.'

The sight of the sandwich, mayonnaise oozing from between the slices, made Susan feel sick but she sipped the hot coffee gratefully.

'And for God's sake, take off those awful glasses and that headscarf,' Karen instructed. 'They make you look like Olga, the Russian spy!'

Susan obediently removed the sunglasses and scarf, smiling in spite of herself. Karen smiled back.

'That's better.' She reached out a hand. 'Oh, Mum, your eyes are all red. You've been crying. Please tell me what's wrong.'

Susan pushed the plate containing the sandwich towards her daughter. 'I can't eat that. You have it.'

'No, I had lunch at school.' Karen took a paper napkin and wrapped the sandwich up. 'Put it in your bag, you can eat it when you get home.' She squeezed Susan's hand. 'Tell me what's wrong, Mum.'

Susan looked at her watch. 'Shouldn't you be getting back to school?'

'I've got a little while yet. Anyway, I'm not moving from here till you tell me what's wrong so if you don't want me to be late … Mum – you're not ill, are you? It isn't something – serious?'

'No, nothing like that.' Defeated, Susan leaned back in her chair. 'It – it's Ted,' she said. 'I thought he was a widower but it seems that his wife is still alive – and in a home.'

'Oh, dear.' Karen shook her head. 'I did warn you, Mum.'

Susan felt the blood rush to her face. 'I don't need you saying *I told you so,* Karen. Not now, if you don't mind. I know what you said and I know I didn't listen, but I don't need my nose rubbing in it, honestly.' She dabbed at her eyes. 'We had such a lovely Christmas together. I rented the allotment next to his and we've had such a wonderful time, gardening together. We had so much in common, Karen. We were going to sell some of our produce at the farmers' market in the summer. We've had some lovely outings to the coast at weekends and – and …'

'And you've been sleeping together?' Karen said gently as she leaned across the table. 'I'm right, aren't I?'

Susan blushed furiously. 'You must think me a silly, gullible old woman,' she said, shredding the damp tissue in her hand.

'Not at all,' Karen said gently. 'And you're not an old woman, Mum. You're a very attractive *mature* woman, with feelings just like anyone else. You've been taken in, that's all.'

'I feel so ashamed.' Susan fumbled in her bag for a fresh tissue. 'You must think I've taken leave of my senses. It was his next-door neighbour who told me – last Sunday. It was almost as though she was waiting for us to get back. It was horrible, Karen. She was so nasty and spiteful. You should have heard what she said – that I was lowering the tone of the neighbourhood; that I was depraved and had no morals. She made me feel like – like some old trollop.' She blew her nose and dabbed at her cheeks.

'So what did Ted say when you told him what she'd said?'

'I didn't give him the chance. I walked out – came straight home.'

'And he hasn't tried to get in touch?'

'Oh, he's tried of course. He keeps on ringing and texting, but I don't reply. I don't want to hear his lame excuses. He lied to me.'

'He definitely told you his wife was dead?'

'Well …' Susan took a deep breath. 'I've been thinking about that. I don't think he actually used those words, but he did say he'd been on his own for some years.' She looked at Karen. 'Devious. That's what he was. He let me believe what I wanted to believe. He thought he could pull the wool over my eyes.'

'Well, I think he owes you an explanation,' Karen said. 'You deserve one.'

'I don't want any more to do with him,' Susan said firmly. 'You said he could be a conman and it turns out you were right.'

Karen looked at her watch. 'I'm sorry, Mum, but I'll have to go. Look, pay for your shopping and I'll run you home on my way back to school.'

'But it isn't on your way,' Susan protested as she got up and took the trolley handle.

Karen took her arm. 'I haven't time to argue with you. Just let's get to the checkout.'

*

In the car on the way home, Susan realized that they'd only talked about her problems and she felt slightly guilty. Turning to Karen she said, 'You must think I'm awful; full of my own woes. I haven't asked how things are going with you. How are Simon and little Peter? And how is that new au pair girl of yours fitting in?'

'Simon and Peter are fine,' Karen told her. 'As for Adrey, I'm letting her go home to Holland for Easter. Her father hasn't been well and she wants to see him and the rest of her family. Seems there are quite a few of them back in Amsterdam. I think she's been feeling a little bit homesick too.'

'Do you think she might want to stay, once she gets back?' Susan asked, but Karen shook her head.

'No. She's promised me she won't do that. She loves it here really and she adores Peter.' She glanced at her mother. 'Why don't you come and stay with us for a few days at Easter?'

Susan hesitated. 'Oh, I don't know.'

'Come on, Mum,' Karen urged. 'You don't want to be stuck in the flat over the holiday feeling sorry for yourself, do you?'

Susan bridled. 'I don't think I've ever been one to wallow in self-pity.'

Karen laughed. 'Come off it, Mum. You know what I mean. It's ages since we've had a real family Easter.'

So finally Susan had agreed.

Once Susan was back in the flat, she felt better. Whether it was Karen's sympathy or just the change of scene, she didn't know or care. She felt stronger; almost ready to try and forget Ted and his duplicity and start making fresh plans for the future. 'I'll be fine,' she told herself as she unpacked her shopping and put it away. 'Maybe that Dutch girl will want to stay at home in Holland once she gets there, then they'll want me to take care of Peter again. Ted Mumford can go and find some other gullible woman to tell his lies to.'

But later that afternoon, as she sat there alone watching afternoon TV, she thought longingly about the allotment. All her little seedlings would be ready for pricking out now. She thought about the little shed on Ted's plot and how they'd brew mugs of tea when they took a break. Often, she'd take home-made cake or scones and jam to eat with it. Ted had loved that. He had loved her home cooking. Her thoughts wandered to the muscles in his strong arms and how they had rippled as he loaded the car boot with their gardening implements – and although she tried hard not to remember – the same strong arms, warm around her later as they lay together in his bed.

She sighed. They could have been so happy together if only – if only …

Chapter Fourteen

'SO I'VE ASKED HER to come and spend Easter with us,' Karen said as she unpacked her briefcase.

'You've *what*?' Simon turned to look at her. 'You might have asked me first.'

Karen stared at him. 'Are you saying that I have to ask your permission to invite my own mother to stay for Easter?'

Simon gave an exasperated snort. 'Bloody hell, Karen! I'd planned for us to have some time to ourselves as Adrey is going home. I even thought we might slip away somewhere for a proper break.'

'I had to ask her, Simon. You should have seen her. She's absolutely devastated about this guy lying to her.'

He frowned. 'Well – she's only got herself to blame really, hasn't she? I mean, some old geezer who goes around picking up women on buses. I ask you!'

'She really liked him though,' Karen said. 'They'd been gardening together. Mum had even got herself an allotment. They'd been on weekend jaunts to the seaside and they had all sorts of plans.'

'You'll be telling me next they'd been sleeping together!' Simon laughed.

'Well – as a matter of fact ...'

'*No!*' He stared at her incredulously. 'At their age? Well, well. He must be a right old dog. Good for him!'

'*Don't*!' Karen said. 'It's not funny. I shouldn't have told you. Don't you dare let on to her that you know.'

Simon pulled a face. 'As if! Right, so they've split up. What happened? Was he two-timing her with some other old ...' He caught the look on Karen's face and stopped. 'What did he lie about anyway?'

'She'd got the impression that he was a widower but it seems that his wife is still alive.'

'But they're not together so...?'

'And in a care home,' Karen finished.

'Oh, I see. Well, I suppose that was pretty devious.'

'Exactly. So now perhaps you can see why she feels let down. She's humiliated and shamed.'

He shrugged. 'That's overreacting a bit, isn't it? It's not her fault. Anyway, I daresay it happens all the time.'

'Not to my mother, it doesn't. She's a different generation, Simon. In her day, that kind of thing was really scandalous. I still think it's a pretty rotten way to behave anyway. Morals have may have reached an all-time low but I'd have expected better from a man of his age.'

Simon smiled indulgently. 'I can see that some of Susan's indignation has rubbed off on you, my love.' He put his arms round her and pulled her close. 'Tell you what, if your mother is coming why don't we slip away, just the two of us and leave Peter with her?'

She pushed him away. 'We can't do that. She'll think I only invited her so that we can use her as a babysitter.'

He shook his head. 'Oh, you and your bloody conscience! OK, so we have to have your mum here and sit listening to her going on and on about her car crash of a love affair.'

'Oh, don't be ridiculous.' Karen sighed. 'You make having Mum for a couple of days sound like a boring chore.'

'Oh, heaven *forbid*!' Simon sneered.

'Why can't you cheerfully do something for someone else for a change? At least we're not going to be saddled with Louise this year. '

Simon turned in the doorway. 'Well, I suppose that's what

you'd call a *small mercy*. Here's to a dreary Easter. I can hardly wait!'

Chapter Fifteen

As we were preparing to break for lunch, Mark sidled across to me.

'Lou, can we talk?'

I was chatting to Phil and I gave Mark my freeze-'em-dead look. 'Can't it wait? I'm busy.'

He looked at Phil. 'Sorry, mate but will you excuse us? It's important.'

To my annoyance Phil grinned and walked away. 'See you, Lou,' he muttered.

I stared at Mark. 'For your information, that was an important conversation. What can you possibly want to talk about?'

'Us.'

I raised an eyebrow. 'I wasn't aware that there was an "us".'

He ignored the remark. 'We can't go on like this,' he said. 'It's too ridiculous for words. All I'm doing is putting a roof over my sister's head for a few weeks and you're hardly speaking to me. Can't you see I had no choice?'

'That's not the point. You let me down, Mark. I'd already told Di – I mean, I'd already given my flat up.'

He frowned. 'But you've found another one. It was only for a few weeks anyway. We'll be going on tour soon.'

'As I just said, that's not the point.'

'Look, you say your flat is quite near, so can we go there and talk? I want to put things right between us, Lou.'

I panicked, remembering the glorified description I'd given him of my so-called flat. I'd die if he saw that it was only a bedsit, and a grotty one at that. I shook my head. 'It's not convenient.'

'Well, the pub, then. It won't be as private but it'll do. Oh, come on, Lou. Get down off that high horse of yours.'

I have to admit that it was nice, being on good terms with Mark again and knowing that he really cared what I was feeling, though I wasn't about to let him think me a pushover. It was true that he'd let me down and he was going to have to work hard for forgiveness. In the end, he offered to treat me to lunch and I accepted. We were just studying the menu when my phone buzzed in my bag. I fished it out.

'Hello.'

'Hello. Is this Louise Delmar?'

'Speaking.'

'This is Daniel from the Salvation Army. I have good news for you.'

I felt my heartbeat quicken as I listened to what he had to say. When I'd spoken to him earlier, he'd laid out the way things worked. He'd warned me that if the person being sought was tracked down, their whereabouts would not be disclosed unless they'd agreed. That was a given and not up for negotiation. I'd explained to him who I was and briefly, the circumstances of losing touch with my mother. I also told him that I was an actress, about to star in a new musical in the West End. I thought that might sway things a little. Now he was on the other end of my phone, telling me that he had found my mother and that she was willing to meet me. I could hardly believe my luck.

'Where?' I asked him breathlessly. 'And when?'

'She suggested meeting in the coffee lounge at Selfridge's – possibly tomorrow afternoon – about three o'clock, but I can give you a telephone number and you can alter the arrangements if you need to. She has reverted to her maiden name by the way. Jean Sowerby.'

'Oh, really? Thanks. I will ring, just in case, so if you give me the number I'll make a note of it. Thank you so much, Daniel.' He gave me the number and I added it to my contacts. As I shoved my phone back in my bag, I looked up at Mark, who was staring at me enquiringly.

'That sounded intriguing,' he said. 'And from the way your eyes are shining, I'd guess it was good news.'

'The best!' I told him what the caller had said and the tentative arrangements that had been made. 'I'll ring her this afternoon,' I told him. 'I have to admit that I feel quite nervous now that it's really happening.'

'Do you want me to come with you?' Mark offered. 'I mean, I'd keep out of the way while you meet and talk, but I could be there for you just in case it doesn't work out.'

'Oh no, everything will be fine. I don't need any moral support,' I told him. Seeing his disappointed face I reached out and touched his arm. 'But thank you, Mark. It's kind of you to offer.' I really did feel touched at his thoughtfulness. I was sorry I'd been so mean to him; after all, I knew he really did care for me and he could be so useful in the months to come, once he got rid of that grasping sister of his.

Our food arrived and we tucked in. Mark told me that he intended to take the Ferrari down to Bournemouth when we opened on tour. 'It'll be much more convenient than catching trains,' he said. 'I was sort of hoping you'd join me. I'd appreciate your company.'

'That would be lovely,' I said, trying not to let him see how excited I was at the thought of being driven everywhere in Mark's glamorous car.

He looked at me with his puppy-dog eyes over his plate of pasta. 'Do I take it that I'm forgiven, then?'

After the exciting phone call I'd just had, I would have forgiven the Devil himself. I leaned across the table and kissed his cheek. 'Of course I forgive you,' I whispered in my sexiest voice. 'How could I not?'

Back at the bedsit I took out my phone and clicked in the number Daniel had given me, holding my breath. In just a few seconds I'd actually be speaking to the mother I'd lost all those years ago.

'Hello.'

I gasped at the sound of her voice. 'Oh – er – hello. Ms Sowerby?'

'Speaking. Who is this?'

'It's Louise. Your – your daughter.'

There was a pause at the other end, then she said, 'Hello, Louise. Fancy hearing from you. It was a real surprise when the chap from the Sally Army rang.'

'Yes, I expect it was.' I swallowed hard, my mouth suddenly dry. 'He said you'd suggested meeting tomorrow afternoon at Selfridge's.'

'That's right, but if it isn't convenient ...'

'No! I mean, yes, it is. I'm busy with rehearsals but I'm free in the afternoon. Shall we say three o'clock?'

'That would be fine. See you there, then.'

'Wait! How will I recognize you?' I asked. 'And you me of course. I've grown a bit since you last saw me.'

'Oh yes, of course. Best if I wait for you in the coffee lounge,' she said. 'I'll be wearing a royal-blue coat.'

I'll wear a black suit,' I told her. 'You used to have dark hair,' I went on tentatively. 'Is it still?'

'Blonde now,' she said quickly. 'You?'

'Blonde too.' I laughed shakily. 'See you at three, then ...' I found suddenly that I had no idea what to call her. I couldn't quite bring myself to call her 'Mum' after all these years.

'Yeah – see you.' She hung up.

I was there at five to three the next afternoon. I went up to the coffee lounge and looked around. She clearly hadn't arrived. I took a corner table and watched from a distance. For some reason, I wanted to see what she looked like before I made myself known to her. People came and went but so far no one in a royal-blue coat. I began to think she'd chickened out. Then at ten past three she arrived. I was shocked. She was nothing like the woman I remembered. She was short – shorter than I remembered her. She was quite plump too. Her hair was longish and bleached a brassy shade of blonde that spoke of home hair-dressing. She looked thoroughly down at heel. Could this really be my mother? But there was no one else around in a royal-blue coat. As I watched she stood looking round nervously. I hadn't the heart to keep her waiting any longer so I took a deep breath and stood up. Catching her eye, I smiled. She hurried towards me.

'Louise?'

I nodded. 'Good to see you. Please, sit down.'

Now that I was closer I could see that life had not been kind to her. Her face was very lined and her upper lip had the creased look of the heavy smoker. I ordered a pot of tea and cakes and when the waitress had gone, I smiled at her.

'Well – I hardly know where to begin.'

'I hear you're a big star,' she said enthusiastically. 'Who'd have thought it?'

I shook my head. 'I'm not quite a star yet but the show I'm about to take the leading part in is destined for success.'

'How wonderful.' She paused. 'How – how's your dad?'

I frowned. Of course, she wouldn't know. 'Dad died a few years ago,' I told her. 'He married again after your divorce. Her name is Susan and they have a daughter, Karen.'

She showed no emotion at all. 'I married again too,' she said. 'But it was a disaster.'

'You weren't happy?'

'Happy?' She gave a short laugh. 'Gotta be joking! He was bad through and through – treated me something rotten but eventually the law caught up with him. He was caught – aggravated burglary, the verdict was. He battered a poor bloke half to death in his own home and this time he went to prison. The only good thing to come out of it was my boy, Steven – Steve.'

I stared at her. 'So – I have a...?'

'Half-brother,' she said. She looked at me. 'I suppose when your dad died, this Susan woman came in for everything, did she?'

'No. He left me the house and everything in it,' I told her. 'He left Susan and Karen provided for, of course, but I think I came in for the bulk of his estate because he felt I'd had a raw deal.'

Her eyes lowered. 'A raw deal, eh? Well, he would say that

– and I suppose it was true in a way.' She looked up at me. 'But believe me, Louise, no one will ever know what it cost me to walk out and leave you behind. But I had no money and nowhere to go so I had no choice. Frank loved you and you had a comfortable home with him. You were better off where you were. I was only thinking of you.'

I shook my head. 'It's all water under the bridge now.'

'I didn't forget you,' she said. 'I sent you birthday cards for a few years but I never got no replies so I reckoned he didn't let you see them. Oh, well…' She applied herself to the cake on her plate. 'Well, now that you've found me, we must stay in touch,' she said at last.

'Tell me about Steve,' I prompted.

She looked away. 'Maybe I will, eventually. We'll see.' I was about to pursue the matter when she suddenly changed the subject. 'When does your show open – which theatre?'

'We're going on tour first so I can't tell you that yet,' I told her. 'At the moment we're still rehearsing at St Mary's Church Hall in Stoke Newington and I've taken a temporary flat nearby, but as soon as I know where we open, I'll send you tickets for the opening night.'

'Ooh!' She smiled. 'Fancy me being the mum of a big star.'

I watched her as she helped herself to another cake. 'Your husband – is he still alive?' I asked.

She shrugged. 'I s'pose so. I've divorced him since he's been banged up and reverted to my maiden name. I daresay he's spittin' blood about that. Still, there's not a lot he can do about it in there.'

I felt a stab of alarm. 'How long is he in for?'

She shrugged. 'He got a seven stretch but they only serve about half o'that, don't they? He'll be due for release in a few months' time, but my Steve'll see I don't come to no harm if he comes round lookin' for revenge.'

'Aren't you worried?'

'No. I'll soon get the police onto him again if he starts threatening me. He won't want to go down for another stretch, will he?' She helped herself to another cake and devoured it with relish. 'Those folks at the Sally Army are really clever, you know,' she remarked through a mouthful of cake. 'Tracking me down through three different names.'

By half past four, all the cakes had disappeared. She looked at me. 'I'm sorry but I'll have to go now, Louise,' she said. 'It's been lovely meeting you. Maybe we could do this again.'

'Perhaps I could come to your home next time,' I suggested.

'Yeah.' She looked flustered. 'Well, maybe sometime, we'll see.'

'I'll give you my mobile number.' I scribbled it down on the corner of a paper napkin and passed it across the table to her.

She took it from me. 'What about your address?' she asked.

'Well, as I said, it's only temporary at the moment. I had to give up my flat in Earl's Court,' I said. 'Do you know Stoke Newington?' To my relief she shook her head. 'Well, as I told you, we're off on the pre-West End tour soon so I took this room in Mason Street. It's not very nice but it's quite close to the rehearsal venue. I won't be there after next week though, but if you give me your address I'll keep in touch.'

She looked shifty. 'Oh – well, they've recently altered the postcode and I can't remember it at the moment but I'll ring and let you know.'

I paid the bill and we travelled down together on the escalator. 'Are you heading for the Undergound like me?' I asked as we stood in the store entrance. She shook her head.

'I've got a couple of things to do yet.' She looked me in the eye. 'Louise, I suppose you couldn't lend me a few quid, could you, love? I'll pay you back next time I see you.'

I was taken aback. 'Oh! Yes, OK. How much?'

'A couple of hundred will do,' she said calmly.

I was stunned. I'd been expecting her to ask for a fiver – a tenner at most; something to tide her over till next pension day and I had to stop myself from gasping with shock. 'I don't carry that much money around with me,' I told her.

'A cheque will do.'

Once more her coolness shook me. I took out my chequebook and managed to write a cheque balancing the book on my handbag. I tore it out and gave it to her. 'I hope I haven't given you the wrong impression,' I told her. 'Dad did leave me well provided for but I've had to put money into this production and it's left me quite short.'

She smiled derisively as though she didn't believe a word of it. 'Yeah, but no doubt you'll get it all back. You'll be quids in once the show opens in the West End, won't you?' She put the cheque away in her handbag and looked across the road. 'Oh, there's my bus. I'll have to run. See you soon, Louise – and thanks.'

I watched as she scuttled across the road. What was I to make of all that? The afternoon had turned out to be far from what I'd expected.

Chapter Sixteen

KAREN PICKED HER MOTHER up in the car on the afternoon of Good Friday.

'Thank goodness we've got a couple of weeks off now,' she said as they drove. 'What with making Easter cards and decorating eggs, it's been pretty hectic these last couple of weeks.'

It didn't sound all that onerous to Susan, who sat beside her daughter quietly. She didn't feel much like talking after yesterday afternoon and she wished Karen would shut up for a minute. Her constant chatter was grating on her nerves like a steel file.

Sensing her mother's unease, Karen glanced at her. 'Everything all right, Mum?'

'Perfectly, thank you.'

'You seem a bit – well, not quite yourself.'

Susan turned to look at her daughter. 'We can't all be chatterboxes.'

'Oh! Is that how I seem? I'm sorry.'

Susan bit her lip. 'Oh, Karen, I'm sorry. I didn't mean that the way it came out.'

'What is it, Mum? I can tell something's upset you.'

Susan shook her head. 'It's nothing. I've got a bit of a headache, that's all.'

'Oh, well, a nice cup of tea and a couple of paracetamol will soon put that right. We're nearly home now.'

'Has your Dutch girl gone home?'

'Yes. She went yesterday. She's only away for a week. Simon's looking after Peter this afternoon.'

'I can't wait to see him. It seems ages.'

Karen flushed guiltily. It was true that she hadn't taken Peter to visit his grandmother for weeks. There just hadn't been the time. 'You always seemed to be so busy with Ted,' she said defensively.

Susan turned to look out of the window so that Karen wouldn't see the quick tears that sprang to her eyes. One spilled over to run down her cheek and she fumbled in her handbag for a tissue. Karen looked at her.

'I knew it! There *is* something wrong, isn't there, Mum?' She pulled the car over and switched off the engine. 'Tell me now, before we get home.'

Susan sniffed and put the tissue away. 'I'm just being silly really,' she muttered. 'It's just that yesterday I had a sudden urge to go and look at the allotment. All the seeds I planted would be needing attention and I ...'

'You saw Ted?' Karen prompted.

Susan swallowed. 'I – came away before he saw me,' she said. 'I couldn't face speaking to him.'

'Oh, Mum. You should have stayed and had it out with him – cleared the air.'

Susan shrugged. 'Maybe.' She forced a smile and changed the subject. 'Let's not talk about it any more. Let's just look forward to Easter.'

Karen said nothing. She wondered what kind of reception Simon's idea of their taking a weekend break and leaving Susan to look after Peter would get, especially as her mother was in this emotional state. She'd tried to talk him out of it but he insisted that they both needed a break and she had to admit that she was looking forward to it.

Susan sat staring out of the window. It had been a shock yesterday afternoon when she'd decided, after a lot of heart searching, to take a walk round to the allotments to see Ted. She'd expected to see him working away on his plot and envisaged his pleasure at seeing her. She'd even taken a couple of slices of newly baked raspberry sponge; his favourite. Over and over in her head she'd rehearsed what she would say and imagined his response. But when she got there, she found him working on what had been her allotment. And he wasn't alone. A woman was helping him. Aged, Susan guessed, in her forties, the woman was tall and slim with dark curly hair. Susan had certainly never seen her before. She could see that all the seedlings she had planted had been pricked out and were now in neat rows, coming along nicely. She watched from a distance as they worked together, hoeing the rows. The woman straightened up, one hand to her back. She said something and they laughed together and set off towards Ted's hut, no doubt to take a break for refreshment. Resentment tore at Susan's breast. It hadn't taken him long to find someone else! Was *she* aware of the situation? she wondered. Did he take *her* back to his bungalow? Had they – were they – more than friends? Unable to bear it any longer she had made a hasty retreat, returning home with her mind in turmoil and wishing she had not given in to her impulse to try to see Ted that afternoon. He obviously wasn't missing her at all!

At they walked in through the front door Peter came running to meet her. 'Granny!' He threw himself into her arms and she lifted him up to hug him. 'Let's see what Granny has brought you,' she said, putting him down again. Opening her bag she produced a bar of chocolate. 'Only a little bit now,' she said. 'And save the rest for after tea.'

She fed Peter his tea and then asked to be allowed to bath him and put him to bed. She read him a story about his favourite Thomas the Tank Engine and came downstairs, savouring the smell of Karen's home-made lasagne. As the three of them sat down to eat, Simon looked across at his mother-in-law.

'Susan,' he said casually, tearing off a piece of his roll and buttering it. 'We were wondering if you would like to do us a little favour while you're here.' He carefully avoided Karen's grimace.

Susan looked up. 'Of course, what is it?'

'Well, we – Karen and I have both been working really hard and we thought – wondered how you would feel about our taking a little break while you're here to be with Peter?'

'Oh.' Susan's heart sank. 'Over Easter, you mean?'

'Well, yes, this weekend is our only chance.'

Susan glanced at Karen, whose eyes were on her plate. 'I don't see why not,' she said. 'It would give me time to spend with Peter – as long as he doesn't miss you too much.'

'Oh, he won't,' Simon said quickly. 'And even if he did it would only be for a couple of days.'

'Where were you thinking of going?' Susan asked.

'Paris actually,' Simon replied. 'It's a nice little hotel quite close to the Eiffel Tower.'

'Oh, I see. You've already booked it, then?' Susan couldn't keep the resentment out of her voice. It was only too clear now that they'd taken her for granted. She had only been invited so that she could babysit while they went away for a break.

Karen broke in. 'We booked before Adrey asked if she could go home,' she said. 'With her father being ill we could hardly refuse.'

'No, of course not,' Susan said, wondering why Karen hadn't mentioned it before. 'And of course I'll be happy to take care of Peter while you're away. It'll be a pleasure. When are you going?'

'Tomorrow,' Simon said, maintaining his casual manner. 'Our flight leaves at ten o'clock so we'll have to be away from here quite early.'

Susan got up early next morning to see Karen and Simon off for their break. She stood at the front door with Peter in her arms, waving as they drove away.

'Where Mummy and Daddy goin'?' Peter asked, his little face bewildered.

'They're going for a little holiday,' Susan told him, giving him a hug. 'But it's only for two days and you and I are going to have such a lovely time while they're away, aren't we?'

The little boy nodded uncertainly. 'Brekspup?' he said hopefully.

Susan laughed. 'Yes, Nana and Peter are going to have breakfast together now. And afterwards we'll go to the park and feed the ducks. Would you like that?'

He smiled. 'Yes, please. Have poddidge now?'

'Yes, porridge coming up.'

After Susan had cleared away the breakfast things and washed up, she got Peter's buggy out from under the stairs. It was a lovely morning, although the wind was still keen, and she wrapped Peter up warmly before they set out. In the park, the snowdrops and crocuses were out and even a few early daffodils. They made her think of the allotment and her heart twisted, reminding her of how much she missed it.

Down at the lake, she broke up the bread they had brought into duck-sized pieces and Peter threw it to them, laughing as they squabbled and splashed, racing one another to get it first. When all the bread had gone they went for a walk, Susan pushing the buggy and holding Peter's hand as they made their way slowly around the lake. She was trying to persuade him to get back into the buggy again when she spotted Ted. Her heart quickened and she tried to hurry away in the opposite direction but Peter complained loudly.

'No – *this* way,' he cried loudly. 'Want to see birdies.'

She had promised him a visit to the aviary where there were peacocks and other rare birds. 'Tomorrow,' she said. 'Nana will take you tomorrow.'

'No! *Now*! Want to go now!'

Susan glanced up and saw to her dismay that Ted had heard Peter's cries and was hurrying towards them. She sighed. 'All right,' she said resignedly. 'We'll go now.'

Ted was breathless as he caught up with them. 'Hello there.' He bent down to Peter. 'Did I hear you say you wanted to see the birdies? Well, suppose I come with you?' he said. 'I was on my way there anyway. Now, are you going to sit in that comfy pushchair again? And will you let me push you?'

'Yes.' Peter obediently climbed back into his buggy and Ted

took the handle from Susan.

'I've tried countless times to ring you,' he said quietly as they walked. 'You never get back to me.'

Susan couldn't look at him. 'There's a good reason for that,' she said.

'Maybe, in your eyes,' he replied. 'But you've never given me the chance to explain.'

'I don't want to listen to any more lies,' Susan said, looking straight ahead.

'You're being very unfair,' Ted said thickly. 'There's a very good reason for everything if only you'd stop and listen.'

They'd reached the aviaries and Peter had climbed out of his buggy to look at the birds. Susan turned to Ted. 'Actually I came up to the allotments yesterday afternoon to see you,' she said.

His eyebrows rose. 'You did? I didn't see you.'

'No, you were too busy laughing with your new lady friend!' She took Peter's hand and bundled him, protesting, back into his buggy. 'We have to go now. Peter is due for his nap.' She set off at a smart pace while Ted gazed after her.

'Susan – wait,' he called. But Susan didn't look back.

When Karen and Simon arrived home, Susan sensed an atmosphere between them.

'Did you have a nice time?' she asked as Karen unpacked her case in the bedroom.

'Yes, thank you. It was lovely,' Karen replied without enthusiasm.

Susan sat down on the bed. 'So – what did you do? Did you go to the Louvre and the Moulin Rouge?'

'Oh yes, we did all the usual touristy things just as you'd expect.'

'You don't sound very enthusiastic.'

Karen sighed and looked at her mother. 'That's probably because we spent most of the time rowing.'

'Oh, no! That's a shame. What did you find to row about?'

Karen hung the last of her clothes in the wardrobe and sat

down beside her mother. 'I broke it to him that I've agreed to be one of the teachers to go on the school trip to Spain during the spring break. He was furious.'

'But surely as head, decisions like that are up to him, aren't they?'

'Neil Harris, who takes P.E. and games, is organizing this trip,' Karen explained. 'He's got carte blanche.'

'I see. So what is Simon's objection?'

Karen shook her head exasperatedly. 'You know how old-fashioned he is. He still hasn't really got his head round the idea of me going back to work. And leaving him and Peter for a whole week is completely beyond the pale.'

'Oh, dear.'

'But that's not all,' Karen said. 'Neil is a very friendly guy and Simon's got it into his head that he ...' She glanced up at her mother. 'That he fancies me and that he's only invited me on the trip for – well, for obvious reasons.'

'And is there any truth in it?' Susan enquired bluntly.

'*Mum*!' Karen looked scandalized. 'Of course there isn't.'

Susan shrugged. 'I only asked.'

'There's something else,' Karen said unhappily. 'Simon keeps on about us having another baby.'

'And you don't want one?'

'It's not that so much, Mum,' Karen said. 'It's just that I feel he's using it as a way to bring me to heel.'

Susan frowned. 'Bring you to heel? What on earth do you mean?'

'He never wanted me to be a working mother. He obviously thinks that I'd be obliged to give up if I had another child.'

'Lots of people do work with more than one. And you have your au pair.'

Karen looked at her mother exasperatedly. 'You don't get it, do you, Mum? Simon wants to *control* me and it's not on. This is the twenty-first century and I'm not his property.'

Susan decided it wasn't up to her to express an opinion or take sides so she just shrugged. 'I'm afraid that's something you're going to have to work out for yourselves,' she said.

'But you do see my point, surely?' Karen pressed.

'It's really none of my business,' Susan said. 'For what it's worth, I think it's something you need to be in complete agreement on.'

'But we've reached stalemate. Simon won't change his mind.'

Susan patted her daughter's hand. 'Then maybe you'll have to come to some kind of compromise,' she advised. 'But don't stop discussing it. Keeping quiet only breeds resentment.' Susan looked at her daughter. 'And while we're on the subject, Karen, why didn't you tell me from the start that you wanted me to babysit Peter while you and Simon went to Paris?'

Karen coloured. 'I didn't know at the time.'

'You did, though. You told me yourself that the trip was already booked.'

'That was a fib, Mum. To begin with I asked you because I thought you needed a break. It was only after I told Simon I'd invited you that he said it would be a good idea for us to take a weekend away.'

'All right, but why didn't you say so? Why lie about it?'

Karen sighed. 'It's what I'm trying to tell you, Mum. He always expects to get his own way. It was never my intention to go away and leave you in charge. I thought it was an imposition but in the end I gave in. Then I had to lie to you to back him up.'

Susan shook her head. 'You're going to have to make up your mind what it is you do want,' she said. 'And then stick to it. It's the only way if you want your marriage to work equally.'

'I know you're right.' Karen sighed. 'I just wish it wasn't so hard, keeping everyone happy.'

Chapter Seventeen

When I got back to the bedsit, I thought a lot about my meeting with Mum. It couldn't have been further from what I'd hoped for. To begin with, I'd stupidly visualized her as looking exactly the same as the last time we'd set eyes on each other. That was more than thirty years ago and the truth of it was that we were strangers. If we'd passed each other in the street we'd have been none the wiser. Getting to know one another all over again was going to take time – a lot of time. The question was, after this disappointing meeting was I prepared to put in the necessary time and effort? I wasn't at all sure.

The fact that she'd asked me for money had definitely put me off. After all, it was our first meeting for more than three decades. Then there was the fact that she'd obviously sunk to the depths; married a criminal and divorced him while he was in prison. If he was as bad as she made out, there were bound to be repercussions. Surely he was going to be fuming when he came out of prison and would be looking for revenge. On the other hand, if she told him that the daughter from her previous marriage had plenty of money (even though I haven't any more) wouldn't that be an incentive for him to get me to share the spoils? I began to wish I hadn't told her so much or given her my telephone number – more, I wished I'd never bothered to look for her.

I was so disappointed and depressed. I'd had some stupid naïve vision of this sweet, gentle woman, overjoyed to find her long-lost daughter – of her being loving and supportive – proud of me even. I'd imagined her as having made good in spite of her difficulties. I'd even visualized her as glamorous. Instead, this shadowy figure from my past, returning to give me the special maternal love I'd missed out on for most of my life, had turned out to be nothing more than a conniving woman on the make.

When I saw Mark at rehearsals the next morning, he was eager to hear all about our meeting.

'Come on, then, how did it go?'

I forced a smile. 'Oh, very well really.'

'Is that all?' he asked looking disappointed. 'Are you seeing her again? Did you hit it off?'

'I couldn't make any arrangements,' I told him. 'I said I'd get back in touch after the tour.'

He pulled a face. 'You don't exactly sound euphoric about it.'

'We hadn't set eyes on each other for over thirty years,' I said. 'It's early days. One hour over tea and cakes isn't going to make up for all that time.' At the mention of cakes, I recalled the greedy way she'd scoffed every single one of them without even offering the plate to me. It was just another of her less than endearing qualities.

Mark was peering at me. 'Oh-oh, it didn't go as well as you expected; you're disappointed?'

'Of course I'm not.'

'Did she fail to live up to your expectations?'

'No. She was – everything was absolutely fine.'

He grinned maddeningly. 'Ah – methinks the lady doth protest too much!'

I nudged him sharply in the ribs. 'Oh, shut up, Mark!'

I was relieved when Paul Fortune arrived at that moment and called for us all to gather round. 'This will be the last rehearsal for us here,' he announced. 'I'm giving you all next week off and then we're going down to Bournemouth where we'll have a week's rehearsal with the full orchestra. There'll be costume fittings as well and we open the following week.'

I felt a flutter of anticipation in my stomach, and Mark and I exchanged excited looks.

After the rehearsal was over, we went to the pub. As we sat down with our drinks I looked at Mark. 'So – we're really on our way!'

'Looks like it.' He took a long pull of his beer and set the glass down with a sigh of satisfaction. 'What are you going to do with your week's holiday?' he asked.

I shrugged. 'Not a lot. I'm practically out of money.'

He looked surprised. 'Really?'

I nodded. 'Down to my last few hundred and I don't suppose I'll be getting any refund on my investment for quite a while.'

'That's true.' He took another swig from his glass and looked at me, one eyebrow raised. 'If you're struggling I can let you have a loan.'

'No, I'll be all right,' I protested. 'Just have to go easily, that's all.'

'There'll be digs to pay for on tour, don't forget,' he reminded me.

'I know. I think I can handle that OK.'

'Well, you know the help is there if you need it.'

I looked at him. Mark was such a good friend – but was he too good to be true? Could I really trust him? I didn't want him thinking he could start calling in favours and I'd no intention of dropping my guard. 'As a matter of fact, I thought I might go home for a few days,' I said. 'Not that it's really what you'd call home, but Karrie and Susan are all the family I've got.'

'I'm sure they'd be more than happy to see you,' Mark said heartily.

I nodded, doubting his optimism. 'How's your sister?' I asked, changing the subject.

'She's fine. She had an interview for a job last week and she heard this morning that she's got it. It's secretary at the local primary school which will fit in nicely with the kids' school holidays. She's seen a little house that she likes too.'

My heart lifted. 'Right, so you'll have the flat to yourself again when we come back from the tour.'

He smiled. 'Are you still interested in moving in?'

Something in me – the contrary side of my nature – resented his taking it for granted that I would jump at the chance. 'Well, we'll see,' I said. 'I might move back in with Di.'

To my satisfaction he looked crestfallen. 'Oh. I thought you'd given your flat up.'

'I did.' I searched my mind for a way out of the corner I'd backed myself into; trying to remember the last thing I told him. 'But Di took on the lease as I told you,' I said as inspiration struck. 'So it's still an option.' Sometimes I amazed myself by my quick thinking.

He drained his glass. 'Well, the offer's there if you change your mind. When are you going home?'

'I'll have to ring and ask when it's convenient,' I told him.

'Well, give me a ring when you get back and we'll arrange to drive down to Bournemouth together.'

Back at the bedsit, I rang Susan's number, but I got the answerphone. I left her a message to ring me and set about making myself a sandwich. I'd just put the kettle on for coffee when my phone rang. I picked it up.

'Hi, Susan, that was quick....'

'It's not Susan,' a male voice said.

'Oh, then who ...'

'This is Steve Harris – your brother.'

My heart gave a jump. 'Oh – er – hello.'

'You don't sound very pleased to hear from me.' His voice was deep and coarse with a strong cockney accent. I had a feeling of foreboding.

'What can I do for you?' I asked stupidly.

He chuckled at the other end of the line. 'Now, there's an offer. I thought we could meet,' he said. 'I think it would be nice to get to know one another, don't you?'

I bit my lip. This was something I had to nip in the bud. 'Forgive me but I can't really see the point,' I said.

'What? You're saying that you don't want to meet your little brother after all these years?' The voice held a mocking tone.

'Until yesterday I didn't even know you existed,' I told him. 'And you're not my brother – only a half-brother.'

'Blood's blood,' he said. 'We share the same mum. She thought you were a bit of all right,' he said. 'In fact it was her idea that you and I got together. She thought we'd have a lot in common.'

'I doubt it,' I told him. 'As a matter of fact, I'm going away tomorrow and after that I'll be away on tour so I can't really see an opportunity for us to meet.'

'Oh dear, that's a pity,' he said smoothly.

'Yes, but there, it can't be helped,' I said, thinking that by the time we came back to London he'd have given up, with any luck. 'Goodbye.' I ended the call quickly and switched my phone off so that he couldn't ring back. 'Damned cheek,' I muttered as I made myself a coffee. No doubt he thought he could get some money out of me. Well, he had another think coming.

Susan didn't ring me back and in the end, I decided to just turn up. Maybe she'd be annoyed but it was only for a few days and I really didn't fancy staying in the grotty bedsit for a whole week with nothing to do. I took the train next morning and arrived in Bridgehampton just before lunch.

Susan didn't look surprised when she opened the door and found me outside.

Her expression was more one of resignation.

'Louise – how nice. Come in.'

'I did ring you yesterday,' I told her as I walked in through the door. 'I left a message. Didn't you get it?'

She shook her head. 'I don't always check,' she said.

I didn't believe a word. 'Is it convenient for me to stay for a few days?' I asked.

She looked at me and then at the small bag I was carrying. 'A few?'

'Just until the weekend,' I said. 'We're off down to Bournemouth next Monday to begin our tour so I'll need time to pack and leave the flat tidy.'

She didn't even try to conceal her relief. 'Oh, well, that's fine then,' she said. 'I'm afraid it will have to be the sofa again.'

'Yes, that's fine.' I unzipped my bag and pulled out the bottle of rosé wine I'd brought as a sweetener. I handed it to her. 'I got you this, Susan,' I said with a smile. 'I know it's one of your favourites.'

She took it from me, looking a bit taken aback, and when she spoke, her voice had softened a little. 'Oh, that was thoughtful of you. I was just going to make a meal. We can have it with that.'

'Better still, why don't you let me take you out to lunch as a thank-you for having me at short notice?' I suggested.

She smiled. 'Oh, thank you, Louise. That would be lovely. I'll get my coat.'

Over lunch I heard all about Karrie and Simon clearing off to Paris at Easter and leaving her literally holding the baby. I thought it was a bit thick of them but I decided to play it cool.

'Well, I know how you love little Peter so I don't suppose you minded,' I said.

We'd drunk a bottle of wine between us and it had loosened her inhibitions somewhat. She frowned.

'Of course I didn't mind having him,' she said. 'We had a lovely time together, the two of us. What I did mind was that Karen felt she had to manipulate me into it. If she'd asked outright I'd still have said yes.'

'Oh well, no doubt she had her reasons,' I said tactfully. 'And what about you? What have you been up to since I was here last?'

Her face clouded. 'Well, to tell the truth, dear, I've had a bit of a disappointment.'

'Over what?'

She looked up at me. 'Over a man actually.'

I raised an eyebrow at her. 'A man, eh? Well, you dark old horse, Susan.' I looked at her, my head on one side. 'So – are you going to tell me about it?'

I got the whole story – about the duplicitous Ted and his double life; about the lovely time she'd enjoyed with him only to discover his deception. When she'd finished pouring it all out to me, she looked up.

'I expect you think I'm a foolish old woman.'

I reached across the table to pat her hand. 'Of course I don't. The truth is, Susan, we women never seem to learn, do we?'

She sighed. 'I suppose not. The truth is I only ever loved one man and that was your father. I'm not used to the kind of games people play nowadays.'

'I met someone I used to be at drama school with,' I told her. 'He's in the same show and I thought we'd get together again. We spent Christmas together and he even asked me to move in with him. Just when I'd decided to say yes, he announced that he'd moved his sister and her two kids in.'

'Oh dear, why was that?' Susan looked concerned.

I shrugged. 'Search me. At a guess, I suppose he regretted asking me and moved them in to make things impossible.'

Susan shook her head. 'How devious of him. And now you have to see him every day – work with him. How uncomfortable.'

'Oh, we're still friends,' I told her lightly. 'We have to be under the circumstances.'

The following morning, Susan went out to the supermarket. She'd asked me to go too but I fancied a lie-in, even though it was only on the sofa. Soon after she'd gone, I got up and made myself some toast and coffee. I was just about to get into the shower when the bell rang. I dragged on my dressing gown and went to answer it. Outside stood a tall, elderly man carrying a large bouquet of flowers. I guessed him to be the notorious Ted.

'Good morning,' I said frostily. 'Can I help you?'

He looked a bit taken aback. 'Oh – is Susan – Mrs Davies in?'

'No, she isn't,' I said. 'I'm Louise, her stepdaughter. I take it you are Ted.'

He nodded. 'That's right.'

'Well, I can tell you for nothing that you're wasting your time,' I told him. 'I've heard all about you. She wants nothing more to do with you and your devious ways so you can take your flowers and give them to the next gullible woman you pick up on the bus!' And with that I slammed the door in his astonished face.

When Susan arrived home I told her about Ted's visit. Her cheeks coloured.

'Oh. What did he say?'

I shrugged. 'Not a lot. I sent him on his way – told him what he could do with his flowers.'

'Flowers?' she enquired weakly.

'Yes. He was carrying this enormous bunch of flowers. Daffodils and irises and those vulgar stripy pink things. I've always hated those, haven't you?'

'Tulips? No. I quite like them actually.' She was unpacking the shopping, her back to me. 'What else did he say?'

'Nothing. I didn't give him the chance,' I told her. 'I don't think he'll be bothering you again though. If he didn't get the message he must be thick!'

'I see.' She turned to me. 'I take it you were rude to him?'

'I only told him you didn't want any more to do with him.'

'Well, I wish you hadn't,' Susan said. 'I wish you'd told him I was out and left it at that.'

I stared at her. 'After what he did? You must be mad if you're even thinking of taking him back after that.'

'But it's my decision, isn't it?' To my astonishment, her eyes filled with tears and she fumbled in her sleeve for a handkerchief. 'What I told you was in confidence and in future, I'd be grateful if you'd mind your own business, Louise.'

'But you said – you were really upset and …' But she'd disappeared into her bedroom and closed the door firmly behind her. I sighed. Really! There was no pleasing some people.

The atmosphere was distinctly frosty after Ted's visit. Susan was distant – hardly speaking to me, and the following afternoon, I decided to go and say hello to Karrie. When I announced my intention to Susan she seemed relieved.

'She should be home from school by four,' she said. 'But she usually picks up some shopping before coming home so I should give her another half-hour if I were you.'

I took myself off to the pictures after lunch, unable to stand Susan's glum face any longer. There was hardly anyone else in the multiplex and the film was mediocre. I came out feeling depressed. Maybe it hadn't been the best idea to come home after all. Even a visit to my bourgeois sister was preferable to going home to Susan. I took the bus to Sunnyside Drive and arrived at Karrie's soon after half past four. I rang the front doorbell and waited but no one came. I rang again – still no reply. But Simon's car was in the drive so I knew that someone had to be at home. I walked round the side of the house and through the back gate. The kitchen looked out over the back garden and I took a peek through the window. What I saw gave me a start. There in the middle of the kitchen, Simon was embracing a blonde girl. She had her head on his shoulder and he was holding her close and murmuring in her ear. I jumped back before either of them spotted me. My God! Simon of all people. And that old cliché: the husband and the au pair! Poor Karrie! Did she even suspect? I tiptoed away as quietly as I could, closing the side gate behind me. Walking down the tree-lined avenue my head was in a whirl. Karrie and Simon. The perfect couple. Who'd have thought it?

I was almost at the end of the road when Karrie's little car came round the corner. She saw me at once and pulled into the kerbside, winding down the window.

'Louise. Mum said you were coming for a visit. Were you coming to see us?'

I opened the passenger door and climbed in. 'I've just been,' I told her. 'Karrie, I'm so sorry but I've got something to tell you.'

'To tell me?' She looked alarmed. 'What are you talking about? Is it Mum?'

'No, she's fine. Look, I don't quite know how to tell you this, but I've just called at yours. No one answered the door so I went round to the back. Karrie – Simon was in the kitchen. He was *kissing* your au pair girl.'

The colour left her face. '*Louise*! What on earth are you talking about?' she said shakily. 'I don't believe you. If this is one of your sick jokes ...'

'It's no joke. I tell you, I saw them with my own eyes,' I told her. 'Do you want me to come back with you? If he denies it, I'll tell him what I've told you.'

'No! Please go now, Louise. I'll handle this on my own. It's between Simon and me.'

'Are you sure?'

'More than sure.'

I was a bit disappointed. I'd been looking forward to seeing the perfect Simon get his comeuppance. 'Well, if you want me you know where I am.'

'I do.' Karen tapped the steering wheel impatiently. 'Please, Louise, if you don't mind.'

'OK, I'm going.' I got out and bent to speak to her through the car window. 'I'm here till the weekend so if you need …' Before I could finish the sentence, she was revving up the car and I had to leap back to avoid the car as it sped off. I stared after it. Well, that was gratitude for you!

Susan and I were having our evening meal when we heard a key grating in the front door. Susan looked up.

'That will be Karen,' she said. 'She's the only other person who has a key. I wonder what she's doing here at this time of the evening.' She got up from the table. 'I hope nothing's wrong.'

Your smug little lives have hit the skids this time, I said inwardly.

Karrie burst into the room before Susan could reach the door. Her face was crimson and she was breathing heavily.

'You *bitch*!' she yelled at me. 'You interfering, trouble-making *bitch*!'

I stared at her, spreading my hands. 'What am I supposed to have done?'

'You put two and two together and made a hundred and four,' she said. 'Can you imagine how I looked, bursting in and accusing Simon and Adrey of adultery?'

'It was no more than they deserved,' I said. I hadn't mentioned the occurrence to Susan and now she was looking from one to the other of us with a shocked expression, at a loss to know what was going on.

'Will someone please tell me what this is about?' she demanded.

Karen looked at her. *'She* ...' She pointed her finger at me. 'She stopped me on my way home this afternoon to tell me that she had seen Simon kissing Adrey in our kitchen.'

Susan gasped. 'Oh my God!'

'He wasn't *kissing* her,' Karen went on. 'He was comforting her because she was upset. She'd just received a telephone call from home to tell her that her father had died. Can you even *begin* to imagine how that made me feel?'

'And you believed him?' I said. 'A likely story if you ask me.'

'Well, I'm not asking you,' Karrie shouted. 'And it's true. She's packing to go home for the funeral as we speak.' She took a step towards me. 'Your trouble, Louise, is that you think everyone is as nasty and devious as you are. Every time you come home there are ructions. The last time you almost had me locked up and now this! It's going to take me a long time to put things right this time. Adrey has already said she won't be coming back and Simon won't even speak to me. And it's all down to you!'

Her hands were clenching and unclenching, and I thought for a moment she was going to hit me until Susan stepped forward and took her arm.

'Calm down, darling,' she said. 'I know you're upset but I'm sure Louise only meant it for the best ...'

'The best?' Karrie screamed. 'Don't try to stand up for her, Mum. She causes chaos wherever she goes. She thrives on it – does it on purpose. If you take my advice you'll kick her out. I know one thing – I'll never speak to her again.' She shook off Susan's restraining hand. 'It's all right. I'm going now. Give me a ring when you've got rid of her. Goodnight, Mum.'

Susan went to the door to see Karrie out. When she came back, her face was grave. 'Why, Louise?' she asked.

I shrugged. 'I know what I saw.'

'And you actually *saw* them kissing?'

'He was holding her,' I hedged. 'She had her face close to his.

They were about to kiss. I'd lay odds on it.'

'So you didn't actually witness a kiss between them?'

I got up from the table. 'Why split hairs? It was obvious what was going on. If Karrie wants to bury her head in the sand; if she wants to let him get away with it, then it's her funeral. I should have ...'

'You should have minded your own business,' Susan finished for me. She began to clear the table. 'I think it might be best if you left first thing in the morning, Louise,' she said quietly. 'Karen was right, unfortunately. There always seems to be trouble when you're around.'

'I'll do better than that,' I told her. 'If this is all the thanks I'm going to get for telling the truth then I'll go now. I can catch the last London train at ten o'clock and be back before midnight.'

Susan didn't argue. 'As you wish,' she said.

It was really late when I got back to Stoke Newington. I wasn't looking forward to going back to my dingy bedsit and as I climbed the stairs wearily, I told myself I'd be glad to be out of the place. I rummaged in my bag for my key but when I went to put it in the lock the door swung open. Inside I found chaos. I'd been burgled. I stood, staring around me in disbelief. The place had been well and truly turned over, drawers pulled out and the contents strewn everywhere. The bed had been stripped; cupboards emptied. I pushed the door closed and it was only then that I saw that it had been forced. I couldn't even secure it for the night. I pulled the chest of drawers across to block the doorway, then sat down on the bed and surveyed the mess. What a horrible week it had been, and now to come back to this. Luckily, there had been nothing of any value in the room. Certainly no money, but the feeling that someone had been here – could possibly come back – gave me the horrors. My first thought was for Mark. I needed him – needed someone kind and sympathetic. I took out my phone and selected his number. After a few rings, a woman's sleepy voice answered.

'Hello, Mark Naylor's flat.'

'I need to speak to Mark,' I said. 'Tell him it's Louise.'

'Isn't it rather late to be ringing?' the voice enquired. 'It's almost one a.m. We were asleep.'

'I'm sorry, but this is an emergency. Who am I speaking to?'

'This is Cathy, Mark's sister,' she said. 'Mark isn't here, he's gone away for a few days. Is there anything I can do?'

My heart sank. 'No, not really. When will he be back?'

'Tomorrow evening. A friend invited him to go up to Scotland for a few days.'

'Oh dear.' I swallowed hard. 'The fact is, I've just come back from a few days away and I've been burgled. The place is in a terrible state and I don't know what to do.'

'Have you rung the police?'

'No. I can't see the point. We're off down to Bournemouth in a couple of days and I'd have to go through all that red tape. Besides, as far as I can see nothing's been stolen.'

'OK, I'll get Mark to come over to you as soon as he gets back. Does he have your address?'

'No. It's room three, fourteen Mason Street, Stoke Newington.'

'Right. I've got that.'

'Thanks.' I switched off my phone and lay down on the rumpled bed, feeling totally sick. I'd been fobbing Mark off about my so-called 'flat' for weeks and now he was going to see this horrible room. But I was past caring. Thank God we were off down to Bournemouth in a few days and leaving it all behind. This had been the worst week of my life. Why had everyone been so bloody awful to me? I'd only done what anyone else would have done under the circumstances, and anyway, it was time they were all jolted out of their little suburban heaven, damn them!

Chapter Eighteen

Karen had given Peter his tea early, bathed him and put him to bed. He'd just about worn her out today, whingeing and crying for Adrey all day. Simon had insisted that she stay at home to look after him. There had been a horrible scene about it last night when she got home from Susan's flat. Karen shuddered at the memory.

By the time she'd arrived home Adrey had gone. Simon had put Peter in his car seat and driven her to the station. He was waiting for her when she got back from her mother's. She found him sitting in the living room with a face like thunder and the moment she got in, he started.

'I hope you're proud of yourself!'

Karen began to take off her coat. 'Well, I'm sorry but Louise was certain you were kissing her,' she said. 'How was I to know...?'

'You didn't wait to find out, did you,' Simon stormed. 'Just came steaming in, throwing accusations around like confetti. That poor girl! It was the last thing she needed, or deserved after the news she'd just had.'

Karen winced. 'I know that now and I'm really sorry.'

Simon snorted. 'Too little, too late. She won't be coming back, thanks to you!'

Karen hung up her coat and came back to sit down opposite Simon. 'Well, I'll write a letter of apology to her.'

'I'm sure that will be a big comfort to her but you still won't get her back,' he said sarcastically. 'She said she'd have to stay anyway, to support her mother.'

'Oh, then it isn't *all* my fault?'

'So – you think that makes it all right then, do you? Bursting in and accusing the poor girl of seducing your husband. And have you even given a thought to how humiliating it was for me?'

'I've said I'm sorry.'

'Have I ever given you any reason not to trust me?'

'No.'

'Then why start now at the worst time possible? And since when have you taken anything Louise said as gospel?'

Karen sighed. 'I don't know. I was tired, I suppose. Adrey is very pretty and just lately …'

'Just lately you've had no time for me or Peter. You're always too tired and preoccupied. You were suffering pangs of guilt, that's it, isn't it?' Karen's shoulders drooped and he went on: 'Incidentally, in case you're wondering, *I* put Peter to bed. You're so self-centred you haven't even enquired about your son!'

'Well, I guessed you'd put him to bed, obviously. And I told you, I'll write to Adrey and apologize. I suppose we'll have to get Peter into a nursery now,' she said, half to herself. 'Either that or ask Mum to—'

'Oh no! We'll do nothing of the sort,' Simon broke in. 'From tomorrow *you'll* be taking care of him.'

Her eyes widened. 'But that's not possible. What about my class?'

'I'll get a supply teacher in until I can replace you. You're packing the job in right away, Karen; at least until Peter goes to school. I've had enough. You're a wife and mother. That should take priority.'

So here she was, stuck in the house all day with a fractious child and a whole heap of ironing that Adrey hadn't had time to do before she left. And all thanks to Louise. Karen ground her teeth at the thought of how much she'd like to wring her sister's neck. Why *had* she believed what she said? She had to admit that Simon had been right. If she faced up to the truth she knew it. She had been neglecting her home and family lately. Once home from school, she had been too tired to play with Peter – too tired to listen to Simon's news when he came home, and at bedtime, too tired to make love. It was weeks since she had shared any intimate moments with Simon and if he had strayed she knew that she would only have had herself to blame.

But to have to stay at home day after mind-numbing day felt like a punishment. Peter was adorable, of course, and she loved him to bits, but to be restricted to the conversation one could have with a two-year-old, or, worse, with the other mothers clustered round the swings in the park, obsessed with which supermarket was the cheapest or which were the best nappies, was enough to drive her mad. A wail from upstairs told her that Peter was awake and demanding attention again. She sighed. Maybe she'd put him in his buggy and walk round to her mother's. It would pass the afternoon and by the time she got home it would be Peter's teatime.

Susan was pleased to see them both and immediately put the kettle on for the inevitable cup of tea.

'What are you doing home at this time of day?' she asked as Karen took off Peter's coat.

'Simon has insisted that I give up my job now that Adrey has left,' Karen told her. 'She's not coming back. Thanks to Louise I made a complete fool of myself yesterday. That poor girl had just received a devastating telephone call and then I burst in with my accusation. Simon was absolutely incandescent with rage when I got back, and that was when he insisted that I give up the job at once and became a full-time mother.'

Susan smiled sympathetically. 'Well, I can't really say I blame him, dear,' she said. 'Peter needs his mum and you know, you never get these lovely baby years back again.'

Privately Karen thought it was just as well but she didn't say so. Like Simon, Susan held the old-fashioned notion that a woman's place was in the home. 'Peter is missing Adrey,' she told her mother. 'He's been really difficult all morning – nearly driven me balmy with his whingeing.'

Susan pulled the little boy onto her knee. 'Poor little chap,' she said, dropping a kiss on the toddler's blond head. 'At this age, stability and routine are important. I'm not surprised he's upset. If you want any help, Karen, you know I'm always willing to lend a hand, don't you?'

Karen smiled. 'Yes, I know you are, Mum, and I'm really grateful. I might take you up on that.'

'It's not as if I'll be doing anything else,' Susan said wistfully. 'Ted came round with a bouquet of flowers while Louise was here one morning. I was out and I'm afraid she told him where to go in no uncertain terms.'

'She had no right to do that,' Karen said. She looked up at her mother. 'Although you weren't going to think of starting the relationship up again, were you?'

Susan sighed. 'I don't know. I really miss him. Maybe I should have given him more of a chance to explain.'

Karen snorted. 'Huh! What's to explain?'

Susan sighed. 'Ah well, I'll never know now, will I?'

'Why can't Louise keep her meddling nose out of other people's business?'

'I think that's what they call the sixty-five thousand dollar question,' Susan said.

Chapter Nineteen

WHEN THERE WAS A ring on my bell on Saturday morning, my heart jumped into my mouth. Ever since the break-in, I'd been really nervy and the landlord had refused to do anything about the broken door, saying that it was down to me to get it repaired.

I picked up the entry-phone. 'Who is it?' I called from behind the chest of drawers that had been securing it ever since the break-in.

'It's me – Mark.'

With huge relief, I buzzed him in and when I heard him outside, I pushed aside the chest and opened the door. He looked at me quizzically. 'What's with all the furniture removal?'

'Being burgled isn't something to joke about,' I told him. 'As I told your sister, I came back on Thursday night to find the place had been thoroughly gone over and the door still isn't fixed.'

He looked around. 'What are you doing in a dump like this anyway?'

'This is where I've been all the time.' I confessed. 'I didn't want you to see it.'

He laughed. 'I'm not surprised.' He looked around. 'Whoever broke in has certainly made a mess of the place.'

'Do you *mind*? I've tidied up since then. You should have seen it when I got back,' I told him.

'And you haven't notified the police?'

I shook my head. 'Nothing's been taken and I can do without the hassle.'

He sighed. 'Just as well we'll be out of here first thing Sunday morning. Are you packed?'

I pointed to the two suitcases standing by the door. 'You bet. I can't wait.'

Mark looked round. 'And you say they didn't get anything?'

I shook my head. 'There wasn't anything worth taking. Certainly no cash.' I looked at him. 'I can't help thinking I know who's behind this.'

'Really – who?'

'Just before I went away, I had a phone call from a guy who said he's my half-brother. He wanted us to meet. I didn't like the sound of him so I said no.'

'That's a bit of a long shot, isn't it? Presumably he doesn't even know where you live.'

I vaguely remembered giving Mum the name of the road. It wouldn't have been that difficult for him to find me. 'He might have found out,' I said.

Mark looked doubtful. 'Why did he want to meet anyway?'

'Said he wanted us to get to know one another.'

'So – is that bad?'

I shook my head. 'What would we have in common? I've never set eyes on the bloke. It seems my mother married again and had another child. His father was a violent man and she divorced him while he was in prison.' I looked at Mark. 'To tell you the truth, I'm beginning to wish I'd never set out to find her.'

He nodded. 'I got the feeling you were less than happy with the meeting.'

'That's not all. She borrowed money from me before we parted,' I told him.

'How much?'

'A couple of hundred.'

He whistled. 'Pheeew! Maybe you told her too much about yourself. I have to agree that it doesn't augur well for re-forging your relationship with her.'

'Or *him*,' I reminded him. 'My so-called half-brother.'

He put his arms round me and gave me a hug 'Poor old sausage. Never mind, once we're out of here you'll have nothing to worry about.' He held me at arm's length and looked at me. 'Would you like to come back and spend the night at the flat?'

I shook my head. 'No. I'll be OK here. After all, it's only for one more night.'

'Well, let's go out for the day, then, spend our last day in London by celebrating.'

I laughed. 'Celebrating what – my burglary?'

'No, the start of our record-breaking success, of course.'

'You hope.'

'I don't hope – *I know*! I've got a really good feeling about *Oh Elizabeth*. We'll start with the London Eye, then lunch at a little place I know in Soho. How does that sound?'

We spent a lovely day in the West End and I felt much better when Mark dropped me off but for some unknown reason I had the nightmare again that night. I hadn't had it for ages but this time it was different; this time *I* was the mother and the one doing the walking out. I couldn't make head or tail of it when I woke but it gave me a dark, disturbed feeling that lasted for ages. It was only when I came to enough to remember that this was the day we were travelling down to Bournemouth, one step nearer to my success as an actress, that I was able to clear my mind and set about getting ready.

Mark picked me up at eight o'clock in the Ferrari. I'd asked him to come early because I didn't want to encounter my landlord again. I'd paid the rent and the man in the room upstairs had fixed the door for me (after a fashion) but I hadn't given him the formal month's notice, and I didn't want an argument on my hands over the extra rent he was bound to demand.

We stopped off for lunch at a nice restaurant in Farnborough and arrived in Bournemouth late that afternoon. Mark did a tour of the town to get our bearings, then parked the car in a multi-storey car park and turned to look at me.

'Right, shall we go and find ourselves somewhere to stay?'

I nodded. 'Got any ideas? I don't know Bournemouth.'

His eyes twinkled. 'How about booking a room at the Royal Bath for a couple of nights? My treat,' he added. 'We can look around for somewhere cheaper once we begin rehearsals and get our bearings.'

We'd passed the Royal Bath on our tour and I'd been well impressed. My heart gave a leap. 'Oh, Mark, that sounds wonderful.'

We took our bags and checked in. A porter took our bags upstairs and ushered us into a wonderful room with a sea view. Once the man had gone, I turned to Mark.

'This is a real treat,' I said. 'I hope it's just a taste of things to come once we're famous.'

We went down to dinner and Mark ordered a bottle of bubbly to celebrate. He held his glass aloft.

'Here's to us,' he toasted. 'Us, the play, full houses and fame and fortune for my favourite leading lady.'

I sipped my champagne with relish. The future sparkled even more than the bubbles in my glass.

We hardly slept that night; partly because we were high on champagne and anticipation of the day to come and partly because our enthusiastic lovemaking kept us busy. Mark was so skilful and practiced that I was swept away on a cloud of sensual pleasure again and again. As I lay in his arms, my head on his chest, I asked him how he came to know so much about how to please a woman. He looked down at me.

'There's no secret,' he said. 'When you're as much in love as I am with you, it comes naturally.'

I didn't really believe him. It was always difficult to know when to take Mark seriously, but I didn't really care. After all, we were staying in the kind of luxurious hotel I'd always seen myself staying in and the future couldn't look brighter. He deserved my appreciation at the very least.

'Do you think you could ever love me back?' he whispered. 'Just a little bit?'

I snuggled closer. 'You are a silly old romantic, Mark Naylor,'

I said. 'Stop getting carried away.'

'You make love as though you mean it,' he said softly.

'Of course I mean it,' I told him. 'You're so damned good at it I'd have to be made of stone not to.' I glanced at the bedside clock. 'Have you any idea what the time is? It's actually getting light. If we don't get some sleep we'll be good for nothing in the morning.'

We were due at the Pavilion Theatre at 10.30 so after a wonderful full English breakfast, Mark and I strolled across the road to the Pavilion. I stood in the forecourt looking up at it.

'Wow! It's so big.'

Mark nodded. 'Restaurants and a massive ballroom as well as a very large theatre. It's a wonderful place to be launching.'

I felt a thrill of excitement. At that moment, several other cast members joined us, led by Carla Dean. She looked at me with her usual disdain.

'Where did you two get to last night?' she asked. 'You weren't on the train.'

'We drove down in Mark's Ferrari,' I told her airily. 'And we're staying at the Royal Bath.'

To my great satisfaction her eyebrows shot up. '*Really*? Get you! Living it up as the leading lady already! Well, we'd better find the stage door, I suppose. Unless you want to stand here boasting all morning.'

We located the stage door and went inside. Carla was first to inspect the dressing rooms.

'Mmm, not bad,' she announced. 'But someone has left their stuff in this one.'

On inspection, the other dressing rooms seemed to be full of other people's belongings as well. Phil shook his head.

'It's too bad,' he said. 'The previous lot should have packed up on Saturday night. They should have moved out yesterday morning at the latest.'

At that moment, music could be heard coming from the direction of the stage above us. We stood speechlessly, staring at each other as a tenor could be heard singing 'The Music of the Night' from *Phantom of the Opera*. When the song came to an end, Carla made her opinion heard in no uncertain terms.

'What the bloody hell is going on?' Without another word, she stormed up the stairs to the stage, the rest of us straggling behind. I arrived just in time to see her striding onto the stage where a rehearsal was clearly in progress.

'Excuse me,' she said, her sonorous voice echoing round the stage. 'May I ask what is going on here?'

A man, presumably the director, stood up from his seat in the stalls and walked down to the stage. 'I might well ask the same question of you. Who are you, anyway?'

Carla swept her hand around in our direction. 'We are the cast of *Oh, Elizabeth*,' she said. 'We are supposed to be rehearsing here from this morning until our opening next week.'

The director looked puzzled. 'My company are opening here next week,' he announced. 'We are the number one tour of *Phantom of the Opera*. If you'd like to go up to the foyer, you'll see our posters and flyers. The forthcoming attractions posters are on display outside too. I'm surprised you didn't notice them.'

We stared at each other. I nudged Mark. 'Maybe we've got the wrong theatre,' I whispered. Mark cleared his throat and spoke up.

'Come along, Carla. We'll go and see the theatre management about it.' He nodded to the director. 'I apologize for the interruption. There's obviously some mistake.'

Appeased, the man nodded. 'That's quite all right. I hope you get it sorted out.'

Carla was furious and complained all the way back to the theatre foyer. 'What a cockeyed arrangement,' she complained. 'Just wait till Paul gets here. He's going to be hopping mad.'

The front-of-house manager knew nothing about us. We all crowded into his office and explained our predicament but he said he'd never heard of Paul Fortune or a show called *Oh Elizabeth*.

'There's another theatre across the road,' Phil pointed out. 'Do you think that's where we're supposed to be?'

We straggled across the road to the smaller Palace Court

theatre, but there were billboards in the foyer advertising a thriller beginning next Monday. Defeated, we all repaired to a café further along the road to try to decide what to do.

'Well, I know what I'm going to do,' I said, getting out my phone. 'You go in. I'm going to ring my agent. I'll join you in a few minutes.' When they'd gone, I switched on my phone and clicked on Harry's number. His secretary answered.

'Sally, it's Louise Delmar,' I said. 'Can you put me through to Harry, please?'

'I'm sorry, Louise,' the girl said. 'But I'm afraid he's in a meeting with his solicitor at the moment. He's asked me not to put any calls through.'

'I can't help that. I must speak to him. He won't mind when you tell him it's me, calling from Bournemouth. Tell him it's really urgent, Sally.'

I waited, tapping my foot impatiently. When Harry came on he sounded upset. 'Louise. I think I know why you're ringing me.'

'It's not good enough, Harry,' I jumped straight in. 'We're all down here in Bournemouth and someone's made a hash of the bookings. There's another company rehearsing in the theatre.'

'I know – I know. Listen, Lou. I'm afraid I've got some rather bad news. There was never going to be a show. It was all a highly elaborate con. The truth is, Paul bloody Fortune has disappeared. He's gone, and taken all our money with him.'

I stood as though rooted to the ground, speechless; poleaxed; the blood freezing in my veins. All the work we'd put in. All those weeks of rehearsal! *All the money I'd invested!* It had to be some kind of horrible nightmare. It *couldn't* be true.

'What – what are you saying, Harry?' I said weakly. 'Paul's gone? Gone where?'

'Anyone's bloody guess! How much cash did you invest, Lou?'

When I told him he gasped. 'Christ! I invested too, but not as much as that. What about the others?'

'I don't know. Harry, look, we have to find him. Has anyone been to his flat?'

Harry laughed dryly at the other end of the line. 'Turned out it wasn't even his flat – borrowed while the owner was abroad. He cleared out of there days ago and no one has seen or heard from him since. He's had a good head start on us.'

'The police?' I suggested. 'Surely you've contacted the police.'

'Of course we have. D'you think we haven't been down every possible avenue? They've had roadblocks in place and they've had men at all the airports and ports. He's gone, Lou. As I said, he's had a week's head start on us. I doubt whether Fortune was even his real name anyway – appropriate though it was. He's probably counting his spoils in some luxury hotel in Monte Carlo as we speak.' He groaned. 'I can't believe I was taken in by him but he was so bloody plausible. I feel such a fool! And with all the years I've been in the business. This'll be the ruin of me when word gets around. God only knows what my other clients are going to say, not to mention the wife. I'm—'

'*Harry*!' I broke in sharply. 'Do you have any idea of the impossible position I'm in? Have you got any suggestions as to how I'm to break this news to the rest of the cast?' Stuff Harry's bloody wife! What about *me*? Here I was, trying to face the fact that all my dreams of fame were down the toilet, not to mention most of my inheritance, and all he was worried about was what his flaming *wife* would say.

'Look, Lou, I'm up to my neck in this as much as you are,' he said. 'I'm afraid you'll have to handle your end of it with your usual tact and diplomacy. I'm going to have to go now. I've got my solicitor here with me. I only hope he can come up with something that'll get us our money back but it doesn't look very hopeful at the moment.'

'I'll be in to see you when we get back, Harry,' I warned him. 'After this, I think you owe me, don't you?' Without waiting for his reply, I punched the red button and stood for a moment, trying to process the unbelievable disaster that was taking place. I'd never felt more like running away but I knew I had to go inside and tell the others. Gathering all my courage, I took a deep breath and pushed open the door of the café.

Inside, the rest of the cast were clustered round two large tables. As I walked in, all their faces turned towards me. I swallowed hard. It all felt so surreal. This had to be the worst day of my life.

'I'm afraid I've got bad news,' I began, my knees shaking. 'It looks as if we've all been taken for a ride. Paul Fortune has skipped the country with all our cash. There is no play. No West End run, no nothing – there never was. It was all a big scam.'

For a moment there was a shocked silence as they all looked helplessly at each other then, Carla sprang to her feet, incandescent with rage. 'Are you seriously telling me that bastard has fucked off with everything?' she shouted. 'That all these weeks he's had us on a bit of string like a bunch of bloody puppets, letting us believe we were going into the West End with his fictitious fucking play while stashing all the cash he'd conned us out of into some sodding Swiss bank?'

I nodded miserably. 'More than likely.'

'But what about those other guys – the choreographer, that useless bloody director?'

I shrugged. 'Either in it with him or being taken for a ride like the rest of us, I suppose. It's irrelevant now anyway, isn't it?'

Carla looked round at the others. 'Well, come on, you dozy lot!' she shouted. 'Are you going to sit there with your mouths open like a lot of fucking goldfish? Are we going to just sit back and let this happen?'

'It's no good, Carla,' I put in. 'He's long gone. Why do you think he gave us a week off?' I related everything that Harry had told me on the phone. When I'd finished, Carla slumped back in her seat.

'So we're all well and truly fucked!'

'Couldn't have put it better myself,' I said ironically. I looked at Mark. 'Shall we go and start packing?'

Mark nodded and got up from his seat. All this time he'd said nothing and his face gave away nothing of what he was feeling.

'It's all right for you two!' Carla accused, staring belligerently at us. 'You and him with his fucking Ferrari. You've obviously got enough cash between you not to worry about the odd fifty k. It might be peanuts to you but what about the rest of us? What are we supposed to do?'

Something in me snapped. I'd had just about enough of Carla. 'Oh, for God's sake, will you *shut up,* Carla!' I shouted. I'd always hated mouthy women who kicked off and threw their weight about. Carla had done it repeatedly all through rehearsals, arguing and turning the air blue with her crude outbursts. She wasn't all that good in her part and I'd often wondered how Paul and Mervyn kept their tempers. *Now I knew.*

'We're all in the same boat here. Some of us have put more than others into Paul Fortune's scam but we've all been well and truly conned, and there's nothing we can do but face it. We're going to have to chalk it up to experience.' I grabbed Mark's arm and left the café with him in tow.

It was only a short walk to the hotel and neither of us spoke until we were in our room. I kicked off my shoes and sank onto the bed, bitter tears of disappointment, anger and frustration streaming down my cheeks. Mark came and sat beside me. He didn't speak, just slipped an arm around my shoulders and pulled my head onto his shoulder, pushing his handkerchief into my hand.

Once I could trust myself to speak again I looked up at him. 'I can't believe it, Mark,' I said. 'How could anyone be so evil?'

He gave me his lopsided, rueful grin. 'The thought of all that money, that's how,' he said. 'I reckon if all the cast members invested even half of what we did he must be the richer by at least half a million.'

I blew my nose. 'I still can't understand it. Why did he need to do it? He's a talented musician – and the play—'

'Face it, Lou. The music wasn't bad but the play was rubbish.' He kissed my forehead. 'He knew how to play a bunch of failed actors, darling, and he pulled us all in like a net full of little fishes. He played on our vanity – our dreams of success. In a way it serves us all right.'

'How can you say that?' I stared at hm. 'And what do you

mean – *failed actors*? It's OK for you. You probably won't even miss the money you invested, whilst I – I put most of what I had into this – this disaster.' The tears began again. 'I was going to be a star. It was my last chance and now – now ...'

'I know,' he pulled me close. 'Hey, for what it's worth, you'll always be a star to me.'

I sniffed and shook my head. 'All my money – I can't believe it's gone.'

'I know,' he said. 'It's awful – for you and for most of the others too. I happen to know that some of the poor devils put their life savings into the project, but we should all have known better. Didn't you ever wonder how he thought he was going to hit the headlines in the West End with no big names in the leading roles? We all had our heads in the sand – me as much as anyone. All we saw were our names up in lights. He held out that tempting bait and we fell for it.'

'And now it's all over,' I said bleakly. 'So I suppose there's nothing for it but to go home with our tails between our legs.'

He hugged me. 'Tell you what, why don't we stay on down here for another few days? Have a little holiday.'

I stared at him. 'A holiday! I've got to get back to London. I've got to go and see Harry and he's got to find me some work. He owes me that much after this fiasco. I'm broke, Mark. I'm homeless too. How could I laze about here and enjoy myself?'

He sighed. 'OK, I see your point. We'll go back this afternoon. And you must come and stay at the flat.'

Relief flooded through me. Thank goodness for Mark.

It was quite late in the evening when Mark drove into the underground car park and hauled our cases out of the boot. My two held all my worldly goods as I'd moved out of the bedsit. We travelled up in the lift and Mark took out his key and let us into the flat. Cathy, his sister, came out of the kitchen. She was small and dark with glasses, not at all as I'd imagined her. She stared at us both in surprise.

'Mark! I wasn't expecting you back again so soon.'

Mark sighed. 'It's a long story. Cathy, this is Louise, by the way. She'll be staying here for a while.'

'Oh?' she said shortly, looking me up and down. She looked at Mark enquiringly. 'So – why are you here? What happened?'

'It's all fallen through, Cath,' he said. 'There's no show after all. We've all been taken for a ride but I'll tell you more later.'

She gave a cynical little laugh. 'Well, I can't say I'm surprised.' She glanced at me. 'If you will play at being the thespian and mix with a bunch of dodgy theatricals, can you wonder? You should know by now that you can't trust them. I always said it would end in tears.' She looked at me again with undisguised distaste. 'I expect you'd like to freshen up,' she said, inferring that I was dishevelled and sweaty. 'And I'd be glad if you'd both keep your voices down. The children are asleep.' She sighed resignedly as she turned back towards the kitchen. 'I suppose I'd better put the kettle on.'

Mark looked at me. 'Put your things in my room. We can sort everything out later.'

In Mark's room I unpacked my overnight things. It was all too clear that Cathy didn't welcome the prospect of having me to stay. I 'freshened up', as she put it, in the en suite and then went back out into the hallway. Outside the half-open kitchen door I hesitated. Mark and his sister were clearly having a disagreement.

'Why are you with her?' she demanded. 'Don't you remember how she messed with your head all those years ago? You never stopped weeping on my shoulder about how badly she treated you.'

'We were just kids back then,' Mark said. 'We're both different people now.'

'Her kind of leopard doesn't change her spots, believe me, Mark. I know her sort. She's got her eye on your money now.'

'Of course she hasn't.'

'Anyway, I can't have her sharing your room.'

'Why not?'

'I'd have thought it was obvious. What kind of example would that be to set Kevin and Sharon?'

'Where else do you expect her to sleep, then?'

'Well, the sofa's available as long as it's just for one night, or do you expect me to wake the children and turn them out for her?'

'Don't be absurd, Cathy.'

'What do you see in her anyway?' Cathy went on. 'What happened to that nice girl you were engaged to – what was her name – Felicity?'

'Francesca.'

'That's right. Your letters were full of her for months. You even sent photos of the two of you together. You were completely besotted, then all at once she was out of the picture.'

I didn't wait to hear any more. Back in Mark's bedroom I waited for him. When he appeared a few minutes later he looked sheepish.

'Lou, darling, there's a slight problem. Cathy ...'

'I heard,' I said. 'I was coming to find you – and I heard.'

He winced. 'I'm so sorry. It's just that Cathy thinks ...'

'Cathy obviously thinks I'm a trollop,' I interrupted. 'I must say I'm surprised at you, Mark. It's *your* flat and your life so why does she think she can call all the shots? Anyway, I thought you said she was moving out.'

'She hasn't exchanged contracts on the house yet.'

'I see, so in the meantime she expects to have all the say in what you do?'

'She's just thinking of the children.'

'Well, *obviously* she and her kids have to come first! I'll stay for tonight,' I told him. 'Unfortunately I don't have a choice. I've nowhere else to go, but don't worry, I'll be out of your way first thing in the morning. I wouldn't want to be an evil influence on innocent children.'

'Don't take it like that.'

'How do you expect me to take it? Your sister obviously thinks I'm not good enough for you – not like the saintly Francesca.'

Mark turned pale. 'She didn't say that.'

'No? What do you call it, then? She even accused me of being after your money.'

He sighed. 'She gets a bit resentful sometimes about my inheritance.'

'Did your uncle leave her out of his will?'

He nodded. 'I told you; he didn't leave a will.' He reached for my hand. 'Let me take you out to dinner. We'll go somewhere nice to make up for everything.'

'No.' I snatched my hand away. I couldn't believe he'd stand there and let his upstart of a sister insult me without a word in my defence. 'I obviously count for very little as far as you're concerned. Anyway, I'm too tired and too disappointed. This has been the day from hell and I just want it to be over.'

I couldn't sleep. Not that the sofa was uncomfortable. Cathy had grudgingly supplied me with a duvet and pillow, but she hadn't offered either Mark or me anything to eat and my stomach wouldn't stop rumbling. I couldn't get Paul Fortune's massive con trick out of my head. My dreamed-of chance of success. All I had ever worked and prayed for, not to mention Dad's legacy – all gone in a puff of smoke. It hurt like a knife twisting in my heart. And as if that wasn't enough, there was Mark's betrayal. If he really loved me, why hadn't he put me first and told his bossy sister where to get off?

At last, in spite of the hunger pains and the disbelief of what was happening to me, I dozed off into a restless sleep in the early hours, only to be rudely awakened by two boisterous children jumping on me.

'*Who are you? What are you doing in our flat? What's your name*?' They shrieked questions at me, whilst bouncing all over the sofa. I sat up and gathered the duvet round me, fleeing in the direction of Mark's room. In the hallway I was stopped by Cathy.

'Where are you going?'

'I need to shower and get dressed,' I told her.

'Not in Mark's room,' she instructed. 'You can use the bathroom.'

'But my clothes – my luggage is in there.'

She sniffed. 'I'll get it for you.' She went into Mark's room and emerged a minute later, holding out my suitcases as though they were something she was putting out with the bins.

'Is Mark awake?' I asked.

She shook her head. 'No, and I'd be obliged if you'd allow him to sleep.' The two brats joined her and she slipped an arm round each of them. The girl put her tongue out at me. 'I'll be taking the children to school by the time you're finished.'

'Really?' I looked at them with distaste. 'Well, let's hope they learn some manners there. They certainly haven't been taught any by you!' I picked up my cases and walked towards the bathroom.

She aimed her final shot at my retreating back. 'I'd quite like you to have gone by the time I get back.'

'With pleasure. I can't *wait* to get away from here!' I slammed the bathroom door and sat on the edge of the bath, despondency sweeping over me afresh. Were things ever going to get better?

I took a cab to Charing Cross Road and climbed the stairs to Harry's office, leaving my suitcases in the lobby. Sally ushered me in and Harry began speaking the moment I walked in.

'Louise – I can't tell you how sorry I am about all this.'

'Not nearly as sorry as I am,' I interrupted. 'We've all been well and truly stuffed and there's nothing we can do about it now, but I'm not here for a post-mortem, Harry. I'm broke. I need a job and I need it *now*.'

He winced. 'I know, love, but look at it from my point of view. It's the end of May. All the summer shows are booked – about to open any day now.'

'Then find me something else,' I demanded. 'A touring play – anything. You know how versatile I am.' I leaned towards him. 'You owe me, Harry. You owe me big-time.'

'I know, love, and I feel for you, honestly.' He assumed a pleading expression and spread his hands. 'But I can't work miracles, can I?'

'Well you'd just better *try*,' I told him. 'You're supposed to be my agent and you've lost me a shed-load of cash and landed me in an unholy mess. If I don't hear from you in twenty-four hours, I'll make sure that everyone in the business knows about this scam you were involved in. And I still haven't made up my mind whether to sue you or not. *Right*?' I got up from my chair and stood glaring at him.

He shrugged resignedly. 'OK – OK.'

I was lugging my suitcases back into the street when my phone rang. I put the cases down and fished it out of my handbag. It was Mark. I switched the phone off in disgust.

Back in Stoke Newington, I tapped on my ex-landlord's door. No, I couldn't have my old room back. He'd already let it. But he did have one on the floor above (sloping ceilings and a dormer window). Only problem was, it wasn't available till the week after next! Did I want it? Its only advantage was that it was cheaper than the one I'd vacated. An offer I was in no position to refuse.

'By the way, you owe me a couple of weeks' rent,' he reminded me with a smile. 'Oh, and I nearly forgot. You had a visitor yesterday,' he added. 'She said she was your mum.'

Chapter Twenty

As Susan put her key in the lock, she heard the telephone ringing. Dumping her shopping on the floor, she rushed to answer it.

'Hello, Susan Davies speaking.'

'Mum, it's me, Karen. I wondered if you could have Peter for me this afternoon?'

'Of course, love. Are you all right?'

'Yes, fine. It's just that I've got an interview.'

'I see. Well, just bring him round when you're ready.'

'Thanks, Mum. You're a lifesaver.' She paused. 'Mum – I had a text from Louise this morning. She had the cheek to ask if she could come and stay for a few days. I told her absolutely no, not after last time. If she rings you—'

'She asked to come home?' Susan interrupted. 'But I thought she was down in Bournemouth, about to open this musical.'

'Well, she's obviously not,' Karen said. 'She didn't go into any details. Anyway it's irrelevant to me. All I care about is that she stays away from us. So, Mum, if she rings you, don't be an old softie again and say yes.'

'I wonder what's gone wrong,' Susan mused.

'Mum! Are you listening? You won't let her come and walk all over you again, will you?'

'What? No – oh, I don't know, Karen. I'm making no promises. We're all she's got and ...'

'Yes, and she plays on the fact that you're a pushover! Promise me, Mum.'

'What time did you say you'd bring Peter round?'

'About two. Did you hear what I said?'

'Two o'clock. Right, I'll be ready. We'll go to the park. He always loves that. See you at two, then, dear. Bye for now.' Susan replaced the receiver firmly. She wouldn't have Karen dictating to her. If and when Louise got in touch, she'd make up her own mind about what to do.

She didn't have long to wait. She was washing up after her lunch when the phone rang. She dried her hands and went to answer it. 'Hello, Susan Davies here.'

'Susan, it's me, Louise.'

'Louise. How are you?'

'Not good, I'm afraid. I'm throwing myself on your mercy, Susan,' Louise said. 'Things are not well with the show and I'm back in London. I gave up my flat to go on tour. I have got another in view but the trouble is, it won't be available till the week after next. I know there were a few hiccups last time I was there and I can't apologize enough for the – misunderstanding, but could you possibly bear to put up with me for a few days? I've nowhere else to go and I'm throwing myself on your mercy.'

'Well, I suppose it's all right,' Susan said. 'But what about that nice friend of yours; Dianne, isn't it?'

'She's moved her boyfriend in.'

'I see, but she's got two bedrooms, hasn't she? And as you said, it's only for a few days.'

There was a slight hesitation at the other end of the line then Louise said, 'There was a bit of a problem when I left last time. We parted on – well, not the best of terms.'

Susan sighed. Clearly Louise had upset her friend. When would she ever learn? 'Well, in that case you'd better come to me,' she said resignedly. 'When do you want to come?'

'Today, if that's all right.'

'Today! When? I'll be out for most of the afternoon but I'll be back about four.'

'OK. Susan – thanks. I do appreciate this. I know last

time things got a bit fraught and I'm sorry, but it was all a misunderstanding.'

'Never mind that now,' Susan said. 'I'll see you later.' As she replaced the receiver she sighed. God only knew what Karen would say. Maybe she needn't find out, she told herself; after all, it was only for a few days. Oh well, she'd just have to deal with that when and if it happened.

Karen arrived dead on the stroke of two and refused Susan's offer of a cup of tea.

'I've left Peter's buggy in the hallway downstairs,' she said, handing over the bulging bag containing all his toys and accessories.

'What is this interview you mentioned?' Susan asked.

Karen hesitated. 'Mum, I'm going mad, staying at home with Peter all day. I love him to bits of course and I wouldn't be without him, but I never visualized being a full-time mum. I saw this advert for teachers willing to do some tutoring. It's an agency.'

'And that's where you're going this afternoon?'

'Yes. I telephoned them and I've got an interview this afternoon.'

'Is Simon all right with this?'

'He doesn't know – yet.' When she saw her mother shaking her head Karen went on defensively, 'Well, it might all come to nothing so what's the point of telling him?'

'And if it does – come to something, I mean?'

'I'll cross that bridge when I come to it.' Karen looked at her watch. 'Look, I'll have to go or I'll miss the interview. Sure you're all right with Peter?'

'Yes, of course. You get off. What time will you be back?'

'I don't really know. By four, I hope.' She bent to kiss Peter. 'See you soon, darling. Be a good boy for Granny. Bye, Mum.' And she whisked out of the door.

Peter loved his outing to the park. They fed the ducks with corn that Susan bought at the park kiosk along with ice cream, and finished up in the children's play area. Peter played happily in the sandpit and then begged to go on the swings. Susan hesitated. He was still very little for the swings. Peter sensed the reason for her anxiety.

'Mummy lets me go on the fwings,' he assured her. 'She pushes me up to the sky.' He raised one chubby arm as high as it would go, treating his grandmother to his most beguiling smile.

'All right,' Susan said. She lifted him and began to ease his fat little legs into the baby swing but he screwed up his face and struggled.

'Not baby fwing,' he protested. 'Big-boy fwing.'

'No, you might fall and hurt yourself.'

'No! Won't – *won't!*' He began to cry and Susan could see that he was going to have a tantrum. Since Adrey had left, she had noticed that he was more prone to them than before.

'All right, then, don't cry,' she said. 'But you must hold on very tightly and not let go.'

Having got his own way, Peter smiled again and allowed himself to be extricated from the barred swing. The next minute he was seated on the swing next door.

'Push, Granny. *Push,*' he shouted as Susan gently moved the swing.

Very tentatively, she pushed him a little higher but it wasn't enough for Peter. He twisted round to shout at her again and the next moment he had let go and slipped from the seat. He let out a loud wail as he hit the ground.

Heart in mouth, Susan picked him up and surveyed the damage. His knees were grazed and bleeding. The moment Peter saw this he let out a scream of fear.

'It hurts! Want Mummy!'

Susan lifted him into her arms and put him into the buggy. There was a little first aid hut not far from the play area with a St John's nurse on duty. Taking him there would be quicker than going home. She glanced at her watch as she hurried along, and was concerned to see that it was a quarter past four. She'd better let Karen know she'd be a bit late getting back. Pausing to take out her mobile phone, she pressed in Karen's number but her phone was switched off. With a sigh of frustration she hurried into the first aid hut, hoping it wouldn't take long.

The kindly middle-aged nurse bathed the grazed knees with antiseptic and applied a couple of plasters. 'There,' she said soothingly. 'You're a very brave boy, aren't you? And now you've got two lovely plasters to show off.' She slipped a chocolate button into Peter's mouth. 'There, no harm done,' she said to Susan. 'Might be as well to put him in the baby swing next time,' she said. 'He's very little to go on the big ones yet.'

'He said his mummy lets him go on the big ones,' Susan said, feeling feeble. The nurse smiled.

'They can be very manipulative, even at this age,' she said. 'Still, no real damage this time.'

When she got back to the flat, Susan found an irate Karen waiting outside the door. 'Oh, Mum, there you are. Where have you been and …' She spotted the plasters on Peter's knees. 'What's happened?'

Susan was puffing a little as she put a struggling Peter down. 'It's nothing, he fell and grazed his knees,' she said. 'I took him to the first aid hut in the park. That's why we're late.'

'I got p'asters,' Peter announced proudly. 'I went on the big-boy fwing and I falled off.'

Karen stared at her mother in horror. 'Mum! Why didn't you put him in the baby swing? Haven't you got any sense at all?'

'He wouldn't go,' Susan said wearily. 'He said you let him go on the big ones.'

'Of course I don't! He's only three, Mum. What were you thinking?'

'I tried to ring you,' Susan said as she put her key in the door, 'but your phone was switched off.'

'Well, of course it was,' Karen snapped. 'I wouldn't leave it on when I was in an interview, would I?'

'Well, Peter's all right, Karen, and please stop shouting at me for all the neighbours to hear.' Inside the door she turned to her daughter. 'After all that, did you get the job?'

Karen's expression brightened a little. 'As a matter of fact, it does look hopeful but they said they'd let me know.' As Susan went into the kitchen to fill the kettle she called out, 'No tea for me, Mum. I'd better get going. Simon will be home soon.'

She was gathering up Peter's things ready to leave when there was a ring at the bell. She called out, 'Are you expecting anyone?'

Suddenly Susan remembered Louise and her heart sank. Before she could say anything Karen was calling, 'It's OK, Mum. I'll get it.'

In the kitchen Susan held her breath – then Karen's outraged voice confirmed her worst fear – Louise had arrived.

'What the *hell* are you doing here? Your bloody nerve after last time! You're not welcome here, Louise, so you can turn right round and get back to where ...'

'It's all right. I said she could stay,' Susan put in calmly, standing behind Karen at the door. 'Let her in, Karen. It's only for a few days.'

Karen stared at her mother. 'You *have* to be joking! Well, on your own head be it!' She stood aside and allowed Louise to walk into the flat. 'What new drama has she cooked up this time?' she asked. 'What tale of woe has she conned you with now, Mum? Well, don't expect any further help from me.'

Susan bridled. 'If you get that job, Karen, I daresay it'll be you, wanting help from *me*.'

Karen opened her mouth to say something then closed it again. Bending down, she picked up Peter and his bag and headed for the door, but as she opened it an irate Simon stood outside, about to ring the bell.

'Oh, so *there* you are!' he said, his voice deep with anger.

'Simon!' Karen looked deflated. 'You'd better come in.'

Simon stepped inside and closed the door behind him. 'You might be interested to hear that you've *got the job*!' he growled. 'The job you applied for behind my back. I'd just got in when they rang on the landline – said they'd been trying your mobile but it was switched off.'

Karen put Peter down. 'I didn't want to tell you until I was sure I'd got it,' she said.

'Can you imagine what a fool I felt, not knowing what they were talking about? And just what do you propose to do with our son while you're doing this job?' he demanded. 'I understand from the agency that you've announced that you're free to work daytimes, as well as out of hours. Did you even *tell* them that you have a child?'

Karen threw a look of appeal towards her mother. 'Mum said she'd babysit.'

'As she's done today, I suppose. So you're in on this too, Susan?'

Susan shook her head. 'Like you, I knew nothing about it until today.'

For the first time Simon looked down at his son. 'Look at him. He's filthy – and what's wrong with his legs?' he demanded.

'He tumbled and grazed his knees,' Karen said defensively. 'You know he's always doing it.'

'Granny let me go on the big-boy fwing.' Peter said proudly. 'I fell off an' I got p'asters!'

Glaring at his wife Simon said, 'What are you thinking about, Karen? You know she's past it. Something always happens when she has him. And here you are, willing to risk our son's safety for some two-bit cramming job. My God! What kind of mother are you?' Suddenly he noticed Louise, who had made herself at home on the sofa. He pointed at her. 'What is *she* doing here?' he demanded. He looked at Karen, who was close to tears. 'Can't you see now what a fool your mother is, letting that devious cow back again after the trouble she always causes?' He picked Peter up. 'I'm out of here!' he shouted. 'You'd better come too, Karen – if you put any value at all on your son and our marriage, which I doubt!'

When they'd gone, the flat suddenly felt to Susan as though all the air had been sucked out of it. She sighed and sank down onto the sofa next to Louise.

'Oh, dear.'

Louise, who had enjoyed every minute of the little drama played out before her, smiled sympathetically at her stepmother. 'Wow! He was in a right strop, wasn't he?'

'He wants Karen to be a full-time mother until Peter goes to school. It's not much to ask really, is it? Although she shouldn't have gone behind his back like that.'

'Personally, I'm on Karrie's side.'

Susan glanced at her. 'Well, you would be, wouldn't you?'

'I can't believe you can be so charitable about Simon after what he said about you!'

'He was angry,' Susan said. 'We all say things we don't mean when we lose our tempers. He's probably regretting it already.'

Louise shrugged. 'Generous of you to be so understanding. He needs dragging into the twenty-first century if you ask me. Other women go back to work after having a child so why should he think his wife is above all that? You wouldn't catch me putting up with that kind of bullying.'

'No, I daresay.' Susan looked at her stepdaughter for the first time since her arrival. 'What brings you here, Louise? You didn't go into details on the phone, except to say that it was something to do with the show.'

Louise had done a lot of thinking on the train coming up. There was no way she could admit to Susan that she'd been conned out of her father's legacy and she had no intention of losing face about her career rise either. 'Just a hitch with the bookings,' she lied. 'There would have been a long gap between the tour and the West End opening so we're taking a bit of time out now instead.'

'Oh, well, that's nice,' Susan said. 'I'm afraid it will have to be the sofa again.'

'That's OK.' *I'm getting used to sofas,* she thought. She looked at Susan. 'I'm sorry I put my foot in it with your – chap,' she said. 'But I expect you've kissed and made up by now, eh?'

Susan shook her head. 'Ted? No. I couldn't keep seeing a man who could be so deceitful.' She glanced at Louise. 'What about you? Are you seeing anyone special at the moment?'

Louise was reminded sharply of Mark and his betrayal. Had he been seeing this other woman all the time and stringing her along? She winced, remembering the previous night and Cathy's obvious loathing of her. She couldn't help thinking that

Mark must have been talking to his sister about her – complaining about the way she'd treated him all those years ago when they were at drama school. Surely she must realize that they'd all grown up since then?

'Louise!' Susan prompted. 'Are you all right?'

'Yes, of course I am,' Louise quickly pushed her negative thoughts away. 'You asked if I was seeing anyone. I was, but like your Ted, he turned out to be devious and two-timing.'

'I'm sorry to hear that. Was it the young chap you were at drama school with?'

Louise remembered now that she'd told Susan about Mark the last time she was here.

'No,' she lied. 'Someone new.'

'Oh dear, you don't have much luck with men, do you? Were you in love with him?'

The question threw Louise slightly. 'No, not really,' she said. 'Whatever love is.'

Susan patted her hand. 'Well, you'll soon be busy with your exciting new play. Once you're away on tour, you'll soon forget him.'

Louise cringed inwardly. If only that were true. She turned to look at her stepmother. 'All that business with Karen and Simon last time I was here. It really was a misunderstanding, you know. I didn't intend to make trouble between them.'

'Well, it's all water under the bridge now,' Susan said. She got up from the sofa. 'I daresay you're hungry and I haven't even offered you a cup of tea. I'll make one now. The kettle has boiled.'

'So – am I forgiven?' Louise asked. 'For Ted, I mean.'

Susan sighed. 'As I said – it's over and done with but I can't say I don't miss him, as I expect you miss your boyfriend too. Come and help me start supper and you can tell me all about this wonderful musical you're going to star in.'

Louise got up and followed her into the kitchen. For the first time she realized that she was going to miss Mark. Underneath the anger she already did – a lot. Did that mean that her feelings for him went deeper than she'd meant them to? *Were you in love*? She'd always considered love to be a myth – something for films and books. Certainly an emotion she'd never experienced. Lust, maybe – infatuation, but love? No, surely not. She watched Susan bustling around, making tea and getting out the biscuit tin. If only her life could be as uncomplicated as hers.

The rest of the week was difficult for Susan. She'd telephoned Karen early the morning after Louise's arrival to ask if she and Simon had made up their quarrel and come to any kind of compromise.

'No, we haven't,' Karen snapped. 'Simon is adamant that I turn down the job and stay at home with Peter until he's five. That's another *two whole years* of sheer drudgery and boredom. If it wasn't for Louise, we'd still have Adrey and I'd still be teaching. And if you hadn't been so careless and stupid yesterday, I'd have been home long before Simon and I'd have dealt with things in the way I'd planned.'

'Please don't speak to me like that, Karen. I am neither stupid nor careless and if you want to know, I think it's a mistake, going behind Simon's back like that.'

'Well, when I want your opinion, Mum, I'll ask for it and I'm sorry but I'm afraid I do think you're careless, putting Peter in danger like that. And as for letting Louise come and stay again after all she's done – well, words fail me!'

'I think there have been times when you've been quite happy to forget my so-called careless stupidity,' Susan said.

'Yes? Well, thanks for rubbing it in! You can safely believe that those days are over,' Karen snapped. 'I won't be asking for your help any more!' And there was an ear-splitting crash as she slammed the receiver down.

She'd only just replaced the receiver when the phone rang. It was Simon.

'Susan – I'm ringing to apologize for the scene I made at your flat yesterday.'

'It's all right, Simon. I understand. You were upset and I do sympathize – with both of you.'

'That's your trouble, Susan,' he replied. 'You're far too understanding. But as I said, I'm sorry. I just lost it and I shouldn't have said what I did.'

'Well, as you've brought the subject up, Simon, you didn't mind leaving Peter with me while you and Karen went to Paris.'

'No, I know and I'm deeply ashamed of letting my temper get the better of me.' There was a pause then he said, 'Look, Susan, don't think I'm interfering but you really shouldn't have allowed Louise back after what happened last time. She is at the root of all this mess we're in now.'

'I know but I couldn't see her without a roof to her head, could I?'

'Well, it's your decision, of course. Anyway, I'm sure you know that I didn't mean any of those things I said and I hope you'll accept my apology.'

'Of course, it's already forgotten. Thank you for ringing, Simon.' She put down the phone with a heavy heart. She'd tried to please everybody and finished up pleasing no one and landing herself in the middle of a row.

In the days that followed, Karen would have nothing to do with her. Every time Susan tried to ring her she'd refused to pick up. As for Louise, she was as untidy and disorganized as ever. Susan gave up trying to tidy up or do any housework. She was a little disturbed by the nightmares that Louise obviously had. On several occasions, she had been awakened in the early hours by her stepdaughter's loud sleep-talking. She wondered if it had anything to do with her broken romance, which she'd already guessed had upset Louise more than she would admit. She'd had a lot of calls on her mobile during the week, but she hadn't taken any of them. Had they been from the disloyal boyfriend? she wondered. Clearly she was upset about something.

It was the night before Louise's visit was over and she and Susan were sitting together in the living room, when Louise suddenly turned to her stepmother. 'Susan – there's something I need to tell you.'

'Yes, dear?' Susan looked up from the little sweater she was knitting for Peter. 'What's that?'

'I found my mother.'

Susan's glasses slipped to the end of her nose as she looked up in surprise. 'You found her – how?'

'The Salvation Army found her for me actually. They arranged a meeting and we met for tea at one of the big stores in London. She told me that she'd married again after she and Dad split up and they had another child; a boy – well, man now. He's called Steven.'

'So you have a half-brother. That's interesting. Is she happy?'

'No. Her husband was – *is* a criminal. He's in prison and she's divorced him. I was disappointed, Susan. Frankly, I didn't like her. I didn't like her son either. He telephoned me and I found his manner quite nasty – almost menacing.'

Susan dropped the knitting into her lap and took off her glasses. 'Oh, Louise!'

'The trouble is that I told her too much. She asked about Dad and when she knew he'd died, she wanted to know what was in his will.'

'And you told her?'

'I got a bit carried away at seeing her again after all the years.' She hesitated. 'I told her about the play too – and my big breakthrough. She obviously got the impression that I'm rolling in cash because before we parted, she asked me for money.'

'Oh dear. I'm afraid you've rushed things a bit too much.'

'You can say that again! The thing is, will she let it go at that or is she going to be constantly on my case? What can I do?'

Susan considered for a moment. Privately, she wondered if all that had been the cause of Louise's troubled dreams. 'Well, you'll be away on tour for some time, won't you? After that, maybe she'll have forgotten.'

Louise sighed. If only it were as simple as that. Susan didn't know the half of it. And because she'd been economical with the truth, she couldn't press it any further.

She left the following morning. Susan saw her off gratefully, longing to get her hands on the hoover and dusters. How could one person make such a shambles of the place?

'Don't forget that I intend to be there on your opening night,' she called as Louise ran down to her waiting taxi. 'I'm going to try to persuade Karen to come too, but I'll be there even if she isn't. Good luck, Louise!'

'Thanks, Susan. Goodbye – and thanks for having me.' Louise forced a smile that vanished as soon as she climbed into the taxi. They'd all have to know about her catastrophe before too long and how would she face them all then?

'Where to, missus?' the taxi driver enquired.

She sighed, recognizing that she didn't even qualify for 'miss' any more.

'To the train station, please,' she told him.

She settled back in her seat. Back to a grotty bedsit in the East End, she reminded herself with a sigh. Back to God only knew what!

Chapter Twenty-One

THE BEDSIT UNDER THE eaves looked more depressing than ever after the cosiness of Susan's flat. The weather had suddenly warmed up and when I unlocked the door, the stuffiness and odours left by the previous tenant hit me like a wave; a mixture of stale sweat, unwashed clothes and several dozen takeaway meals. By the stench that still hung about the room like a fog, most of them had been curries. I hurriedly crossed the room and threw the window open wide, making a mental note to buy a can of air freshener next time I went to the shops.

It was only when I turned back to close the door that I saw the note that had obviously been pushed under the door. I slipped a finger under the envelope flap and pulled out a single sheet of lined paper, obviously torn from a cheap notebook. My heart plummeted. It clearly wasn't from Mark, not that I expected it, after refusing all his calls. He'd surely have to give up soon.

There was no address on the note and the handwriting was barely readable. The spelling was atrocious too but the message was only too plain.

Deer Louise. Sorry you an me coodnt meet up. Yor landlord sais you are cumin bak so Ill giv you a bell soon. Yor luvvin bruther, Steve.

I ran downstairs and tapped on the landlord's door. He answered it looking as though he'd only just got up. He wore jeans and a grubby singlet and his greasy hair was tied back in a ponytail. He blinked at me blearily. 'Yeah?'

I held out the envelope. 'How did this come to be pushed under my door?' I asked him. 'I've been away so how did he manage to get in?'

He peered at the envelope. 'Oh, yeah,' he said. 'That was a couple of days ago. When he couldn't get you, he rang my bell and asked me to give you that.'

'Right. Well, if he turns up here again don't let him in, OK?' How on earth had he found out where I lived? My stomach lurched with apprehension as I turned back towards the staircase.

'That's all right,' he called after me. 'Don't *thank* me, will you. I'm not your bleedin' butler – snotty cow!'

I carried on up the stairs without a backward glance. What did he do for the rent he charged anyway?

I hardly slept at all that night. The bed was lumpy and lop-sided and the sun burning down on the roof all day had made the heat and the overpowering smells unbearable. The thought that Steve Harris had tracked me down kept me awake too. If only I hadn't chosen to stay on at the same address!

I was up early, down the street to the supermarket to stock up with food and air freshener. By the time I got back I reckoned that Harry would be in his office so I took out my phone and clicked on his number. To my surprise he answered himself.

'Harry Clay Theatrical Agency.'

'Harry, it's Louise. I've just got back from a week away. I've been expecting a call from you. What have you got for me?'

At the other end of the line I heard him sigh. 'I've tried, Lou, but I told you, there's nothing going in your line at the moment. The summer shows are all booked and there's nothing else suitable.'

'So what do you suggest I do? I'm sure I don't have to remind you that I've been conned out of a job and most of my money. You owe it to me to find me something, Harry.'

'Look, Lou, I'll be frank with you. I'm winding the agency

up. I'm past retiring age and I've had enough. I've already let Sally go.'

'I see, so the rat is leaving the sinking ship, is he?' I snapped. 'I've a bloody good mind to sue you.'

'Go ahead,' he said wearily. 'You can't get blood out of a stone. I'm sure I don't have to remind you that I'm in the same position as you. Look, I know it's tough but my blood pressure has gone sky high and my doctor says I'm heading for a heart attack. My wife has put her foot down. This business with Fortune has just about finished me. It's time to pack it in and that's what I'm doing.'

'It's all right for you, Harry, but—' Before I could complete the sentence he'd hung up, leaving me listening to the dialling tone. I hit the red button in disgust. What the hell was I going to do now? I'd soon get through the little money I had left. I had to earn some cash somehow. I clicked on my list of contacts. There was only one name on the list that would sympathize. I highlighted Mark's number but my pride refused to let me press the call button.

For a long time I sat on the bed, despair washing over me like an all-engulfing tide. Why couldn't something go right for me? What had I done to deserve all this bad luck? I got up eventually and opened the suitcase containing most of my clothes, as yet unpacked. It was hardly an haute couture collection. It made me angry to think of the lovely things I could have bought, if only I hadn't blown all my inheritance on Paul bloody Fortune and his godforsaken play. I hung everything up and examined each garment critically. Nothing looked fashionable or smart. I had to admit that even the Chanel suit was beginning to look a bit shabby as I pulled at a loose button. Suddenly I made up my mind. If I was going to get myself a new agent, I was going to have to look a bit less down at heel. I'd go up west today and buy myself a few nice things to wear with some of the cash I'd got left. They wouldn't be designer but I'd always had a good eye and the summer sales were just beginning. I might find some bargains. I'd get my hair done too. It would be a good investment.

First, I trawled my list of agents. As I'd expected, none of them would see me there and then, so I left a copy of my photos and profile along with my mobile number, then I took the Tube to Oxford Street. I bought a classic suit, a casual skirt, a jacket and three tops. I also found some really elegant shoes and a handbag, all at reduced prices, then I sat in the best hairdresser's I could afford for most of the afternoon, having my roots done and my hair cut and blow-dried.

It was while I was walking to the Tube station that I passed a bridal shop and a tiny card in the corner of the window caught my eye.

Assistant wanted.

I stood looking into the window at the beautiful designer bridal gown displayed for some time. It was unusual for a shop of this calibre to advertise in this way. I was intrigued. Eventually, I pushed the door and was immediately surprised by the tinny notes of Mendelssohn's Wedding March. As I walked into the shop I was trying not to laugh. A stylishly dressed woman of about fifty approached me.

'Can I help you?'

I took a deep breath. 'I've just noticed your advertisement in the window,' I said. 'It so happens that I'm looking for work.'

She looked me up and down critically. 'Do you have retail experience?'

I treated her to my best smile. 'I'm actually an actress,' I told her. 'But I have often taken retail work when I've been between engagements as I am now.' It wasn't strictly true of course but I hoped she'd swallow it.

'In that case you'll have references.'

Trying not to look taken aback, I shook my head. 'All the jobs were temporary, of course. But they were all in the fashion trade,' I added quickly. 'I do have quite a flair for fashion; being an actress it's all part of the training.'

'This is a designer boutique,' she said. 'I design most of the gowns, although I do stock a few low-budget dresses.' She appraised me again. 'As it happens, I only need temporary help at the moment, so this might very well suit us both.' She eyed my outfit doubtfully. 'I take it you own a smart black dress or suit?'

'Naturally.' I made a mental note to give the Chanel suit a good sponge and press and get the loose button firmly sewn on. I looked at her. 'What salary are you offering?'

She named a figure that was ludicrous. I shook my head. 'I'm afraid I'd need twice that,' I told her. 'London rents don't come cheap and if you want me to look smart ...'

'All right, I agree,' she said, throwing me completely. 'When can you start?'

'Well – tomorrow if you like.'

She nodded. 'As you can see, I'm on my own here at the moment so I really do need help – and quickly,' she said. 'Which is why I put the notice in the window. Normally I'd advertise in the usual way.'

'That would suit me. What time would you like me to be here?' I asked.

'Be here at eight, then I'll be able to show you around and explain our routine.'

I walked out of the shop feeling really cheered up. I'd get back onto my feet in no time, I told myself.

But on the Tube on the way back to the horrible bedsit, the spectre of Steve Harris and his threat to get back in touch reared its ugly head. I couldn't stay on in that room like a sitting duck, just waiting for him to come and find me again. Besides, Stoke Newington was an awfully long way from the new job. A sudden thought hit me and I took out my phone and called Dianne. She'd be home from work by now. She could always say no.

She answered the call almost immediately. Clearly she'd deleted my number from her phone because she didn't know who was calling.

'Hello?'

'Di, it's me, Lou,' I said. 'Long time no see. How are you?'

'I'm fine – and you?'

She sounded a tiny bit frosty so I turned on the charm. 'I've really missed you, Di,' I said. 'I hated the way we parted company last time. Any chance I could pop in and see you some time soon?'

She hesitated. 'Aren't you busy with the new play? I'd have thought you'd be on tour by now.'

'It's a long story,' I told her. 'Actually it all fell through in quite a spectacular way. I'd love to tell you all about it. Look, I'm working in the West End at the moment. Any chance I could come and see you after work tomorrow?'

'Yes, OK then,' she said. I knew I'd have aroused her curiosity. 'Though I won't be able to make an evening of it. I have to go out later.'

'That's all right.' I paused. 'How is Mike?'

'Fine – I suppose.' There was a long pause at the other end then she said. 'I'll tell you all my news when we meet. Around six, OK?'

'Fine, look forward to it.'

I returned to the dreadful bedsit that evening, feeling much better. I was just tucking into my microwave meal for one when my phone rang. It was an unfamiliar number. Could it be a call from one of the agents already? Full of optimism, I clicked the green button.

'Hello. Louise Delmar speaking.'

'Hello, darlin'. Little brother Steve here. How are you – all right?'

'Leave me alone,' I snapped, my good mood evaporating instantly. 'Look, there's no point in you ringing me. I don't want to meet you, so please don't ring me or write any more notes.'

'Oh, now is that nice?'

'Frankly, I don't care whether you think it's nice or not,' I said. 'Just go away!'

'I only want to meet you, sis,' he said smoothly. 'I only want to get to know you. After all, blood's thicker than water.'

'I told you. If you don't piss off, I'm calling the police,' I told him. 'This is harassment.'

There was a short pause and then he asked, 'Louise – have you seen last week's edition of *The Stage*?'

'No, why?' My stomach lurched. What did he know that I didn't?

'It was Mum who spotted it,' he said. 'She's so proud of you, our mum. She bought the magazine to see if there was anything in it about you. Imagine how upset she was when she saw an article with the headline: The Show That Never Was. She's really upset for you, Louise.'

My blood ran cold. I'd no idea that the news was buzzing around the business already, but at least it was only the trade paper. 'I don't know what you're talking about,' I bluffed.

'Oh, I think you do,' he said softly. 'I bet your family in Bridgehampton are pissed off for you. Or maybe you ain't broke the news to them yet, eh?'

I caught my breath. 'Look, just leave my family out of it,' I said. 'If you were thinking of blackmailing me, forget it. I've got no money left, and after tomorrow I won't be here so don't bother trying to ring again.'

I cut him off before he could reply and switched off my phone. Something about that voice of his chilled me to the marrow. Maybe I should buy another mobile and throw this one away. I only hoped I'd be able to persuade Di to take pity on me tomorrow. I looked at the remains of my meal, congealing in its plastic tray, and tossed it into the bin, my stomach churning. For the second night I hardly slept, tossing and turning as I tried to think of ways to stop the man who called himself my 'brother' from persecuting me.

The bridal boutique was called Camilla and I soon discovered that it was the name of the boutique owner and designer. I arrived bright and early and she looked me up and down.

'It's a nice suit,' she said.

I nodded. 'It's Chanel.'

'Mmm.' She pursed her lips. 'The best thing about Chanel is that their suits keep their shape – however old they are.'

That was me told! She showed me around, drawing back the velvet curtains on the rails. The gowns were beautiful but some of the price tags took my breath away – anything from a 'humble' two grand as she put it, to over £5,000. Camilla, as she asked me to call her, caught my expression and smiled.

'You get what you pay for, I always say,' she said smoothly. 'All the highest-priced dresses are unique – one-off, so the bride who wears them can be sure that no one else can upstage her.' She fingered the material. 'Nothing but the best fabrics, French lace and silk from manufacturers in Belgium where I have a standing order. Here, feel for yourself.'

I touched the material reverently, feeling that it might mark if I as much as looked at it. 'It's lovely,' I said. 'And you design them all yourself?'

'All the best ones, yes. In my studio upstairs.'

'And do you do the actual sewing?'

'No. I have four expert out-workers,' she told me. 'And two embroiderers, one of them a young man.' She turned away. 'Follow me and I'll show you the staffroom where you can take your breaks.'

After the showroom and the luxurious fitting rooms with their mirror-lined walls and little gilt chairs, I was surprised at the so-called staffroom. It consisted of a sink unit with a kettle and toaster, a Formica-topped table and two wooden chairs. The floor was covered in cracked vinyl. I looked at her. 'Isn't there a microwave?'

She looked down her nose. 'No. If you want a hot lunch, you could always go out. I'm told there's a McDonald's in one of the side streets. Personally, I like to watch my figure.'

My first customer arrived when Camilla had popped out to the bank. A very pretty young girl and her mother stood looking around them self-consciously as I approached.

'Good morning. Can I help you?'

The mother spoke first. 'We're looking for a wedding dress,' she said.

'I love the one in the window,' the girl said. 'But there isn't a price tag on it and we ...' She looked at her mother uncertainly and I guessed that Camilla's prices were going to frighten the pants off them.

'That one is very expensive,' I told them. 'Camilla herself

designs all the expensive dresses and they cost a lot because they are unique. Would you like to try something on?'

The young girl's cheeks flushed. 'Oh, could I?'

'How much *is* the one in the window?' the mother insisted. She cast a warning glance in her daughter's direction. 'Before we get too carried away.'

'I'll just check.' I drew aside the curtain at the back of the window and peered at the price label, concealed inside the back of the neck. Holding my breath, I stepped out again and turned to look at them. 'It's £3,500,' I told them. 'But of course, as I said, it is ...'

'I think we'll leave it, thank you.' The mother grasped her daughter's arm and began to hustle her towards the door.

'That is one of our most expensive dresses,' I said quickly. 'We do have some lower-priced gowns, if you'd like to come this way.'

At the back of the showroom were two rails of what Camilla called 'budget dresses'. I drew one or two out and the faces of mother and daughter relaxed a little. The girl picked out a couple to try on and eventually chose one. As I packed it carefully in tissue paper and one of the distinctive black and pink *Camilla* boxes, I was thrilled to think I'd made a sale and when Camilla returned, I couldn't wait to tell her. She looked pleased until I told her which dress it was.

'I told you to push the *designer* gowns,' she said, looking cross.

'I did but they were obviously out of their price range.'

'What does that matter? You'll find that if you push in the right way they usually give in. After all, they can always economize on something else.'

'How do you mean, push in the right way?'

Camilla sighed. 'Flatter them, of course. Tell them the dress was made for them – that they have the perfect figure for it. Point out that they'll regret it for the rest of their lives if they don't get the very best – that sort of thing.' She glared at me. 'Use your two and a half brain cells for once, *dear*!'

I opened my mouth to give her both barrels but I bit my

tongue just in time. For two pins, I could have walked out there and then but like it or not I needed this job and after all, she was paying me what I asked. But as she turned away, I promised myself that the minute one of the agents I'd contacted came up with a half-decent job, I'd be out of here in a flash.

As the day went by and more customers came in, I observed Camilla's sales technique. She really did go over the top with her flattery and oiliness. The amazing thing was that it seemed to work. I wondered how much mark-up there was on a dress designed by her and reckoned it couldn't be far off eighty per cent. No wonder she pushed so hard. After each sale she was impossible; so conceited and overconfident that I longed to bring her down a peg.

Di had only just got in from work when I arrived at the flat. She looked tired and took a bottle of wine out of the fridge and poured us both a glass. I'd had nothing to eat since breakfast and I secretly hoped she'd ask me to stay for supper.

'So …' she said as she handed me my glass. 'Tell me your news. What's this West End job you've landed?'

'Not what you think.' Perched on one of her kitchen stools at the breakfast bar, I told her about Paul Fortune's treachery. I went on to describe my first day as sales assistant at Camilla's. As the wine relaxed me I found myself camping it up a bit – imitating Camilla's voice and mannerisms – and soon Di was in fits of laughter.

'Oh, Lou, you are priceless,' she said. 'I'd love to have been a fly on the wall. Looks as if you've met your match in the formidable Camilla.'

I wasn't sure what she meant by that. Somehow it didn't feel like a compliment, but I decided to let it go. 'Look, Di, I might as well cut to the chase and tell you why I'm here,' I said. 'I've got this horrific bedsit in Stoke Newington. It takes ages to get up to the West End in the rush hour and I wondered …'

'If I could put you up,' she finished for me.

'It would only be temporary,' I assured her. 'I'm expecting an offer from my new agent any day now.'

'Your new agent? What about Harry Clay?'

'He's decided to retire,' I told her. 'He put money into the show too and it's just about finished him. Of course, if Mike isn't happy with the situation, I'd look elsewhere. I don't want to ruin your relationship.'

'Mike and I aren't together any more,' Di said.

I'd suspected as much but I feigned surprise. 'Oh, Di – I'm sorry to hear that.'

'I found out that he was only using me and my flat as a stop-gap until he found somewhere else to live,' she said with more than a hint of bitterness. 'Plus the fact that he met someone else.' She shook her head. 'I won't go into details but it wasn't the happiest of partings. It'll be a long time before I trust another man!'

'Well – I know the feeling.' I left a respectful gap in the conversation then I said tentatively, 'Does that mean you'd be willing to put up with me for a week or two?'

She sighed. 'OK, I suppose so, as long as you pay your way, Lou. I can't afford to let you stay rent free.'

'Of course, I wouldn't dream of putting on you.'

'And as long as you try to keep the place tidy,' she went on. 'No dirty laundry hanging around the place. And you take your turns with the shopping and the chores.'

'But of course.' I frowned. 'Didn't I always?'

'Not always, no.'

'You make me sound like a slut.'

'Precisely,' she said, looking me straight in the eye.

For a moment we stared at each other, then we both burst out laughing.

'Oh, Lou,' Di said at last. 'Slut or no slut, it's good to have you back again.' She looked thoughtful. 'You know, from what you've told me about this Paul Fortune crook, I'm surprised you haven't thought of selling your story to one of the tabloids.'

Her words hit me like a ton of bricks. *What a brilliant idea*! Why hadn't I thought of that? But had the thought crossed the minds of any of the other members of the cast? Had I missed the boat? It was certainly worth finding out.

'Dianne,' I said. 'You are a genius!'

Using Di's computer and at her suggestion, I emailed three of the most popular tabloids there and then while she rustled up a quick supper.

'I'll text you if there are any replies,' she promised. We arranged that I'd move in with all my worldly goods the following week and by the time I got back to the bedsit that night I was feeling a lot more optimistic.

It was a couple of days later at Camilla's that it happened. It was halfway through the morning and business had been slow. I was in the grotty little staffroom making coffee when I heard the shop doorbell chime out its naff ringtone. Camilla rushed into the showroom like she had a wasp in her knickers.

'Good morning. Welcome to Camilla's. How can I help you?'

Hearing her dulcet Estuary English tones, I peeked through the curtain and got the shock of my life. There in the centre of the shop was Cathy, Mark's stroppy sister. She wasn't alone though, the woman she had with her was about thirty, very pretty and quite well dressed. I eavesdropped shamelessly. Cathy's friend was getting married and wanted a wedding dress immediately.

'My fiancé wants us to be married as quickly as possible,' she explained. 'He's given me *carte blanche* on the dress and I want to look spectacular.'

'I'm sure you'd look that, dear, even without the help of one of my creations,' Camilla oiled. She must be feeling as though all her birthdays had come at once as she reached for her most expensive creations.

'This would be perfect for you,' she simpered. 'This style is just right for your lovely figure.' I turned back to my coffee, a wave of nausea washing over me, then I heard something that stopped the breath in my throat and I almost choked as Cathy said, 'Oh, yes. Do try it on, Franny. I'm sure you'll look lovely in it.'

Franny! Surely that was short for Frances – Mark's ex? Was he going to marry this girl he'd been engaged to after all? My first thought was that he must be on the rebound. Then another thought occurred to me: had *I* been the one on the rebound after *her*? Had he never really loved me? Had I lost him for good? I held my breath as Franny disappeared into one of the changing rooms, hoping against hope that the dress wouldn't require any alterations. If it did, Camilla would be sure to ask me to help her with the pinning. The thought of facing Cathy made my stomach churn. I couldn't bear the thought of her taking the news of my humiliation triumphantly back to Mark, and them laughing about it together.

Luckily, the dress fitted perfectly and Franny decided to buy it. She paid the exorbitant sum with her credit card and she and Cathy went off together in high spirits, chattering away excitedly. Camilla came back into the staffroom, flushed with the pleasure of success, only to glower at me as she tasted her coffee.

'Stone cold!' she pushed the cup at me. 'Make me another. And please do not peep round the curtain in that vulgar way. Don't think I didn't see you. If you want to learn how to conduct yourself, just come into the shop and help me as any sensible person would do.'

'I was on my break,' I pointed out to her.

'Business comes before breaks in my establishment, as you'll soon learn,' she said.

Who the hell did she think she was, speaking to me like that? I was a mature woman, not some spotty teenager. I longed to pour the cooling coffee over her elegant coiffeur but I controlled myself. My turn would come, I promised myself with gritted teeth. At the first sniff of a job, I'd be out of here like a rat up a drainpipe.

Whilst Camilla was out at lunch, I switched on my phone. There were three missed texts from Di. I went into 'messages' and read them. They all said the same.

Editor of the Daily Sphere *wants you to get in touch ASAP. Good luck, Di.* A phone number followed.

Praying that no customers would come in, I tapped in the number and waited with bated breath, hoping he wasn't out at lunch. His secretary put me through at once. Yes, he'd heard rumours about Paul Fortune's scam, and yes, he was certainly interested in my story. I tentatively asked what the paper would pay, pointing out that I'd lost my entire savings. He was sympathetic. Could I go in and talk about it?

Yes, I could!

This afternoon?

That threw me for a moment. Camilla would never agree to letting me have the afternoon off so I'd have to wangle something. One thing was for sure – I wasn't going to pass up a chance like this, whatever it took.

'Is that a problem for you?' he asked.

'No! Not at all,' I assured him. 'Just say a time and I'll be there.'

The meeting was scheduled for four o'clock so I was going to have to think quickly. As it happened the fates were with me. The answer dropped into my lap minutes before Camilla returned from lunch and I saw at once that this was my chance. Two people came into the shop, one young, the other, I guessed, around fifty-something. Mother and daughter, I guessed, but it soon emerged that the customer was not the daughter but the mother.

'I'm getting married for the second time,' she simpered. 'My first wedding was a rushed affair in a register office so I want this one to make up for what I missed the first time round. I'm planning the full works.'

I looked at her. She was short and on the tubby side, with a figure that I guessed owed much to pies and cakes. She wore too much badly applied make-up and her hair was an unlikely auburn with magenta highlights. I drew out one of Camilla's most expensive gowns.

'Oh, that's lovely but it's a bit plain and …' She took one look at the price tag and gasped.

'Good God! I wasn't thinking of paying that kind of money,' she said. I saw the daughter flinch.

'Mum – you wouldn't wear a dress like that anyway, would you?'

The mother bridled. 'Why not?'

'Isn't it a bit – well – *young*?'

'Everyone says I look *years* younger than I am!'

'We do have a budget range,' I put in. 'Shall we see if there's anything there that you'd like?'

She cheered up at once – just as Camilla walked in. Seeing that I had a customer, she went to the back of the shop and disappeared into the staffroom, where I guessed she had her ear to the gap in the curtain, if not her eye. In the rail of budget dresses was one particularly hideous dress. I'd spotted it on my first day and asked Camilla about it. She told me it had been foisted on her by a sales rep. In return for taking it off his hands, he'd given her a good deal on six other dresses. It was a gypsy-style gown in fuchsia pink with a ruched skirt and the lowest neckline I'd ever seen. It was generously decorated with black lace and diamante and as I'd guessed she would, the woman fell in love with it on sight.

'Oh! I *do* like that!'

I saw her daughter wince but ignored it. Adopting my best Camilla manner I went into my act. 'This would look *perfect* on you,' I gushed. 'You have just the right figure for it. When your groom turns and sees you coming down the aisle in this, he's going to go weak at the knees.'

'He's eighty so he's weak at the knees already,' the daughter muttered. I tried hard not to laugh.

'Try it on,' I invited. 'And just you see if I'm not right.'

The woman emerged a few minutes later, flushed with delight. The dress was too tight for her. She'd obviously had trouble with the zip and her bosom was spilling over in the most alarming way, but at least she had it on. She looked at herself in one of the full-length mirrors, smiling as she turned this way and that. 'Ooh, I have to have this,' she said. Once again the daughter winced.

Casting me an apologetic look she said, 'Mum! You look like something out of a pantomime.'

Her mother rounded on her. 'Shut up, Norma! You've got no fashion sense and anyway, you're just jealous!' She turned to me. 'Every bloke she gets dumps her after a fortnight,' she said nastily. The daughter flounced out of the shop. I could almost feel Camilla's eyes burning into the back of my neck and a few moments later, she emerged from the staffroom, a false smile plastered onto her face.

'I'm sure we can do better than that for you, madam,' she said.

The woman shook her head. 'Oh no, I've set my heart on this one,' she said. 'And don't you worry: I shall recommend your shop to all my friends.'

She changed back into her own clothes as I packed up the hideous dress, feeling Camilla's fury just waiting to erupt the moment the customer left the shop.

As the tinkling notes of the Wedding March door chime died, she started spitting venom.

'What on *earth* did you think you were doing?' she stormed. 'That – that awful woman will tell everyone she bought the dress here. My reputation will be ruined!'

I raised my eyebrows. 'Well, it *was* for sale. And she was a very satisfied customer.'

'Do you honestly think I want *her* kind of customer?' Her eyes flashed. 'If I'd been here I would have put her off. I was going to take that wretched thing home tonight,' she said. 'I wish I'd done it weeks ago. I should have *burnt it*!'

'Well, you didn't, and now it's sold and you've got the money for it,' I told her blandly.

Her eyes narrowed. 'You did it on purpose, didn't you?'

I smiled at her. 'Well, you did ask for it,' I said. 'You and your high-handed ways. If you think you can speak to me as though I'm your inferior and get away with it, you've got another think coming!'

'How *dare* you!' Her face turned a peculiar shade of puce and as she spoke, flecks of spit landed on my jacket. 'Get out of my shop this minute. You're dismissed! Do you hear me? *Get out*!'

'Not until you've paid me,' I said, standing my ground. 'If you refuse, I shall take it further and we don't want to ruin your reputation even more, do we?'

Practically fizzing with anger, she went behind the desk and opened the till, snatching a handful of notes she almost threw at me. She was flexing her fingers as she glared at me, and I looked at the long scarlet talons and decided that it was time to make my exit before she scratched my eyes out. As I left the shop I glanced at my watch. If I caught a bus now I'd be right on time for my date with the *Daily Sphere*.

The meeting was more than successful. I made the most of Paul Fortune's scam and the mess he'd landed me and everyone else in. The editor was enthusiastic and offered me a full page spread in the Sunday edition, complete with photograph. It meant that everyone I knew would get to hear of my humiliation but the fee I managed to negotiate more than made up for that. When I got back to the bedsit, I texted Di.

Thanks to your brill idea I'm on the up again. See you soon – Lou.

After I'd pressed 'send' I sat looking at my list of contacts then, on impulse, I highlighted Mark's number. He'd given up trying to contact me and now I knew why.

Congratulations, Mark! I tapped in. *Love, Lou.*

At least he'd know I was still thinking of him.

Chapter Twenty-Two

SUSAN WAS WASHING UP her breakfast things on Sunday morning when the phone rang. She pulled off her rubber gloves and went into the living room to answer it.

'Hi, Mum, it's me, Karen. Have you seen the paper this morning?'

'No, I haven't been up very long. I'm still in my dressing gown. Why?'

'You do have the *Sphere*, don't you?'

'On Sundays, yes, I have to confess that I like the sensational stories they publish.'

'Well, there's none as sensational as the one on page three,' Karen went on. She sounded excited. 'Just go and get it, Mum. Have a look now while I'm on the phone.'

Mystified, Susan picked up the paper from the coffee table and unfolded it, Spreading it out, she turned the pages. There, on page three, the face of her stepdaughter smiled up at her. *Louise*! But what...? She scanned the headline and the story beneath and gasped.

'Well – have you found it?' Karen sounded impatient at the other end of the line.

'Yes, I've found it. It says that the show she was supposed to be opening in turned out to be a huge confidence trick.' Susan was still running one forefinger down the page. 'Oh my God! It also says that she put – it says here – her life savings into the project and that she's lost the lot!'

'Have you been holding out on us, Mum? Did she tell you any of this last time she was with you?'

'No. Just that there was a problem and that there'd be a delay with the opening.'

'She's such a liar, Mum.'

Susan was shaking her head. 'It must mean that she's lost everything that her father left her. She wouldn't want to admit it, would she? Poor Louise. What a terrible blow for her.'

'Mum! You can't be serious. She richly deserved this. She's a nightmare. You know she is.'

'I know she has an unfortunate habit of upsetting people but you have to admit that this is something you wouldn't wish on your worst enemy.'

'Want to bet?' Karen said, half to herself. 'Well, I just hope that she isn't going to descend on you again, taking advantage of your good nature. Promise me you'll say no if she rings and asks you to take her in again.'

Susan bridled. 'I'm promising you nothing of the sort, Karen. I shall do as I think fair and right when and if it arises.'

'Well, on your own head be it. Don't say you haven't been warned, Mum.'

'I won't.' Susan waited a moment then asked, 'Is everything all right there – with you and Simon, I mean?'

'Of course. Why do you ask?'

'Don't be naïve, Karen. You know perfectly well why I'm asking. Have you resigned yourself to staying at home with Peter?'

'No, not completely. I've said I'll do the odd spot of tutoring. Not full-time, of course. Just when they're stuck for someone.'

'And is Simon all right with that?'

'Mum – I'm not a slave. Simon is my husband, not my keeper. I do wish you'd update to twenty-first century thinking.'

'Perhaps it's not me you need to convince,' Susan said. 'Anyway, thanks for ringing, Karen. I hope you feel that Louise's stroke of disastrous bad luck has vindicated you in some way. I'll have to go now, I need to get dressed.'

Susan sat thinking for a few minutes after hanging up. Louise had been so convinced that this play was her big breakthrough. She must be devastated. On impulse, she lifted the receiver again and dialled Louise's mobile number. After a few minutes Louise answered.

'Hi, Susan.'

'I've just seen the article in the *Sunday Sphere*,' Susan said. 'How awful for you and all the rest of the cast. You must be so upset and disappointed. Did you know about this when you visited last time?'

'Well, yes, I did actually.'

'And you never said a thing. How are you managing, dear? Did you lose all your money?'

'Quite a lot of it, yes, but don't worry, Susan. I had a temporary job and the paper paid me well for the article. Of course, everyone will know now that I ended up with the proverbial egg on my face, but never mind.'

'Have they caught this man – this Fortune person?'

'No and I'm not holding my breath that they will. He's left the country and covered his tracks pretty well.'

'Do you have a place to stay? Will you be able to find another job?'

'I'm staying with Dianne at the moment. We've made up our little quarrel. My agent, Harry, put money into the show too and it's put him out of business so I have to find a new agent.'

Susan sighed. 'Oh dear, what a disaster for you all! Louise – have you heard any more from your birth mother?'

'No and I don't want to. That's a closed chapter as far as I'm concerned. Anyway, if she's read the article, she will have lost interest in me by now. Thanks for ringing, Susan. I appreciate your concern.'

'Not at all. You know where I am, don't you, if you need me?'

'Yes, and thanks again. It's nice to know there's someone on my side.'

'What about your young man – the one you told me about? He was in the show too, wasn't he?'

'Yes, he was, but that's all over, Susan.'

'Oh. I'm sorry to hear that, dear.'

Susan hung up with a sigh. Whatever Louise had done in the past, it was certainly catching up with her now. She thought briefly of the rebellious child she had taken on all those years ago when she married Frank. Louise hadn't been easy to bring up, especially once she reached puberty. There was a time when it seemed that she would never have a normal life, but somehow she'd put all the distress she had suffered in her youth aside and made a life for herself. It was true that she'd become a difficult and unpredictable woman, but she'd suffered so much in her young life that Susan tried to make allowances. Maybe the trauma and the underlying sense of loss would never leave her.

It was two days later that Karen rang again.

'Mum – can you do me the most amazing favour?' She rattled on before Susan had time to reply. 'The things is, this agency, you know the one that offered me the job, need me to do some work for them today.'

'And you have no one to babysit?'

'Not at such short notice. It would only be for the morning, Mum. I'd be so grateful.'

'Does Simon know?'

'What's that got to do with anything?'

'I'll take that as a no, then.'

'What he doesn't know can't hurt him. I really haven't got time to discuss it now. Can you have Peter or can't you, Mum?'

'I suppose so. When and where do you want me?'

'I'll pop him round to you on my way. Have to go now. See you soon.'

Karen's visit was swift and brief. She handed a bewildered-looking Peter over unceremoniously. 'I've left the buggy downstairs by the front entrance,' she said breathlessly. 'It's a lovely morning so if you take him to the park I'll meet you there – say, by the café at one o'clock. OK?' And before Susan could confirm that this was convenient for her Karen was already halfway down the stairs.

Peter popped his thumb into his mouth and looked up at his

grandmother. 'Mummy gone.'

Susan bent to pick him up. 'Yes, darling. Mummy's gone but she'll be back again soon. Have you had any breakfast?'

His little face brightened. 'Poddidge?'

'Yes, Granny'll make you some porridge and then we'll go to the park, shall we?'

As Karen had said, it was a lovely morning and Susan let Peter walk beside the buggy down to the lake. She'd brought stale bread and they bought a bag of corn which Peter delightedly threw to the ducks. They went to the playground and Peter went on a swing – a baby one this time – and sat on Susan's lap for a gentle ride on the roundabout. At the café they each had an ice cream, after which Peter began to look decidedly sleepy. Susan put him into the buggy and tucked his blanket round him, and by the time they had walked across the park to the bowling green, he was fast asleep. Grateful for five minutes' respite, she sat down on one of the benches to watch the elderly men playing their sedate game. The warm sunshine made her drowsy too and her eyelids had closed when suddenly she heard someone say her name.

'Susan.'

She opened her eyes to find Ted standing in front of her. 'Good morning, sleepyhead.'

Her heart leapt and she felt the warm colour stain her cheeks. 'I wasn't asleep,' she said. 'It's just the sun. It's very warm and – and dazzling.'

'Of course it is.' He chuckled and sat down beside her. 'Young Peter looks cosy.'

'Yes.' She sat up straight. 'I – we were just going, as a matter of fact.'

'Back to the flat? Mind if I walk with you?'

Susan glanced surreptitiously at her watch and sighed. It was almost half past twelve. Karen would be here in half an hour. She couldn't go now. 'No, not back to the flat,' she told him. 'Karen is picking Peter up at the café at one o'clock. I was going to walk across there in a minute.'

'Then I'll come with you.' He looked at her searchingly. 'That is if you've no objections.'

'There's really no need,' she said stiffly.

He laid a hand on her arm. 'Susan, please. Surely at our age there's no need for us to keep up this ...' He shook his head. 'Whatever you want to call it.'

'No – well.'

'I've missed you very much these last weeks.'

She turned to look at him. 'I – thought you'd found a new – companion.'

He smiled. 'The lady you saw helping me on the allotment was a fellow gardener's wife,' he explained. 'They saw me struggling to keep your plot going as well as mine and offered to help.'

Susan bit her lip. 'Oh. I see.'

'There's a lot you don't see, my dear,' he said. 'And that's all down to me. I should have been more upfront with you. My only excuse is that I didn't want to frighten you away. If you'll just let me ...'

'How did you know I'd be here today?' she asked him.

'I didn't.' He smiled. 'Believe it or not, I come here every day in the hope of running into you.'

Susan stood up. 'I'm sorry, Ted, but the time's getting on. I have to go. Karen will be waiting.'

He stood and faced her. 'Then may I?'

'I suppose so – if you must.'

'Oh yes, I must,' he said with a smile. 'I *really* must. If you only knew how much courage it took to speak to you just now.'

Susan said nothing as she rose and began to push the buggy, but she was slightly mollified by Ted's humility as he walked silently beside her.

Karen was already waiting. Her cheeks were glowing and she looked happy.

'Thanks so much, Mum,' she said as she took the buggy's handle. 'I've enjoyed this morning so much. Tutoring is so rewarding. They've asked me to continue. It's for a boy who's broken his leg in an accident and as he's taking his GCSEs next year, they don't want him to fall behind.' She kissed her mother and hurried off. Ted had been standing to one side and now, Susan felt him looking at her. He touched her arm.

'Lunch? Please say yes, Susan.'

She relented. 'Well – I normally only have a sandwich. They do quite nice ones here at the café.'

He nodded. 'Then a sandwich it shall be – for both of us. You pick a table out here in the sunshine and I'll go and get them.'

He returned with an assortment of sandwiches and coffee for them both. She looked at him.

'I never thanked you for the flowers you brought me,' she said. 'And I owe you an apology for my stepdaughter's outspoken remarks.'

He shook his head. 'Don't give it another thought.' For a few minutes they ate in silence then he said, 'Susan, I owe you an explanation and I'm determined that you shall hear it. It's all a bit convoluted and you'll have to bear with me but—'

'You owe me nothing, Ted,' she interrupted.

'Oh, but I do,' he insisted. 'There are things I have to tell you, if only for my own peace of mind.'

'All right.' She looked around. The tables were filling up now and two elderly women at the next table were clearly listening to their conversation with interest. 'But let's finish our lunch first and find a quiet spot.'

They finished their sandwiches in silence, then got up and walked slowly down to the lake. She waited for him to begin and it was obvious that he was nervous and hesitant.

'First, I must confess that it's true that I'm still married,' he said at last.

She stiffened. 'You gave me the impression that you were a widower.'

He looked at her. 'I'm sure I never actually said so.'

'Maybe not, but …'

'Meg and I have been married for more than forty years. We were both very young when we married, especially Meg, and I'm sorry to say that she cheated on me from very early on in the marriage. She had an endless stream of affairs, none of which lasted for long, and like a fool I always forgave her and took her back. But eventually she met someone and fell seriously in love. They ran off together – went to live abroad. I filed for divorce when it became clear that she had no intention of coming back to me but for some reason she refused to cooperate. I thought it must be because she was still unsure about her new relationship, but eventually that thought petered out and I began to pick up the pieces and make a new life for myself. I thought that eventually she'd want to marry her new partner and agree to a divorce, but the years went by and it never happened.'

'Do you know where she is now?' Susan asked.

He gave her a wry smile. 'Oh yes. I know where she is. Four years ago, Meg's partner got in touch with me quite out of the blue. He told me that she had developed dementia and he could no longer have – as he put it – the *responsibility* of her. He informed me that as I was still her husband that duty now fell to me.'

'Oh no!' Susan stared at him, appalled. 'But – how many years had you been apart?'

'Almost thirty.'

'But surely, isn't there something about a marriage being null and void after a certain period of desertion?'

'Desertion is grounds for divorce, yes, but I never applied to divorce her on those grounds.' He shook his head. 'And I could hardly do so at that stage.'

Susan shook her head. 'So what happened?'

'Her new partner brought her back to me and just left. He handed her over like an unwanted pet and disappeared over the horizon. Naturally, she didn't understand. She didn't even remember who I was. Life was sheer hell – not only for me but for her as well. Eventually, my only alternative was to get her into a care home. I chose the best I could afford and she's still there.'

Susan felt chastened. 'Do you visit her?'

He shrugged. 'Occasionally, though she doesn't recognize me.'

'Oh, Ted! How awful. I'm so sorry.'

'The woman next door knows some of this but not all. She

had no right to speak to you as she did.'

For a while, Susan was silent as she tried to take in all that Ted had told her. At last she looked at him. 'I wish you'd told me this in the beginning.'

He smiled wryly. 'I wish I had too, but it isn't a happy story and it doesn't make me much of a prospect, does it? Apart from that, you might have thought I was to blame.'

'How could you be to blame?'

'For all you knew I might have driven her away. I might have been a bad husband – might have been violent or abusive.'

'Knowing you, I'm certain none of those things applied to you.'

'But back then you *didn't* know me. I would have told you eventually, Susan. I had no intention of keeping you in the dark. But when we first met I knew at once that you were going to be someone special and I couldn't risk losing you.' He sighed. 'Unfortunately Mrs Freeman forestalled me.'

'I should never have listened to her,' Susan said. 'I should have given you the chance to explain.'

He shook his head. 'It was understandable that you felt shocked and let down, believed that I'd misled you.' He looked at her. 'Under the circumstances I'll understand if you don't want to see me again. I just wanted you to know the truth. I couldn't bear the thought of you thinking me a liar and a – philanderer.'

Very tentatively Susan reached for his hand. 'Oh, Ted, what a sad life you've had,' she said. 'Now that I know all this how could you imagine that I would want to end our friendship? I've missed you too – so much.'

His eyes lit up. 'Are you saying you'd be happy for us to start seeing each other again?'

'Of course, though it might be better for us to meet at my flat in future.'

'Whatever you say.' Ted stood up and held out his hand to her. 'Shall we walk back to the cafe? I don't know about you but I could murder a cup of tea.'

Susan laughed and took his hand. 'Me too.'

Walking back through the park in the sunshine with Ted, Susan's heart lifted. Although she felt desperately sorry for the poor woman in the care home she felt that Ted deserved some happiness and contentment at last for all that he had suffered. As for her, she couldn't remember a time when she'd felt so happy.

Chapter Twenty-Three

I COULDN'T BELIEVE IT. THE very next day after the article had appeared in the *Sunday Sphere,* I had a call from one of the agents I'd left my details with. I called at once and an interview was arranged for the following day.

Di was thrilled. 'There! What did I tell you?' she said. 'I knew it was a good idea to put yourself out there.'

I pulled a face at her. 'Making myself look like a gullible idiot, you mean,' I said. 'Maybe they're just looking for a cleaning woman.'

'Get away with you,' Di said. 'This could be the making of you.'

Dressed in the Chanel suit, carefully sponged and pressed, I made my way to the agent's office at the appointed time and sat nervously in reception. Looking around the room, I was impressed by the signed photographs of many well-known celebrities displayed on the walls.

The door opened and a man came out. A moment after he left, the receptionist's phone rang. She listened briefly, then replaced the receiver and looked across at me. 'Mr Jason will see you now.'

He was middle-aged with silver-grey hair and an attractive, warm smile. He rose and offered his hand. 'Good morning,

Miss Delmar. I'm Patrick Jason. Please have a seat.' When we faced each other across his desk he said, 'I read the article in the *Sunday Sphere* about your bad experience. The name rang a bell and I looked in my in-tray and found the details you left at the office a few days ago.' He looked up at me. 'Surely you had an agent before all this?'

I nodded. 'Yes, Harry Clay. Unfortunately he put money into the project too and it has put him out of business.'

'I see. I had heard he was retiring, but I had no idea that he was another victim of this terrible business.' He looked at me speculatively. 'This Fortune man fooled us all. It was brave of you to go to the national press with your story,' he said.

'I don't know about that.' I smiled. 'It might well turn out to be the end of my career, but frankly it was a case of desperation. I needed the money. It was as simple as that.' I looked at him and decided I might as well lay my cards on the table. 'I couldn't believe my luck when I got the leading role in this new musical. It seemed like the big break I'd been longing for, and when I was asked to put money into it I was only too eager. I was supposed to get my money back, plus generous interest once the show was up and running.'

'I can't begin to imagine how you all felt when you found yourselves stranded high and dry in Bournemouth.'

I gave him a wry smile. 'It was a blow to say the least. Then when Harry told me he was closing the agency, it looked as if I was going to be out of work for some time. I had to do something.'

He shook his head. 'So you decided to go public?'

'Yes.' I sighed. 'The price I paid being that now everyone will know how vain and gullible I was.'

'So, you've had no work since?'

'Not in the business. I did have a job at a West End wedding-dress boutique,' I told him. 'It only lasted a few days though. The owner was a twenty-four-carat cow.'

He laughed. 'I liked the sound of you in the article,' he said. 'It showed me that you have character, the ability to laugh at yourself.'

I decided to ask him point blank where all this was going. 'So – why did you ask me to come in? Are you offering to represent me?'

He pursed his lips. 'I do have one or two things in mind that might suit you.' He looked at me. 'Meantime, would you be willing to do some commercial TV?'

'Anything to keep the wolf from the door.'

'Right.' He made a note on his pad. 'Have you done any TV work before?'

I opened my mouth to tell him I had and then closed it again. The time had come to be honest. If I lied about this he'd be bound to find out and that could ruin any future chances I might have. 'No,' I said. 'But it's always been an ambition of mine.'

'OK. If anything comes up I'll give you a ring.' He leaned back in his chair and eyed me for a moment. 'I've just had a thought. One of the TV soaps is auditioning next week,' he said at last. 'Have you done character?'

I shook my head, remembering my decision to be truthful. 'Not really, though I admit I'm getting close to that age. What's the part?'

'A middle-aged motherly type,' he said. 'The kind of sympathetic woman everyone turns to in times of trouble. Do you watch *King's Reach*?'

I nodded. I'd seen it a few times when I was staying with Susan. *Susan*! She was just the kind of woman he'd just described to me. I knew Susan well enough to use her as a role model. 'I quite like the sound of that,' I told him.

'You wouldn't mind playing older than your age?'

'Not at all.' At that particular moment I wasn't at all sure about playing older but beggars certainly couldn't be choosers. And I reminded myself that some of those soap stars had been playing the same part for years. A guaranteed income sounded pretty good to me, playing older or not.

He opened a file on his desk and took out a sheet of paper. 'Here's the character description. She's called Amy Armstrong. Take it home and have a read. The audition is next Thursday. I'll text you the address of the venue when they let me have it.

Meanwhile, I'll ring you if a commercial opportunity comes along.' He paused. 'Are you still at the address you left me?'

'Oh, no. What a good job you thought to ask. I'll be staying with a friend for the foreseeable future.' I scribbled down Di's address and passed it to him. 'But you can always get me on my mobile.' I stood up. 'Thank you so much for seeing me, Mr Jason.'

He smiled. 'Patrick, please. Let's hope it all works out for you.'

The following morning I had two calls; the first was from Patrick Jason, giving me the address of the audition venue. I was so excited when I clicked the call off that I went to the fridge and poured myself a celebratory glass of wine. I wished Di could have been with me to share the excitement. I was sipping my wine and studying the character description once more when my phone rang again.

'Hello.'

'Hello, Louise.'

I recognized the voice at once and my heart plummeted. 'What do you want?' I asked bluntly.

'Oh, come on Louise,' my mother said. 'I'm just ringing to say how sorry I was to read about your disappointment. It was such a lovely photo in the paper too.'

'So now you and your son will realize that I'm actually broke,' I said. 'If you think I'm rolling in cash and a soft touch, you're going to have to think again.'

'You're very suspicious, Louise. I can't think where you get that from.' Her voice had a hard note to it now.

'Neither can I. So we might as well call it a day now,' I told her. 'I'll be frank. I don't want a relationship with you. It's too late and I don't think we have anything in common anyway.'

'I'm sorry to hear that you think that, Louise,' she said. 'But before we part company, there's something you should know about yourself. Something important that no one knows about but me.'

'I don't think you know anything that I don't.' I held my

finger over the 'end call' button but what she said next stopped me.

'Don't hang up, Louise! This is something you really should know – for your own sake. I'm not joking.'

There was something chilling about her tone and I began to be apprehensive. 'Then tell me now.'

'Not on the phone,' she said. 'It's not trivial, Louise, and I'm not kidding. We really have to meet for me to tell you – even if it is for the last time.'

She'd got me now. In spite of myself I was curious. 'All right,' I said. 'Where and when?'

'King's Cross Station,' she said. 'In the café – Thursday, at one o'clock.'

'Thursday?' It was the day of the audition. 'Can you make it any other day?'

'Thursday would be best.'

I thought about it. I should be able to make it by one o'clock. If not she'd have to wait. But why meet at a railway station?

'King's Cross? Why there?' I asked.

'I'm meeting someone off a train at two. It's convenient.'

'All right, I'll try to be there.'

'It's to your advantage to be there, Louise. See you on Thursday, then.'

After the call, I sat there for a long time, my excitement about the audition temporarily forgotten as I wondered what she was about to tell me about myself. Could it be that I was a carrier for some horrible disease? Or was it just some devious trick? I decided not to think about it for the moment. Concentrating on the audition was my priority.

I arrived at the studio in good time and was dismayed to find about twelve other actresses waiting, not least of which was none other but Carla Dean. To my dismay she soon spotted me and came across.

'Fancy seeing *you* here, darling,' she said. 'I saw the article all about you in the *Sunday Sphere*. Anyone would think you were the only one to be cheated.' She sniffed. 'A very flattering photograph, I thought. When was that taken – ten years ago?'

'It was taken by the paper's own photographer,' I told her.

'Then they must have airbrushed it,' she said. 'So how did you hear about this?'

'My new agent, Patrick Jason.'

Her finely plucked eyebrows rose. 'Jason, eh? Personally I was tipped off by a friend and it seemed like the sort of thing that would suit me down to the ground.' She smiled smugly. 'I know a couple of the production crew actually. Just between you and me the rest of you might as well go home now. I think it's pretty much a foregone conclusion.'

'Well, good luck, then.'

We were called in one by one. When it was my turn I pulled out all the stops, reading the test piece with Susan very much in mind. When everyone had auditioned, the production assistant came out and told us we'd be notified in a few days' time. Carla looked at me. 'Coming for a drink?'

I shook my head. 'No. I have another appointment,' I said. No way was I going to sit in some wine bar being bombarded with personal questions by Carla. We parted company in the street outside and she wandered off. I looked at my watch. It was a quarter to one. If I was going to make King's Cross on time I was going to have to hail a taxi.

I knew there was more than one café or coffee shop at King's Cross station and my 'mother' hadn't said which one, but I soon spotted her, sitting at one of the tables outside. She saw me and waved me over.

'You're late. I began to think you weren't coming.'

I hoisted myself onto the high stool opposite her. 'The traffic was bad.'

'Oh well, you're here now. Do you want to go and get a coffee?'

'No. I just want you to get to the point,' I said. 'What is this you need to tell me?'

She took a leisurely sip of her own coffee, looking at me speculatively over the rim of the mug. 'Do you remember the night I left?' she asked. 'Or were you too young at the time?'

'I remember it as though it was yesterday,' I told her. 'In fact, I've been having nightmares about it ever since.'

She snorted disbelievingly. 'Well now, aren't you the drama queen!'

'So – it's something about the night you left,' I said. 'I thought you said it was about me.'

'It is.' She picked up her spoon and began to swirl what was left of her coffee. 'Did your dad ever tell you what we rowed about?'

'Of course not. Look …' I was fast losing patience with her. 'Just get to the point. How does this concern me?'

She looked up at me with a hint of triumph in her eyes. 'Your dad and I rowed because I told him he wasn't your father.'

I stared at her. 'You *what*?'

'I told him the truth: that he wasn't the father of my child!'

It was as though a chill hand clutched my heart. 'I don't believe you.'

She smiled maddeningly. 'What do you want – a DNA test? It's a bit late for that!'

My mouth dried. Suddenly I had trouble breathing. 'But – he kept me – brought me up. He was my dad and I loved him. If what you say is true, why didn't you take me with you?'

She shrugged. 'I was young. I wanted to be free.'

I winced. 'If Frank Davies wasn't my father, then who was?'

'Could be one of several,' she said casually. 'I was a good-time girl back then. I played the field, as they say.'

I felt sick. Getting down from my stool, I took one last look at the woman who had given birth to me. '*You bitch*!' I said. 'I hope I never have to see you again.'

'Likewise, I'm sure,' she said with a laugh. 'You and your boasting about being the big wealthy star. You're nothing but a small-time extra – *if that*!' As I walked away she called after me, *'You'll never amount to anything – you're bloody useless – just like that fool you called Dad!'*

I made a beeline for the ladies' and locked myself in a cubicle where I was wretchedly sick. My heart was thumping and the tears ran unchecked down my cheeks. I came out of the cubicle feeling weak and stood shakily, clutching one of the wash-basins. A concerned-looking woman asked me if I was all right. I shook my head.

'I'll be fine in a minute,' I said 'It's a stomach upset – something I ate.' I dashed some water on my face and hurriedly made my escape.

As luck would have it, Di was out all evening. One of her colleagues was having a hen night. I made myself a sandwich but it was like sawdust in my mouth and eventually I threw it in the bin and took myself off to bed. I heard Di come in around midnight but she was quiet and I didn't call out to her. I fell at last into a fitful sleep but the dream seemed to begin almost at once. A crowd of people were laughing derisively at me and in the centre of them was my mother, her face a mask of hate as she pointed and jeered. When I looked down I held a baby in my arms. In deep shock, I threw the baby from me and heard it scream as it fell. The screams and the scornful laughter grew louder and louder until they became unbearable. I put my hands over my ears and I heard myself shouting, *'Stop it! Stop it!'*

'Lou – *Lou*! Wake up!'

I opened my eyes to see Di's concerned face looking down at me. Still shaking, I hoisted myself into a sitting position, my heart thudding. 'Was – was I talking in my sleep?'

'Shouting more like,' she said. 'It must have been a horrible dream. Do you want to tell me about it?'

Suddenly everything crowded in on me. I was overwhelmed by a terrible feeling of grief and I burst into tears. Di put her arms round me and held me close.

'What is it? I've never seen you cry before.'

'I – I had some bad news,' I stammered, swallowing hard.

'What was it, Lou – the audition?'

The audition! I could almost have laughed. It seemed like a million years ago. So trivial that I'd completely forgotten about it. I shook my head. 'No. That was OK, they're letting us know. There were dozens of applicants so I don't suppose I've got a chance.' I looked at her. 'Afterwards I went to meet my mother

– birth mother, I mean.'

'So you found her? You didn't tell me.'

I shook my head. 'It was some time ago and it wasn't what you'd call a roaring success. In fact it was a disaster. She was nothing like I'd imagined her. She was only interested in meeting me because she thought I was going to be a star and have lots of cash. Then I made the mistake of telling her I'd inherited Dad's money. She asked me for cash the first time we met but when she realized I wasn't going to be a never-ending source of easy money, she didn't want to know.'

'So why did you meet her again today?'

'She rang me to say she had some important information about me – something I really should know.'

'And did she?'

'Yes.' I bit my lip to stop myself from getting emotional again. 'She told me that Dad wasn't my real father. Worse – she'd cheated on him several times and she didn't really know who my father was.'

Di looked shocked. 'Oh, Lou. But is it true, do you think? Was she just being spiteful?'

'I don't know. I've been thinking – the one person who might know is Susan. I'll have to go and see her.'

Di squeezed my hand. 'I'm so sorry, love.'

'Me too.' I looked at her. 'Can you imagine, Di, how it feels, not knowing who you really are?'

'But you're *you* – Louise Davies. You're your own person.'

I sighed. 'I even dropped Dad's name. It's ironic, isn't it? I know I haven't been a nice person. I treated poor Mark horrendously when we were at drama school. No wonder his sister hates me. I've taken advantage of people – you included. I thought the world and everyone in it owed me when really I was just a pain in the backside. I'm a mess, Di. Maybe I take after my father – whoever he might be. Now there's no way I'll ever know.'

Di patted my shoulder. 'My advice is to get on a train first thing tomorrow; go and see Susan and have a good talk. She probably knows more than you realize.'

Chapter Twenty-Four

'MUM! YOU CAN'T BE serious. It's no time at all since Louise came to stay with you. What's her excuse this time?'

'She says she needs to talk,' Susan said. 'She's somehow managed to locate her birth mother and she seemed so upset on the phone. I couldn't say no to her.'

'But you can say no to me!' Karen protested. 'You know it'll all be some trivial nonsense she's dreamed up.'

'I don't think so. She ...'

'So you're saying no to having Peter for me?'

'I'm afraid I can't,' Susan said firmly. 'Not this time. Anyway, Simon's at home now for the holidays. Can't he ...' She paused. 'I take it you have told him – about the tutoring?'

'Not exactly.'

'Well, don't you think you should?'

'I told him I had to be out for the morning and he's gone off somewhere in a huff so I can't; at least not at the moment.'

'Well, you know my feelings on that matter. You should start talking to Simon,' Susan said. 'Just be honest. No marriage can survive all this subterfuge. And I don't want to be a party to it either, Karen.'

Karen slammed the telephone down so hard that Susan was almost deafened. She hung up with a sigh. If Karen wasn't careful she'd ruin everything between herself and Simon. She was just making a shopping list ready for Louise's visit when there was a ring at her bell. She opened the door to find Simon standing outside.

'Can I come in, Susan?'

'Of course. Is something wrong? Can I get you a drink – coffee?'

He ran a hand through his hair. 'Have you got anything stronger?'

'I think there might be some whisky left over from Christmas,' she said. 'Will that do?'

'Yes, fine, thanks.'

Susan fetched the whisky and a glass and watched as he threw the drink back in one gulp. 'I think you'd better tell me what's wrong,' she said as he put the glass down.

'It's Karen,' he said. 'I don't know what she's up to, forever going off somewhere. I'm beginning to think she's – that there's someone else. I know she's lying to me and I'm really not happy about the way things are between us at the moment.'

Susan sighed. Things were getting to a ridiculously complicated stage between those two. Maybe it was time she put Simon straight. At the risk of being classed as an interfering mother-in-law, she took a deep breath and said, 'It's nothing like that, Simon. She's doing a bit of private tutoring, that's all. She didn't want to tell you because she thought you'd be angry.'

He sighed. 'Oh, for God's sake! Not that again.'

'She's a bright girl, Simon, and she hates wasting her education and ability. Surely a little part-time tutoring can't hurt? And I'm happy to have Peter when I can.'

He sprang to his feet. 'What does she take me for – some kind of ogre? I asked her outright this morning when she was making some excuse about being out again tomorrow morning but she always hedges.' He glared at her. 'And you knew about it all along!'

'You must admit you've been rather inflexible in the past.'

'Surely a man has the right to say what he wants in his own house!'

Susan smiled. 'Simon, can you hear yourself? You sound like some Victorian patriarch. Times have changed. Young women are no longer satisfied with a life of undiluted domesticity. Why can't you cut her some slack?'

'It was only going to be till Peter goes to school.' He sat down again, slightly calmer. 'She only had to ask.'

'Ask?'

He frowned. 'You know what I mean. We could have discussed it.'

'I remember the last time you *discussed* it,' she said. 'Right here in my living room. Look, why don't you go home now and try to be more reasonable with her? Offer to stay with Peter tomorrow while she does her tutoring. What she's doing is important to that family. It's very worthwhile.'

He shrugged. 'I suppose so.' He looked at her. 'Has she asked you to have Peter tomorrow?'

'Yes, but I had to refuse because Louise is coming to stay for a couple of days.'

He stared at her. 'Louise! *Again*? You have to be joking! After the trouble she always causes. You must be mad, Susan!'

She stood up. 'Never mind whether I'm mad or not, Simon. Just take care of your own problems and let me take care of mine. Go home and get it over with. Make things right between you.'

Louise arrived just after eleven. When Susan opened the door to her, she was shocked by her appearance. She looked pale and drawn.

'Come in, dear,' she said. 'You're looking tired. I'll put the kettle on.'

Louise came in. She had a small bag with her, Susan noticed. Only an overnight case, so she obviously didn't intend to stay long. Susan bustled about the kitchen, putting cups on a tray and getting out the biscuit tin.

'Did you have a good journey?' she called out. 'You must have made an early start. Have you had any breakfast?' Louise didn't reply so she gave up in the end and when she carried the tray through to the living room, she found her sitting on the sofa looking miserable. She sat down beside her. 'What's wrong, dear? You said on the telephone that you wanted to talk to me. Has something happened?'

'Yes.' Louise took the cup that Susan handed her and took a long drink. 'Thanks. I needed that. Yes, Susan, something has happened and I might as well come straight to the point. You know I told you I'd found my mother?'

'I do. And you were worried that she might not leave you alone.'

'After the newspaper article about the collapse of the play, there wasn't much fear of that. She got the message at last that there was no money.' Louise looked at her stepmother. 'I'm sure you saw the article too.' Susan nodded. 'So now you realize what a damned liar I am – if you didn't already know.'

Susan looked up in alarm. 'Louise!'

'Oh, I'm well aware of what a pain I've been to you all,' Louise said. 'There's no need to pretend. The thing is, she, my – the woman who gave birth to me – asked to see me one more time. She said she had something to tell me about myself and it was vitally important.'

'So you met her again?' Susan said cautiously.

'Yes. And what she gleefully told me was that Frank Davies wasn't my father. That was what they rowed about on the night she walked out. She didn't take me with her because she didn't want me and Frank wasn't my real father.' She looked at Susan, her eyes full of pain. 'Can you imagine how that made me feel?'

Susan laid a hand on her arm. 'Oh my dear, of course I can. How horrible for you.'

'So – the reason I'm here, Susan, is to ask you if you knew about it.'

Susan leaned back in her seat with a sigh. 'Yes, I have to confess that I did know. Frank told me when we were first married. Not that he ever really believed it. He always insisted that you were his daughter and nothing would dissuade him from that belief.'

'I'll never *ever* think of him as anything else,' Louise said.

'He loved you very dearly,' Susan said. 'But being abandoned by your mother like that at such an early age had a very profound impact on you. It damaged something deep inside you and although maybe I shouldn't say it, it made you a very difficult child to handle.'

'I know. I remember how awful I was to you.'

'In your early teens things got worse. You rebelled – stayed out late – mixed with a crowd of young people that were – well, a disastrous influence on you.'

Louise frowned. 'I don't remember much about that.'

'No, you wouldn't. You ended up getting into real trouble and having a kind of breakdown. You spent quite a long time in hospital.'

Louise looked at her. 'Is that a nice way of saying I got mixed up with drugs and had to go into rehab?'

Susan sighed. 'I'm afraid it is. But you got better, that's the main thing. It took some time but you got better and you came home with very little memory of what had happened. I think that may have had something to do with the treatment. The psychiatrist warned us about that.'

Louise's eyes widened. 'Psychiatrist! I was *that* sick?'

'I'm afraid so. When you came home you had this deep desire – almost an obsession – to go to drama school and become an actress. I was doubtful at the time. I wondered how you would cope, being away from home, but Frank was only too pleased that you had something you really wanted to do – a goal in life. He was happy to make the necessary sacrifices to pay for you to go.'

'I never knew that.' Louise's eyes filled with tears. 'You both stuck by me – and look how I repaid you.'

Susan patted her hand. 'Never mind that now. It's all water under the bridge. Frank always loved you unconditionally. As for whether he was your true father or not, we'll probably never know for sure, but one thing I can tell you and that is that he was a father to you in every possible way he could be: a father to be proud of.'

'I know, and I am proud of him. And of you too, Susan.'

'Louise ...' Susan paused and took a breath. 'There's something else you should know,' she said. 'And I hope this is the right time for me to tell you. But you must prepare yourself for a shock.'

'Why couldn't you be straight with me, for God's sake?'

Simon and Karen were standing in the kitchen. Karen was wiping down the worktops, her back to him. Now she spun round to face him. 'Because you're always so bloody unreasonable if you must know.'

'Don't swear at me.' He looked round. 'Where's Peter?'

'He's fine. He's playing in the front garden.'

'So – about this tutoring. How did you think you could manage to do it in the holidays without me knowing?'

'I was going to tell you.'

'Oh yes – when?'

Karen threw the dishcloth she'd been using into the sink with a splash. 'When I could work up the courage. I knew you'd say I'd have to give it up. You're so controlling – always making me out to be a bad mother and wife. Do you have any idea how that makes me feel?'

He sighed. 'All I ask is that you stay at home with our son till he's old enough for school,' he said exasperatedly. 'He's three now so that's only another couple of years. Can't you even wait that long to shake off the shackles of motherhood?'

Karen snorted. 'Oh, will you *listen* to yourself? *The shackles of motherhood*! You sound like a character out of a Victorian novel. I ask you – how can it hurt for me to be away from him for a couple of hours a week?'

Simon seethed. For the second time that morning he'd been accused of being 'Victorian'. He'd always considered himself to be a forward-thinking man. In favour of female equality at home and in the workplace. But surely when a woman had a child …

'The poor kid's a bundle of nerves,' he lashed out. 'When you're not here he's constantly asking for you. You're damaging him – making him into an anxious, neurotic little wreck!'

Karen laughed. 'I've never heard of anything so ridiculous. Peter's a very well-adjusted child. Every …' Suddenly, through the open front door came the sound of squealing brakes, a bang and a loud shout. They stared at each other for a stunned second, then made a dash for the door.

'Peter!'

The front gate was open and a car stood at the kerb; the driver was kneeling over a tiny prone body in the road in front of his car. He looked up, white-faced, as Karen and Simon came running out.

'Christ! I'm so sorry,' he said. 'He ran out. I couldn't stop in time – didn't have a chance!'

Karen screamed, a hand over her mouth. Simon rushed to kneel by the lifeless child. He felt for a pulse and turned to Karen. 'He's alive. Quick, ring for an ambulance!'

'You were fourteen when you went off the rails and started getting into trouble,' Susan said. 'Your dad and I were already worried but the final blow fell when we discovered you were pregnant. We never found out who was responsible. It was soon after that you had your breakdown.'

Louise was staring at her stepmother, open-mouthed. 'I had a baby? But what happened? Why don't I remember any of it?'

'You were very ill. The doctors thought you might lose the child or that it might be affected by the drugs. They kept you in hospital throughout the pregnancy.'

'And the – baby?'

'Thankfully the baby was fine – taken straight from you at birth for adoption.'

'Without my consent?'

'You were in no fit state to know what you wanted – you never asked once about the baby and afterwards you seemed to have blanked it completely from your mind.' Susan laid a hand on her arm. 'Louise – we – your dad and I adopted her.'

'*You* did? ' Louise took in the implication of what she had just heard. She stared at Susan. 'You don't – you *can't* mean…?'

'Yes. Karen is yours. She was four months old when you came out of hospital and we thought at first that seeing her would trigger your memory, but it didn't. You never took any interest in her at all. So we decided to let sleeping dogs lie and bring Karen up as our own. You went back to school – a new school, then later to drama school.'

Louise was deeply shocked. It seemed bizarre. It was all so hard to take in. Then she remembered the dreams – about her mother leaving, then later, about rejecting a baby. Somewhere, deep in her subconscious she'd buried it all. She looked at Susan.

'I'm so glad you told me,' she said. 'I take it that Karrie doesn't know?'

Susan shook her head. 'She still thinks of Frank and me as her parents, although we did tell her a long time ago that she was adopted.' She looked appealingly at Louise. 'I really think that after all this time it's probably better to leave it at that.'

Louise was silent for a moment then she slowly nodded. 'I'm sure you're right.'

Susan took her hand and squeezed it. 'It's not too late to have a child of your own,' she said, but Louise shook her head.

'I've never been the maternal type. Karen was lucky to have you and Dad. Better not to stir things up now.' She looked at Susan. 'I owe you so much, Susan; far more than I ever dreamed. I'll never be able to make it all up to you.'

Susan smiled. 'No need, my dear. I'm so sorry about your disappointment over the play. I only wish things could work out for you.'

The telephone began to ring and Susan got up to answer it. Louise watched as her stepmother's face suddenly drained of colour.

'Oh my God!' she cried. 'Yes. We'll get there as soon as we can. Thank you for ringing me, Simon. Goodbye.'

In the family room at the hospital, Susan and Louise found a shocked-looking Karen and Simon sitting silently together. Simon was holding Karen's hand. Susan sat down beside Karen.

'Any news?'

'Not yet.' It was Simon who spoke. Karen wept silently into Susan's shoulder.

'It was my fault,' she sobbed. 'If only we hadn't been rowing ...'

Simon squeezed her hand. 'Don't. It was my fault as much as anyone's. I think it's time we got our priorities straight.'

Louise reached out to touch Karen's shoulder. 'Karrie – darling, I'm so sorry.' When Karen did not respond, Louise looked at Susan. 'Shall I go and get some coffees?'

Susan nodded. 'That sounds like a good idea.'

When she'd left the room Simon looked up angrily. 'Why on earth did you bring *her* with you? She's the last person we want with us at a time like this.'

'Something quite monumental has just happened in her life,' Susan said. 'I think you'll find she's going to be very different from now on.'

Simon grunted. 'Huh! I'll believe that when I see it.'

Louise returned with the coffees on a tray and they sat in silence as they drank them. When the door opened and a tall young man walked in, all four looked up expectantly. He introduced himself.

'Good morning. I'm Paul Grainger, senior paediatrician here at St Mary's and I've been looking after your small son.'

Karen was on her feet. 'How is he?'

'He has a hairline fracture of the skull but apart from a few bumps and bruises, that's all.'

'A skull fracture!' Karen cried, leaping to her feet. 'But that's serious, isn't it?'

'A hairline fracture in a child of his age heals very quickly,' the consultant told them. 'We'll keep him in for a couple of days just to be on the safe side.' He looked at Karen. 'You can stay here with him if you like. After that there's no reason why you can't take him home.'

Simon stood up and put his arms round Karen, who was weeping with relief. He looked at the consultant. 'Thank you so much.'

'Not at all.' On his way out the consultant said, 'By the way, the driver of the car is in the waiting room. I think he'd appreciate some reassurance.'

'I'll go in a minute.' Simon kissed the top of Karen's head. 'Don't cry, darling. Everything's going to be all right.'

'It's my fault,' she sobbed. 'If we hadn't been arguing about my wanting to work, it would never have happened.'

'Never mind that now.'

Susan gestured to Louise that they should give them some time alone and they quietly left the room.

As they made their way back to Susan's flat Louise said, 'Maybe I should go back to London. You'll be wanting to help Karrie and Simon and I'll just be in the way.'

Susan smiled. 'It's considerate of you to suggest that but don't go tonight. Leave it at least till the morning.' She looked at her watch. 'I must ring Ted and let him know what's happened.'

'Ted?' Louise looked at her. 'You're back together, then?'

Susan smiled. 'We met and he explained everything. I'll tell you about it over supper.'

Chapter Twenty-Five

I LET MYSELF INTO THE empty flat and stood for a moment in the silence. In spite of Susan's revelations and what had happened over the last few days, I felt more estranged from my family now than ever before. They didn't need me. I was Karen's mother and Susan's stepdaughter. Another thought hit me: I was Peter's *grandmother,* for heaven's sake! And yet none of them needed or wanted me; in fact quite the opposite. Susan was kind and good. She always had been and no doubt she always would be. But could I ever begin to mend all the fences I'd ridden roughshod over in the past? They had all put up with so much from me. Perhaps now it was time to give them all a much deserved rest; to stand on my own feet and try to turn over a new leaf. It was a really strange feeling. It could be an end and yet if I really made up my mind to it, it could be a beginning – a fresh start. This could be make-or-break time and I realized that the outcome was up to me.

I put my case in my room and went into the kitchen to put the kettle on. As I waited for it to boil, my thoughts turned to Mark! Dear, patient Mark. I'd been so horrible to him, yet he'd given me nothing but love and consideration. But now it was over. He'd given up trying to contact me and he was about to marry someone else. I'd lost him. Anyhow, I'd never be able to face him now. He'd be sure to see me differently when he knew the truth. I knew I owed him that but I shied away from telling him about my past – seeing the look of disgust and disillusionment on his face.

I found cold meat and salad in the fridge and as I set about putting a snack lunch together, it occurred to me that in the past I'd have taken Di's food without a thought. Now I promised myself that I'd go out later and replace what I'd taken. When I'd eaten, I went out to the supermarket and it was as I was letting myself back into the flat that my phone began to ring. I put down the bags of shopping and fished my phone out of my bag.

'Hello.'

'Louise. It's Patrick Jason. I've got some good news,' he said.

'Oh yes?'

'The audition you did – for the soap, *King's Reach* – I'm delighted to tell you that you've got the part. Congratulations!'

I stared speechlessly at the phone. Had he just said what I thought he'd said?

'Louise – are you there?'

'Yes – yes. Are you sure?' I asked, my knees trembling.

He laughed. 'Of course I'm sure. You start rehearsing next week, so if you'd like to come into the office tomorrow and sign the contract …'

'Oh, yes of course,' I said quickly. 'I'm still trying to take it in. I can't thank you enough, Patrick. What time would you like me there?'

'Ten would be fine,' he said. 'I'm glad you're pleased.'

'*Pleased*! I'm over the moon,' I told him. 'It couldn't have come at a better time. See you in the morning, then.'

As I unpacked the shopping, my hands shook so much that I kept dropping things. As soon as I'd put the last item away I went straight back out again – to the off-licence on the corner to buy a bottle of champagne.

Di was late getting home and I kept looking at the clock. I couldn't wait to tell her my news. She looked tired when she got in. I'd already cooked and dinner was waiting in the oven – the champagne chilling in the fridge.

She seemed pleased to see me. 'Hi. How was your home visit?'

'Traumatic,' I told her. 'Peter, my little nephew, was involved in an accident and rushed to hospital.'

'Oh no! Is he OK?'

'He's got a hairline skull fracture but apart from that he's all right. The doctor said that skull fractures in young children are fairly quick to heal.'

Di hung up her coat. 'How did it happen?'

'Seems that Karrie and Simon were having a row about Karrie working. No one was watching Peter. He was playing in the front garden and he got out onto the road and ran in front of a car.'

'They must feel so guilty.'

'I think they do. Maybe they'll realize now that they have to come to some kind of compromise over Karrie's work.'

'So what else happened?'

'Quite a bit. Susan and I had a long talk. But something happened after I got back and I can't wait to tell you that first.'

She smiled. 'I knew there was something. I can feel you fizzing from here!'

'Di – I got the part,' I told her. 'The part I auditioned for, in *King's Reach*.'

Her face lit up and she reached out to hug me. 'Wow! That's fantastic! I'm so happy for you.'

'I've got champagne and I've cooked us a special meal,' I said excitedly. 'Shall I open the bottle now?'

'What do you think?'

The cork popped and we toasted each other. I served the meal I'd cooked and we chatted excitedly. Di wanted to hear all the details. We were having coffee when she asked, 'So what about Susan? Was she able to throw any light on what your mother told you?'

'She knew about it of course – said that my dad never accepted that I wasn't his. It was something they didn't talk about.' I put my cup down, suddenly serious as I came down to earth. 'She told me a lot about my childhood, Di. I must have been a nightmare for them. Apparently I went off the rails big-time in my early teens – drugs. I ended up in rehab and later I had a bad breakdown.'

'Oh, Lou.'

I didn't tell her I'd given birth to a daughter and that Susan and Frank had adopted her. I'd already decided to keep it to myself in case somehow it got back to Karrie. Di was looking at me aghast.

'I've just had a thought – those nightmares!'

'I know. It all makes sense now. It says a lot about the kind of person I became too,' I added. 'Somewhere at the back of my mind I felt I needed to pay someone or something back. God knows why. Susan and Dad were wonderful to me. I've been so lucky and I've got a hell of a lot of making-up to do. One thing I do know, and that is that I wish I'd never tried to get in touch with my mother. If I'd known all this sooner there's no way I would have wanted to know her.'

'My advice is to put it all behind you,' Di said. 'It's in the past and there's nothing you can do about it. Just concentrate on this new challenge and look forward to a fresh start.'

I smiled. 'Yes, I will. I can't wait to begin.'

'And what about Mark – surely you want to ring him with your news?'

'I don't think he'll be interested. I'm afraid I've blown it with him.'

'Why do you say that?'

'I think he's finally written me off. Anyway he's getting married.'

'Married! Who to?'

'A girl he used to know.'

'How do you know this?'

'It was when I was working at the bridal shop. His sister came in with her to choose the dress.'

Di looked crestfallen. 'It didn't take him long, did it? Maybe it was on the rebound. But even so, he'll want to hear about the new job, surely?'

I shrugged. 'I doubt it.'

'Well, you should at least try.' Di shook my arm exasperatedly. 'Go on, ring him now. I'll give you some space while I'm doing the washing-up.'

She disappeared into the kitchen and I took out my phone and sat looking at it. Suddenly, I had cold feet. Maybe he'd be out with his fiancée. He wouldn't be interested in what I was doing any more. It would be nice to tell him my news, yet I dreaded hearing the indifference in his voice as he tried to sound interested. I knew Di wouldn't let me get away with feeble excuses so I quickly clicked on his number, hoping that he'd be out or that the number would be engaged.

'Hello, Mark Naylor.'

My heart missed a beat. 'Oh – Mark. It's Louise.'

'Lou! I've been trying to come to terms with the fact that you'd had enough of me. I'd even deleted your name from my phone. How are you?'

'I'm fine – you?'

'Yes, fine. Cathy and the kids moved out a couple of days ago and I'm getting used to having the place to myself again.'

'Oh. You must miss them.'

'I do, but in a good way. You can have too much of being woken up by two little monsters jumping all over you at seven in the morning.' There was a pause and then he said, 'Is everything all right, Lou? You sound a bit – odd.'

'Do I?'

'A bit. Do you have a special reason for ringing me?'

'Do I need one?'

'Well – you tell me.'

'Actually I'm ringing because I've got some news,' I told him. 'Quite a lot of news in fact. I've—'

'*No!*' he broke in. 'Don't tell me now. Let's meet. Are you doing anything tomorrow?'

'I have to go to Patrick Jason's office in the morning. He's my new agent.'

'Right – what time?'

'Ten o'clock.'

'OK if I pick you up from there – say ten thirty?'

'That sounds fine.'

'OK, see you then. And Lou ...'

'Yes?'

'I've missed you.'

I swallowed hard. 'Me too.'

Patrick had the contract all ready for me to sign the following morning. Putting my name to it felt good. The first really important contract I'd signed in my whole career and when I saw the salary Patrick had negotiated for me, my heart gave a leap. It was more money than I'd ever earned.

Patrick looked down at my signature. 'You've signed, Louise Davies.'

'Yes. That's my real name and it's how I want to be known from now on,' I told him. 'Louise Delmar is dead and buried.'

He laughed. 'Right. I'll make a note of it.'

It was all over and done with by a quarter past ten and I sat in reception waiting for Mark to arrive.

When he walked in, I was surprised at the receptionist's reaction. She looked up with a smile. 'Mark! What brings you out of the woodwork so early in the morning?'

He leant across her desk to give her ear a tweak. 'Less of your cheek, young Sharon. I'm here to escort a lovely lady to lunch.' He turned to me. 'It's terrible the disrespectful way these receptionist treat you nowadays, isn't it?' he quipped. 'You can't get the staff, you know.'

On the way downstairs I asked him how he knew Patrick Jason's receptionist. And he grinned.

'He's my agent too,' he said.

It suddenly occurred to me that he hadn't sounded surprised when I said I had to see Patrick this morning. 'He's *yours* – but ...' I looked at him with narrowed eyes. 'I suppose his contacting me wouldn't have had anything to do with you, would it?'

He frowned. 'Come to think about it, I suppose I *might* just have mentioned you in passing.'

'Then it wasn't down to the article in the *Sunday Sphere*!'

'Oh, that!' He laughed. 'I had a good laugh at that. Good on you!'

'I didn't do it for a laugh,' I told him. 'I did it because I needed the money. You didn't have to put in a word for me with Patrick, but thanks all the same.'

'Well, it seems to have worked out. Has he come up with anything for you?'

'He has as a matter of fact. That was what I was going to tell you last night. I auditioned for a part in a BBC soap last week.'

He looked at me. 'And…?'

'And – I got the part.'

'Great!' He slipped an arm through mine. 'So we've got something to celebrate. What are we waiting for?'

We had lunch at a small, intimate restaurant quite close to the Savoy, overlooking the Thames. As we were having coffee, Mark looked at me.

'You said you had a lot of news,' he reminded me. 'What else has happened?'

I came down to earth. After the excitement of signing the contract the memory of Susan's revelations had been pushed to the back of my mind. 'The rest of it is a bit more serious.' I looked at him. 'But before I tell you, I think you have some news for me.'

He looked bemused. 'Me? No, nothing springs to mind.'

'Not the little matter of your forthcoming nuptials?'

He burst out laughing. 'That'll be the day! Where did you get that idea from?'

'I did a short stint working in a bridal boutique,' I told him. 'Your sister came in with your fiancée.'

He shook his head. 'My *what*?'

Slightly irritated I went on. 'Come off it, Mark. The name Franny ring any bells?'

His face cleared. 'Oh! Franny! Francesca Barratt. She's an old school friend of Cathy's. She's getting married next month and Cathy has been helping her with the preparations.'

'Oh.' I bit my lip, feeling slightly foolish. Mark laughed softly.

'You didn't actually think it was me, getting married, did you?'

'Well, I ...'

'And were you at all upset by the news?' I shook my head and he leaned towards me. 'What, not even a little bit?'

'It was an easy mistake to make,' I blustered. 'Was I upset? Not really, no. You deserve to be happy.'

'But we don't always get what we deserve, do we?' he teased.

I looked up at him. 'Will you please stop baiting me?'

'Not until you tell me how you *really* felt when you thought I was marrying someone else.'

'OK, I was ...' I searched my mind for the right word. 'I was sad,' I said at last. 'Sad and – OK – a bit jealous.'

His eyes danced. 'Why would you be jealous? *You* don't love me, do you?'

'Mark – I've got something quite serious to tell you. It's about me – things I only discovered a few days ago – things that might make a difference about how you see me.'

'It all sounds very solemn.'

'It is.' I looked around. 'Could we go somewhere quiet where I could tell you?'

'Of course.' The smile left his face as he beckoned the waiter for the bill. 'You go and get your coat. I'll take you back to the flat.'

'So that's my background, Mark. Not very inspiring, is it?'

We were seated opposite each other in Mark's sunlit living room. It had been painful, pouring out my past to him. It had taken all my courage and strength and he obviously saw that.

He poured a large glass of wine and put it in my hand. 'It's all so long ago,' he said gently. 'You were still a kid, Lou. It's in the past and what's done can't be undone. Why should it make me see you any other way than I see you now?'

I took a sip of the wine and looked at him. 'Not even the fact that I had a baby at barely fourteen years old?'

He shook his head. 'That was then, Lou. So you made some mistakes – who hasn't? It must all have been sheer hell for you.

It's no wonder your mind refused to retain any of it.'

Suddenly I decided to tell him the truth about Karrie. No one else must know but I felt I owed Mark not just some, but all of the truth. 'My baby daughter was adopted by Susan and Frank,' I said slowly. 'Karen is – was my daughter and neither of us has ever been aware of the fact. Susan has asked that it remains a secret and I feel bound to honour that wish, so please, Mark, you are the only other person to know this and it mustn't go beyond these walls.'

He moved to sit beside me. 'I feel flattered that you're prepared to trust me with a secret like that and of course it goes no further.' He slipped an arm round my shoulders. 'Thank you for what you've just confided in me, Lou. It makes no difference to the way I feel about you. You know how much I love you. It's a love that has lasted for years so it's not likely to stop now. I understand so much more about you now and if anything it makes me love you even more.' He searched my eyes. 'Please will you tell me truthfully how you feel, because if your feelings don't match mine this must be goodbye. I couldn't go on, knowing that what I feel will never be returned.'

I put my arms round him and held him close. 'Of course I love you, Mark. I've loved you for years without recognizing the fact. I've been a total bitch to you in the past. I can't understand why you kept on loving me.'

He kissed me. 'Maybe I can make you understand.'

For a long time neither of us spoke and the next time I glanced at the time it was almost five o'clock. I sat up.

'I must go. Di will be home.' I looked at him. 'That's another thing, Mark. It was always Di's flat – never mine. I lied about it. God knows why. I've lied about so many things in the past, but never again. All that ends from today onwards and that's a promise.'

He winced. 'If that's not tempting providence I don't know what is!'

'Well, I believe it anyway.' I looked around the room. 'You asked me once if I would move in. Does that invitation still stand?'

He grinned. 'What do you think?'

I stood up and gathered my coat and bag together. 'Then will you take me to Di's to collect my things, please?'

Chapter Twenty-Six

As Susan rang the bell of Karen and Simon's house, she caught Ted's hand and gave it a squeeze. He returned it, looking down at her.

'You are sure they invited me too?'

Susan smiled. 'Of course. I want you all to get to know one another and this is the perfect opportunity.'

Karen opened the door. She looked tanned and relaxed from the recent holiday the three of them had just enjoyed.

'Come in, both of you. Simon is just reading Peter a story. He'll be down in a minute and then we can all have a drink.' She led them through to the living room. Turning to Ted she said, 'I'm so glad you could come. I hope you like steak.'

Ted grinned. 'What man doesn't? It's very kind of you to invite me.'

Simon appeared. 'Phew! Getting away from that young man is like tearing off a plaster.' He smiled and held out his hand to Ted. 'Hi! I'm Simon, as I expect you've guessed. What will you have to drink?'

Karen turned towards the door. 'I'll have mine in the kitchen while I put the finishing touches to dinner. We want to have the meal over and done with ready for the big event, don't we?'

Simon inclined his head towards Ted. 'Whisky?' Ted nodded and Simon poured. He glanced at his mother-in-law. 'Don't need to ask you or Karrie. G and T. Ice and a slice – right?'

Susan laughed. 'I'll take mine and Karen's into the kitchen and see what I can do to help. I'm sure the two of you are dying to talk football.' She took the glasses from Simon and he opened the door for her. Outside in the hall, she paused for a moment to listen but she was soon reassured by the sound of the two men talking. It looked as though they were going to get along. In the kitchen she found Karen tossing new potatoes in parsley and butter. 'I'll put your drink on the worktop, shall I?' she said. 'Now – what can I do?'

'You could put the dressing on the salad, Mum.'

As they worked Susan looked at her daughter. 'You enjoyed your holiday, then? You certainly look well.'

Karen nodded. 'Simon and I have come to a compromise. I can do my tutoring as long as Peter doesn't suffer in any way.'

Susan's eyebrows rose. 'Suffer! In what way?'

Karen blushed. 'Well, you know, being passed round.'

'Being cared for by me, you mean?'

'Not at all! It's just that it's better for him to be in his own home until he's older.'

'So he's not even going to playgroup?'

Karen shook her head. 'Yes, of course he is but …'

'I should stop now, dear,' Susan said with a smile. 'You know what they say – when you're in a hole, stop digging.' They both laughed and Susan went on, 'Actually it's as well you're not going to be asking me to have him very often. Ted and I are going to be really busy, harvesting all the produce from the allotments and manning our stall at the farmers' market.'

Karen smiled. 'Oh, Mum, I'm so glad you're really back together again.'

'Are you? He is still a married man, you know,' Susan pointed out.

'Yes, but it's only a platonic relationship you have with him, isn't it?'

'Mmm.' Susan bent her head to take a sip of her G and T. 'I can't wait to see Louise in this show,' she said, changing the subject.

'Neither can I,' Karen agreed. 'Since she landed that job she's been a different person. In fact you'd hardly know her these days.'

'Yes, and of course she's finally admitted that she loves that delightful man, Mark. She couldn't have a better partner. He keeps her feet on the ground and she might need that if she's going to be a big hit in this show. I understand that the part she's playing is a pivotal one.'

'Yes, so I believe.' Karen handed her mother the potatoes and salad bowl. 'Take these through, will you, Mum? The steaks are done and we'd better get on if we don't want to miss the programme.'

At eight o'clock the four of them were seated in the living room, their eyes on the TV screen. They watched as the initial captions rolled and then it was the opening scene and Louise was seen in the character of Amy, cooking breakfast in her kitchen.

Susan was surprised to see that she was made up to appear at least ten years older than she actually was and as the scene progressed, she could see for the first time what a good actress her stepdaughter was. The character she was playing was just about as different from her own personality as it was possible to be. Watching the scene she soon lost herself in the story, almost forgetting that she was watching her own stepdaughter.

Ted was watching Susan. Her expressions went from surprise to enjoyment and then to pride in Louise's achievement. She was certainly very good in the part. In fact, he thought he could see something of Susan in Louise's development of the character. He looked again at Susan's face and smiled inwardly. She was such a lovely woman. He told himself daily how lucky he was to have met her. He didn't dare to think too far into the future but it was his dearest hope that one day he would be able to ask her to marry him.

Karen watched her sister's performance with interest. Clearly she had worked hard for this and her work had paid off. She was good; very good indeed. This was what Louise had wanted so badly. It had always been her ambition to land an important part on stage or TV and now here she was, fulfilling her dream. Now she could understand some of the reason for her sister's past recklessness, her spiteful, hard-to-forgive actions. They must have sprung from frustration and although they still rankled with Karen, she felt that now she could put them to one side and wish her sister well.

Simon watched the programme with his tongue firmly in his cheek. He didn't like soap operas anyway and he felt that Louise had always belonged in something tacky. She had found her niche and he wished her well of it. Hopefully it would keep her too busy to come and visit and happy enough not to want to cause any more problems.

Giving the programme more attention, he was suddenly aware of something interesting. Louise's portrayal of the motherly Amy was very much like his mother-in-law, Susan. How like Louise to steal someone else's personality for her own gain. He hoped that Susan wouldn't recognize the fact. Surely she would not see it as a compliment. He glanced around the room. Karen and Susan wore rapt expressions and Ted – well, Ted only had eyes for Susan. Good luck to him!

I watched my first episode of *King's Reach* full of apprehension and self-doubt. Did I really come across as I'd intended? Did I look right – *sound* right? And that wig! I hadn't been too sure about the make-up or costumes that wardrobe had chosen for me, but now, looking at them from the other side of the screen I could see that they were right.

'So – proud of yourself, Miss Davies?' Mark handed me a glass of champagne and I took it from him, wrinkling my nose.

'Not really. Do you think they'll ring tomorrow and say they're terminating my contract?'

He laughed and sat down beside me. 'As if! You're the best thing that's happened to that show in a long time. You bring it to life.'

I leaned across to kiss his cheek. 'I suppose you wouldn't be the teeniest bit prejudiced, would you?'

He looked wounded. 'I hope I'm too honest for that. I'm an actor, remember? And I know good acting when I see it.'

I looked at him. 'What about your career, Mark? Has Patrick come up with anything for you?'

He shook his head. 'To be honest, darling, I'm not that bothered. I was never the actor that you are.' I made to protest but he held up his hand. 'No. I mean it.' He slipped an arm around my shoulders and nestled closer. 'I'm quite happy to bathe in your reflected glory; to chauffeur you hither and thither and be known as the celebrated *Mr* Louise Davies.'

'Don't say that!'

'I will say it because whether you recognized it or not, my darling, that was my fumbling way of asking you to marry me.'

I put down my glass and wound my arms around his neck. 'I don't deserve you, Mark Naylor, and you will never *ever* be *Mr* Louise Davies.'

'Well, OK, but the question still is, will you be Mrs Mark Naylor?'

I kissed him hard. 'I thought you'd never ask,' I whispered. 'And – just for the record, you've got yourself a deal!'